ADVANCED LIGHT MICROSCOPY

1. PRINCIPLES AND BASIC PROPERTIES

Advanced Light Microscopy

Advanced Light Microscopy

Volume 1
Principles and Basic Properties

Maksymilian Pluta

Professor of Applied Optics
Head of the Physical Optics Department
Central Optical Laboratory, Warszawa

Elsevier

Amsterdam—Oxford—New York—Tokyo

PWN—Polish Scientific Publishers

Warszawa

1988

Cover design by *Zygmunt Ziemka*

Distribution of this book is being handled by the following publishers:

for the USA and Canada
ELSEVIER SCIENCE PUBLISHING CO., INC.
52, Vanderbilt Avenue
New York, N.Y. 10017

for Albania, Bulgaria, Cuba, Czechoslovakia, German Democratic Republic, Hungary, Korean
People's Democratic Republic, Mongolia, People's Republic of China, Poland, Romania, the
USSR, Vietnam and Yugoslavia
ARS POLONA
Krakowskie Przedmieście 7, 00-068 Warszawa 1, Poland

for all remaining areas
ELSEVIER SCIENCE PUBLISHERS
25 Sara Burgerhartstraat
P.O. Box 211, 1000 AE Amsterdam, The Netherlands

Library of Congress Cataloging in Publication Data

Pluta. Maksymilian.
    Advanced light microscopy.

    Translated from the Polish manuscript.
    Bibliography: p.
    Includes index.
    Contents: v. 1. Principles and basic properties.
    I.   Microscope and microscopy — Technique —
Collected works.   I.   Title.
QH207.P54   1988   502′.8′2   87–24605

ISBN 0-444-98939-0 (Vol. 1)
ISBN 0-444-98940-4 (Series)

Copyright © by PWN-Polish Scientific Publishers—Warszawa 1988

Printed in Poland by D.R.P.

Preface

This book covers, in principle, all fields of modern light microscopy although advanced methods have been given priority over routine techniques. Special emphasis is given to quantitative methods which are at present developing rapidly and are becoming instrumental in further advances in science and technology.

The work is divided into three volumes. Volume 1 deals with the physical fundamentals of light microscopy, geometrical and diffraction theory of image formation in the microscope, construction of modern microscopes and their components, different systems of bright-field microscopy, and methods of quality assessment of microscopical imaging.

Volume 2 is devoted to special qualitative and descriptive methods, with particular regard to phase contrast, differential interference contrast, amplitude contrast, dispersion staining, modulation contrast, reflection contrast, fluorescence, UV and IR microscopy, holographic microscopy, and other new techniques.

Volume 3 contains quantitative and measuring methods, with special emphasis on micrometry, microphotometry, polarizing microscopy, microinterferometry, optical diffractometry, Fourier transform microscopy, and automatic image analysis.

Microscopy is employed by biologists, medical researchers, chemists, physicists, textile engineers, optical fibre manufacturers, electronics specialists, geologists, criminologists, and many others. This book is addressed to all these potential users, but primarily to cell biologists, chemists, physicists, textile and optical fibre researchers, specialists in thin film technology, metallographers, all general users of microscopy, both professional and amateur. For most of these, the microscope is a basic instrument of research, although they may have had little instruction in the use of more advanced microscopy during their school or university education. The result is that experience in specialized microscopy must be acquired on training courses or by individual study and practice. This book

aims to be of help in such self-education and to promote more advanced knowledge of modern light microscopy. It is assumed, however, that the reader is
familiar with secondary school mathematics, and has some knowledge of higher
mathematics. The latter is nevertheless applied only when absolutely necessary.

The book is extensively illustrated by drawings and photomicrographs. The
author is indebted to a number of scientists, researchers, and manufacturers
for providing some of these illustrations. Extensive references to the literature
are given so that, where desired, the reader can pursue his studies in greater detail.

The general conception of this book, which was hazy at first, manifest itself
clearly to the author from the moment of realization that the microscope is an
instrument which encompasses and successfully displays what are perhaps the
most fascinating phenomena of physical optics, namely, the diffraction, interference and polarization of light. This strong connection with physics in microscopy
attracted the present writer to the microscope, and a result of that attraction is
this book. In spite of its considerable bulk, it cannot cover the topic fully; the
field of light microscopy is now so vast that no book can satisfy all users of the
microscope. The author will therefore be grateful for critical remarks and sugges-
tions oriented towards possible improvements in a future edition.

Warszawa
June 1987   *M. Pluta*

Contents of Volume 1

Contents of Volume 2: *Specialized Methods*

Contents of Volume 3: *Measuring Techniques*

Introduction

In recent times, studies of biological specimens, surfaces of metallographic samples, textile fibres, powder of different materials, and many other microscopic objects have aimed at a quantitative analysis of their nature, physical properties and structures. Such studies cannot be effectively performed without a knowledge of the physical principles of light microscopy. Thus Chapter 1 of this volume covers the interaction of light with matter, interference, coherence, diffraction, polarization of light, and other phenomena which reside and operate constructively or destructively in the microscope. An indispensable presentation of some special microscope components (polarizers, retarders, interference filters, birefringent beam-splitters, etc.) is also included. It is hoped that this largely introductory chapter will suffice to give the reader a suitable acquaintance with the basic physical aspects of light microscopy and related topics, and to prepare the ground for the perusal of this and further volumes without the need to dip into advanced textbooks on physical optics.

The principles of microscopical imaging within the scope of geometrical optics are based on refraction and specular reflection of rays as well as on the ability of lenses and mirrors to focus and change the convergence or divergence of light beams. This geometrical approach is, however, insufficient if, for example, some fundamental problems relating to resolution and contrast transfer in microscopical imaging are to be clearly explained. A satisfactory explanation is possible only on the basis of the diffraction theory of image formation in the microscope. These two approaches, geometrical and physical, are presented in Chapters 2 and 3.

Although the basic principles of the diffraction theory were formulated by Ernst Abbe in 1873 and refined by a Polish physicist, Mieczysław Wolfke, in 1911-1920, the theory's real potential was only revealed some decades later when phase contrast was discovered by Frits Zernike (1935), and Fourier optics was developed by P.M.Duffieux (1946). Today, the deepest insight into the microscopical imaging process is provided by Fourier optics, which is also outlined in Chapter 3. A certain amount of higher mathematics was consequently unavoidable because this branch of optics involves quantitative relationships which can only be expressed by integrals, convolutions, and other rather complicated functions. However, no more than a necessary minimum of higher mathematical procedures were included. Moreover, if a reader feels that higher mathematics is not for him, he can still study the descriptive text, which is backed by many illustrative line drawings and photomicrographs. Nevertheless, Chapter 3 is addressed to all who want to understand fully the origins of microscopical imaging and the fundamentals of many special microscope systems discussed in Volumes 2 and 3.

The development of light microscopes themselves and their applications has accelerated greatly during the last two decades. This is due both to gradual improvements in the industrial technology of microscopes and to the use of computer-optimization in the design of optical systems. Today, high quality images can be provided for large fields of view (up to 32 mm). Most present-day microscopes are much easier to handle than their predecessors of twenty years ago. Due to rapid progress in mechanical engineering and electronics, all leading manufacturers offer a great number of different types of microscopes for both ordinary and specialized use. A bird's eye view of these modern microscopes, starting with the standard type and ending with the latest most sophisticated instruments, is also given in Chapter 2. Exhaustive descriptions of standard microscopes are omitted, as detailed information on them can be better obtained from manufacturers' booklets or catalogues. Routine bright-field microscopy and common stereoscopic microscopy are merely outlined, as they are largely covered by numerous other books. Here, attention is mainly paid to instruments and unconventional techniques for advanced light microscopy.

Present-day microscopes and their high performance components need some sensitive methods and tools for quality control. The accurate measurement of the parameters of a microscope system is a specialized operation, but a qualitative evaluation of the performance of the microscope used is a relatively easy procedure. However, the user should be able to recognize different aberrations and to evaluate their magnitude. Qualitative procedures and more sophisticated quantitative methods for an assessment of the performance of a microscope optical system, and especially of microscope objectives, are discussed in Chapter 4, which refers closely to the subjects of the preceding two chapters and brings this volume to an end.

It is evident that the theoretical and practical considerations outlined here refer primarily to bright-field microscopy, but they are also valid for other more specialized microscope systems, methods and techniques dealt with in the next two volumes.

1. Phenomena of Physical Optics in Light Microscopy

The light microscope is an instrument which displays almost the entire range of the phenomena of optical physics. Though the principles of the ordinary microscope may be sufficiently well explained by geometrical optics, the majority of advanced or specialized microscopical methods can only be understood on the basis of wave optics and Fourier optics. Therefore, two principal phenomena of physical optics, i.e., diffraction and interference of light, as well as properties of light coherency must be taken into account. Such an approach is needed, in particular, when the phase contrast method, differential interference contrast, microinterferometry, holographic microscopy, and Fourier transform microscopy are discussed. Moreover, in order to deal with polarized light microscopy, reflection contrast, some special phase contrast and interference systems based on anisotropic optical elements, and even typical incident light microscopy, one needs to be familiar with light polarization, whereas for fluorescence microscopy some knowledge is necessary of the interaction of photons with matter at an atomic level.

In short, to make rational use of different light microscopes, it is useful to familiarize oneself beforehand with some aspects of physical optics. Therefore, a concise recapitulation of some basic phenomena of physical optics is supplied at the beginning of this book.

1.1. The spectral range of light microscopy and its background

Maxwell was the first to propose that light is an electromagnetic wave motion. Light waves differ from radio waves only in *wavelength* λ, and they propagate in a vacuum with the same velocity $c = 3 \times 10^8$ m/s.[1] Röntgen (X) rays and gamma (γ) radiation are also part of the electromagnetic spectrum.

[1] According to recent official data, $c = 299\ 792\ 458 \pm 1.2$ m/s (*La Recherche*, **9** (1978), 643).

In the whole of the spectrum of electromagnetic waves, extending from $\lambda \approx 10^{-16}$ to 10^{10} m, *visible light* (VIS) occupies the insignificantly small range of $\lambda \approx 0.4 \times 10^{-6}$ to 0.76×10^{-6} m (Fig. 1.1). The wavelengths of this range are correlated with the colours we perceive as a gradual change in hue from violet

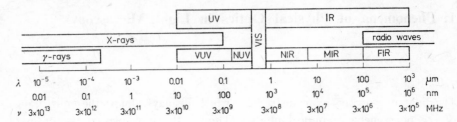

Fig. 1.1. Spectral ranges of light waves. VIS—visible light: V—violet, I—indigo, B—blue, G—green, Y—yellow, O—orange, R—red; UV—ultraviolet: NUV—near ultraviolet, VUV—vacuum ultraviolet; IR—infrared: NIR—near infrared, MIR—middle infrared, FIR—far infrared λ—wavelength, ν—frequency).

at one end of the visible spectrum through indigo, blue, green, yellow, orange, and red at the other end (see Fig. Ia).[2] There is, of course, no exact correspondence between the light wavelength and the sensation of a given colour. Nevertheless, it is useful to divide the *visible spectrum* into several (6–9) hues. In various textbooks slightly different wavelength limits are given for the colours mentioned above, but the values listed in Table 1.1 are roughly in agreement with the data of many authors.

Beyond the violet and red ends of the visible spectrum we have *invisible light*: *ultraviolet* (UV) and *infrared* (IR), respectively. Visible light together with the UV and IR regions constitutes the optical range of light microscopy. In general, it covers in theory a wavelength spectrum from 10×10^{-9} to 10^{-3} m, but in fact a range between $\lambda \approx 0.2 \times 10^{-6}$ and 10×10^{-6} m (200 to 1000 nm) is the practical field of activity of light microscopists.

[2] Figures denoted with Roman numerals are assembled as plates at the end of the volume.

TABLE 1.1

Correspondence between light wavelengths and colour sensation

Colours	Wavelength ranges [nm]
Violet	400–445
Indigo	445–465
Blue	465–500
Green	500–578
Yellow	578–595
Orange	595–620
Red	620–750

In microscopy light wavelengths are usually expressed in micrometres (1 μm = 10^{-6} m) or nanometres (1 nm = 10^{-3} μm = 10^{-9} m),[3] or angstroms (1 Å = 0.1 nm = 10^{-4} μm = 10^{-10} m). In accordance with current practice, light wavelengths will be expressed in nanometres or micrometres throughout the present book.

The microscope is employed to see fine objects and structures which are not visible to the naked eye. However, the primary function of any microscope is not to obtain an extremely large magnification but to yield resolved images of fine details. The *resolving power limit* ϱ of the microscope is dependent on the wavelength λ of waves used, namely it is proportional to λ. For modern light microscopes equipped with high-power objectives, ϱ is comparable to 0.5λ. As the wavelength λ_e of matter waves associated with electrons accelerated by very high voltages (V) is much shorter than that of visible light, considerable gain in resolution can be achieved by means of electron microscopes. A resolving power, for instance, $\varrho_e \approx 0.5$ nm if $\lambda_e \approx 0.005$ nm (at $V = 60$ kilovolts). In contrast to light microscopy, the effective resolving power of the type of electron microscope mentioned above is considerably greater than λ_e ($\varrho_e \approx 100\lambda_e$). The reason for this discrepancy lies in the optical imperfection of electron lenses and their low numerical aperture in comparison with the glass lenses of light microscopes.

X-rays also offer far better resolution than light microscopes, but at the present time no X-ray microscope is available commercially. On the other hand, an acoustic or ultrasound microscope is likely to be produced in the foreseeable

[3] There are some books in which millimicrometres (mμ) are employed instead of nanometres; this is obsolete today.

future; a commercial prototype has finally been built in the E. Leitz factory (Wetzlar). The resolving power of the latter instrument is somewhat worse than that of a high power light microscope, but it appears theoretically possible to produce ultrasound waves with a wavelength comparable to or even smaller than that of visible light; in that case the resolution of the acoustic microscope will be as good or even better than that of the light microscope.

At the present time, however, the microstructure of nature is mainly explored by means of light and electron microscopes. The light microscope, developed over the last 300 years, has revealed the basic microarchitecture of plants and animals. By contrast, the electron microscope became a useful research instrument in the 1940's, and showed up ultra-fine structures in cells, minerals, and metals. Simultaneoulsy, the potentialities of light microscopy in research were greatly extended by the application of the phase contrast and interference methods; this made possible for the first time the study of living cells and tissues. In short, these two instruments of microscopy—the light microscope and the electron microscope—are complementary to one another.

1.2. Propagation of light waves

According to our requirement, light may be regarded as *rays*, as *wave motion*, or as *photons* (discrete packets of energy). The ray is the principal concept of geometrical optics, while one of the basic concepts of wave optics is the *wave-front*. The ray is defined by a line along which a continuous stream of luminous energy propagates, whereas the wavefront is a continuous surface determined by points where a light wave has the same phase. In isotropic media rays are perpendicular to the wavefront, therefore knowing the latter one can determine the rays, and vice versa.

1.2.1. Light motion in transparent isotropic media

Light waves are regarded as rapid variations of an electromagnetic field. Their mathematical description is given in *Maxwell's equations* which express the relations between basic electric and magnetic quantities: *electric field intensity* E, *magnetic field intensity* H, *magnetic induction* B, *electric displacement* (or *electric induction*) D, *electric current density* j, *charge density* ϱ, *specific conductivity* σ, *dielectric constant* (or *permittivity*) ε_0 and *magnetic permeability* μ_0 of a vacuum, *relative dielectric permittivity* ε and *relative magnetic permeability* μ

of matter. In general, **E**, **H**, **B**, **D** and **j** are vectorial quantities, whereas ϱ, μ_0, and ε_0 are scalars, and σ, μ, and ε may be scalars or vectors. Substances for which $\sigma = 0$ (or is very small) are called *dielectrics* or *insulators*. Otherwise, when σ is not negligibly small, substances are *conductors* and *semiconductors*. A medium free from electric charges and currents ($\varrho = 0$, $j = 0$) is called a *free space*. Substances for which $\mu \neq 0$ belong to *magnetics*; they hold a negligible place in microscopical research.

It is not necessary to present Maxwell's equations here (see Ref. [1.1] or [1.2] for details), and only some basic facts resulting from them will be summarized. First of all, a time variation of the electric field and current density create a whirl of the magnetic field, and vice versa, time variation of the magnetic field produces an opposite whirl of the electric field. Both whirls are coupled to each other and that is what constitutes an electromagnetic wave. Its *electric component* and *magnetic component* can, however, be separately described by two analogical *wave equations*. For isotropic homogeneous media free from charges and current, the said equations take the form

$$\nabla^2 \mathbf{E} - \varepsilon_0\,\varepsilon\mu_0\,\mu\,\frac{\partial^2 \mathbf{E}}{\partial t^2} = 0, \tag{1.1a}$$

$$\nabla^2 \mathbf{H} - \varepsilon_0\,\varepsilon\mu_0\,\mu\,\frac{\partial^2 \mathbf{H}}{\partial t^2} = 0, \tag{1.1b}$$

where ∇^2 is the Laplace's operator and t denotes the time.[4] The first of these equations describes the propagation of the electric component and the other that of the magnetic component of an electromagnetic wave. The velocity of both components is the same and equal to $v = 1/\sqrt{\varepsilon_0\varepsilon\mu_0\mu}$. If an electromagnetic wave propagates in a vacuum, its velocity $c = 1/\sqrt{\varepsilon_0\mu_0}$ because for the vacuum $\varepsilon = 1$ and $\mu = 1$. The ratio of velocities c and v is simply the *absolute refractive index* (n) of a material medium:

$$n = \frac{c}{v} = \sqrt{\varepsilon\mu}. \tag{1.2a}$$

For most substances the magnetic permeability μ is practically equal to 1, thus

$$n = \sqrt{\varepsilon}. \tag{1.2b}$$

At a fixed moment t the electromagnetic wave moving in a vacuum or a free space can be sketched as shown in Fig. 1.2, where the vectors **E** and **H** denote

---

[4] Bold-faced type as usual signifies the vectorial quantities.

the intensities of the electric and magnetic fields, respectively, while E_0 and H_0 are their *amplitudes*. The wave moves in the $+z$ direction. The vectors **E** and **H** vibrate with the same phase, they are perpendicular to each other and to the direction z, in which the *wave energy* propagates. This direction is fixed by the

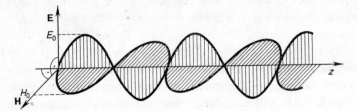

Fig. 1.2. Electromagnetic wave.

Poynting vector **P**, which expresses the quantity of wave energy that flows in a unit time through a unit area perpendicular to the direction of wave propagation. This vector is defined by the vectorial product of the electric and magnetic vectors: $\mathbf{P} = \mathbf{E} \times \mathbf{H}$. According to the SI of units, its value is expressed in watts per square metre (W/m^2). Let E, H, and P denote the values (lengths) of the vectors **E**, **H**, and **P**, respectively, i.e., $E = |\mathbf{E}|$, $H = |\mathbf{H}|$, and $P = |\mathbf{P}|$. Then it may be shown that

$$H\sqrt{\mu_0 \mu} = E\sqrt{\varepsilon_0 \varepsilon}, \tag{1.3}$$

and

$$P = \sqrt{\frac{\varepsilon_0 \varepsilon}{\mu_0 \mu}} E^2, \tag{1.4a}$$

or

$$P = \sqrt{\frac{\mu_0 \mu}{\varepsilon_0 \varepsilon}} H^2. \tag{1.4b}$$

Referring to geometrical optics, it can be stated that light rays are traced by Poynting vectors.

In general, one accepts that only the electric component of the optical electromagnetic waves is of direct importance for the phenomenon of light, since the magnetic component does not act on the retina of the eye, photographic emulsion or other photodetectors. Besides, the possible effects of the magnetic component occurring in some optical phenomena are so weak as to be negligible in comparison with the effects of the electric component. Consequently, a light wave is usually represented as a transverse electromagnetic wave motion the magnetic component of which is ignored. Thus in the optical spectral domain of electro-magnetic waves the *electric vector* **E** is also called the *light vector*.[5] Its wave motion is described only by Eq. (1.1a).

Let us suppose that the electric (light) vector vibrates in a fixed plane. This kind of vibration is called *linearly polarized,* and in this case Eq. (1.1a) reduces simply to

$$\frac{\partial^2 E}{\partial z^2} - \frac{1}{v^2}\frac{\partial^2 E}{\partial t^2} = 0, \tag{1.5}$$

where v is the *velocity of light*, z denotes the direction of light propagation, and t represents the time involved. One of the particular solutions of the above equation is a sine function, which describes the *plane light wave*, or the *collimated (parallel) beam of rays*, advancing in a vacuum along the z-axis. In general, this function may be written as

$$E = a\sin(\omega t - kz + \psi_0), \tag{1.6}$$

where E is the instant value of electric field intensity, a represents the *wave ampli-tude* (i.e., the maximum value of E), ω denotes the *angular frequency* of wave motion,[6] k is the *wave number*, and ψ_0 represents a possible *initial phase* of wave motion. The wave number[7] is defined as

$$k = \frac{2\pi}{\lambda}, \tag{1.7}$$

where λ is the wavelength. The angular frequency is related to the *wave period T, frequency v* $(v = 1/T)$, velocity v, wavelength λ, and wave number k as follows

$$\omega = \frac{2\pi}{T} = 2\pi v = \frac{2\pi v}{\lambda} = kv. \tag{1.8}$$

Equation (1.6) expresses a simple *harmonic wave motion* both at any single point z_0 of the wave path (z-axis) during time t, and at every point of this path for a fixed value of time (t_0). This kind of wave motion can be illustrated sche-

---

[5] In fact, there are some situations it is the electric induction vector **D** that should be considered as the light vector (**D** $= \varepsilon_0\varepsilon$**E**). However, in most microscopical research the vectors **E** and **D** are found to act identically, and thus the electric vector **E** will generally be treated as the light vector.

[6] Instead of angular frequency the terms *circular frequency, pulsation* or *pulsatance* are also in use.

[7] In spectroscopy the wave number is defined as $1/\lambda$, therefore the parameter $2\pi/\lambda$ is also called the *cyclic wave number*. In integrated optics and waveguide optics this parameter is usually called the *propagation constant*.

matically by means of two identical sine graphs, as shown in Figs. 1.3a and b, their periods being equal to T and λ, respectively. In the second case one can speak of the *spatial period* in contrast to the *temporal period T*. The graphs in Fig. 1.3 are a static representation of harmonic wave motion. A real dynamic process may be demonstrated by the continuous shift of an infinite sine curve along the z-axis in Fig. 1.3b. The initial phase ψ_0 in Eq. (1.6) is meaningless if a single wave is considered, but it is of basic importance when the superposition of two or more light waves is taken into account.

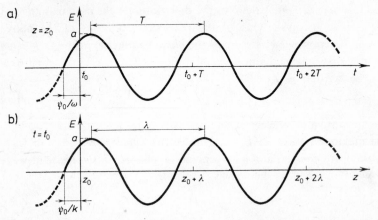

Fig. 1.3. Harmonic wave motion: a) at a single point of space, b) for a fixed value of time.

Let us leave aside the initial phase for the present. Then Eq. (1.6) may be written as

$$E = a\sin(\omega t - kz) = a\sin(\omega t - \psi), \qquad (1.9)$$

where

$$\psi = kz = \frac{2\pi}{\lambda}z, \qquad (1.10)$$

and represents the *phase* of the wave at a distance z measured from an initial point z_0 of the wave path. Expressions (1.6) and (1.9) represent the mathematical description of a *harmonic plane wave* which moves in the positive direction of the z-axis (Fig. 1.3b). The same wave motion travelling in the opposite direction, i.e., in the negative direction of the z-axis, is described by similar expressions, except that a plus sign instead of a minus sign is inserted before the phase ψ. Hitherto it has been assumed that a plane light wave moves in a vacuum. However, if it propagates in a material medium (isotropic and transparent for this wave), then an *optical path* (s) instead of a geometrical one (z) must be inserted into expressions (1.6) and (1.9). The optical path is defined as the product of the geometrical path and the refractive index of the medium, i.e., $s = nz$. Thus the wave phase ψ at a distance z (measured from an initial distance z_0) is not expressed by Eq. (1.10), but is equal to

$$\psi = \frac{2\pi}{\lambda} nz. \tag{1.11}$$

Now, it is interesting to note what energy is transferred by a plane light wave. To evaluate this quantity, it is sufficient to introduce Eq. (1.9) into Eq. (1.4a). Then

$$P = \sqrt{\frac{\varepsilon_0 \varepsilon}{\mu_0 \mu}}\, a^2 \sin^2(\omega t - kz). \tag{1.12}$$

The eye and other common photodetectors do not record P but rather a quantity called the *intensity* or *irradiance*,[8] denoted by I. This is defined as the time-averaged value of light energy reaching an area A perpendicular to the direction of wave propagation in a time unit. This quantity is then expressed by the following formula:

$$I = \frac{1}{T} A \int_0^T P\,dt = \frac{1}{T} A \sqrt{\frac{\varepsilon_0 \varepsilon}{\mu_0 \mu}}\, a^2 \int_0^T \sin^2(\omega t - kz)\,dt. \tag{1.13a}$$

Since the integral is equal to $T/2$, the intensity of light

$$I = A \sqrt{\frac{\varepsilon_0 \varepsilon}{\mu_0 \mu}}\, \frac{a^2}{2} = Ca^2, \tag{1.13b}$$

where $C = (A/2)\sqrt{\varepsilon_0 \varepsilon/\mu_0 \mu}$ is a constant factor. It appears, therefore, that light intensity is directly proportional to the square of the amplitude of the light wave. In most optical considerations the constant factor C is omitted and one assumes that

$$I = a^2. \tag{1.13c}$$

It is well known that many mathematical operations become simpler when they are expressed in exponential instead of trigonometrical functions. There-

---

[8] *Intensity* is now being replaced in optics by *irradiance* (see, e.g., Ref. [1.21]).

fore, it is useful to introduce an exponential representation into the wave theory of light by using *complex numbers* of the form

$$w = x+iy = ae^{i\alpha} = a(\cos\alpha+i\sin\alpha), \tag{1.14}$$

in which a and α are the *modulus* (or *amplitude*) and the *argument*[9] (or *phase*) of the complex number, respectively, and i is the imaginary unit $(i = \sqrt{-1})$. The first part $x = a\cos\alpha$ and the second $y = a\sin\alpha$ are called the *real* and *imaginary parts* (or *components*) of the complex number w. They are denoted by the symbols Re and Im. Thus the above formula may also be written as $w = \text{Re}\{w\}+i\text{Im}\{w\}$. It is worth noting that w is simply a real number if $y = \sin\alpha = 0$ (i.e., if the argument $\alpha = 0, \pm\pi, \pm 2\pi, ...$) and w is an imaginary number when $x = \cos\alpha = 0$ (i.e., if $\alpha = \pm\pi/2, \pm 3\pi/2, ...$).

According to the formula (1.14), the motion of a plane light wave, described previously by Eq. (1.6), can now be written as

$$\mathscr{E} = ae^{i(\omega t-kz+\psi_0)}. \tag{1.15}$$

It is understood that only the real component,

$$\text{Re}\{\mathscr{E}\} = E = a\cos(\omega t-kz+\psi_0), \tag{1.16}$$

of the expression (1.15) has physical significance and represents the wave motion under consideration, while the imaginary component has a solely formal (mathematical) denotation. Besides, the latter can easily be separated, if necessary, from the former, because when adding two or more complex quantities one adds their real and imaginary components separately; in other words, the real component of a sum is equal to the sum of real individual components. Hence to obtain the resultant of a number of superposed waves, one can add their complex expressions and then take the real component of the sum. Such a summation frequently takes the form of integration, which can be performed more easily if complex quantities are used. As can be seen, Eq. (1.16) is a cosine function, whereas Eqs. (1.6) and (1.9) are sine functions. This difference has no significance because it is enough to assume an initial phase $\psi_0+\pi/2$ instead of ψ_0 and at once a sine function is transformed into a cosine function, and vice versa.

A great advantage of complex representation is that the spatial and time parts of a wave motion can be split, since Eq. (1.15) may also be written in the form

$$\mathscr{E} = ae^{-i(kz-\psi_0)}e^{i\omega t} = \mathscr{A}e^{i\omega t}, \tag{1.17}$$

[9] The modulus $|w|$ of the complex number $w = x+iy$ is by definition equal to $\sqrt{x^2+y^2}$ and the argument of w (arg w) is defined as $\arctan(y/x)$. If w is given as $w = ae^{i\alpha}$, we have $|w| = a$, and arg $w = \alpha$.

in which the quantity

$$\mathscr{A} = a\mathrm{e}^{-\mathrm{i}(kz-\psi_0)}$$ (1.18)

represents the *complex amplitude* and $\mathrm{e}^{\mathrm{i}\omega t}$ is a harmonic *time factor*. The modulus and argument of the complex amplitude express the amplitude and phase of the light wave, i.e., $|\mathscr{A}| = a$ and $\arg\mathscr{A} = kz-\psi_0$. When a number of light vibrations of the same frequency ω is added, the time factor $\mathrm{e}^{\mathrm{i}\omega t}$ is a common quantity and can be omitted in the operation of summation. Thus the resultant complex amplitude is simply the sum of the constitutive complex amplitudes, and the modulus and argument of the resultant complex amplitude give, respectively, the amplitude and phase of the resultant vibration. This property of the complex amplitude is very useful when phenomena of light interference are considered.

It is often sufficient to know the intensity of light, i.e., the squared amplitude of the light wave. This quantity is usually given by the product of the complex amplitude $\mathscr{A}$ and its *conjugate value*,[10] i.e.,

$$I = \mathscr{A}\mathscr{A}^* = |\mathscr{A}|^2 = a^2.$$ (1.19)

Apart from plane light waves, *spherical waves* are of fundamental importance in the analysis of optical phenomena and different light processes. In particular, the spherical light wave is radiated by a single point source. Wave energy propagates within a full solid angle, and so the intensity of the electric field is a function of the radius r with the origin at the source. If the time dependence is of the form $\mathrm{e}^{\mathrm{i}\omega t}$, the spherical light wave expanding from the point source can be expressed as

$$\mathscr{E} = \frac{a_0}{r}\,\mathrm{e}^{\mathrm{i}(\omega t-kr)},$$ (1.20)

where a_0 represents the amplitude at the origin of the wave motion. The factor $1/r$ reduces the light intensity I of such a wave as $1/r^2$. Expression (1.20) describes an outgoing spherical wave. However, if the minus sign in the argument of this expression is replaced by a plus sign, then Eq. (1.20) will describe an incoming spherical wave. Its light intensity I will, of course, increase as r decreases.

---

[10] The complex conjugate of the complex number w is denoted by w^* and is obtained by replacing each i with $-\mathrm{i}$ in Eq. (1.14). Hence, if $w = x+\mathrm{i}y = a\mathrm{e}^{\mathrm{i}\alpha}$, then $w^* = x-\mathrm{i}y = a\mathrm{e}^{-\mathrm{i}\alpha}$. It is easy to show that $ww^* = x^2+y^2 = |w|^2 = a^2$. For more details about the algebra of complex numbers, the reader is referred to other books, e.g., to [1.3].

1.2.2. *Passage of light through an absorbing isotropic medium*

Hitherto it has been assumed that light waves propagate in a vacuum or in an isotropic and transparent material medium. Now, it is interesting to see how light waves propagate in an absorbing medium. Let the medium obey *Lambert's law of absorption*:[11]

$$I = I_0 e^{-Kz}, \tag{1.21a}$$

where I_0 and I are the intensities of a parallel light beam which first enters the medium and then traverses a path of length z, whereas K is a constant and is called the *absorption coefficient*. It is sometimes more convenient to express the above relation in the form

$$I = I_0 e^{-(4\pi/\lambda_M)\varkappa z}, \tag{1.21b}$$

or

$$I = I_0 e^{-(4\pi/\lambda)\varkappa n z}, \tag{1.21c}$$

where λ_M and λ are the wavelengths of light in the medium and in vacuum, respectively, $\varkappa$ is the *extinction coefficient*,[12]

$$\varkappa = K \frac{\lambda_M}{4\pi}, \tag{1.22}$$

and n is the *refractive index* of the medium,

$$n = \frac{\lambda}{\lambda_M}. \tag{1.23}$$

The expression (1.23) results directly from Eq. (1.2a), but needs a commentary. When light enters into a material medium, the light velocity changes, but the light frequency ν remains constant because the frequency of a light wave is a property determined by the light source and is thus unaltered by the medium through which the wave travels. Light velocity in a vacuum can be expressed, of course, as

$$c = \frac{\lambda}{T} = \lambda\nu, \tag{1.24a}$$

---

[11] This law was discovered by Bouguer in 1729, and next rediscovered by Lambert. Consequently, it should rather be called *Bouguer's law*, but the name "Lambert's law of absorption" prevails.

[12] There is no general agreement as regards the names of different parameters employed for the description of light absorption phenomena. It is worth noting that $\varkappa$ is also called the *attenuation index* or *absorption index*, whereas it is the quantity $\mathscr{K} = n\varkappa$ that is said to be the absorption index. The latter quantity is, however, also called the absorption coefficient.

and because $1/T = v$ is a constant parameter, light velocity v in a material medium must be expressed as

$$v = \frac{\lambda_M}{T} = \lambda_M v, \tag{1.24b}$$

where λ_M is the light wavelength altered in proportion to the changed velocity of light. When Eqs. (1.24a) and (1.24b) are inserted into Eq. (1.2a), the relation (1.23) is obtained.

Leaving aside the initial phase ψ_0 and using Eqs. (1.7), (1.8), (1.23) and (1.24), the expression (1.15) may be written as

$$\mathscr{E} = a e^{i\omega[t-(n/c)z]}. \tag{1.25}$$

If, according to Lambert's law, the medium absorbs the light, the amplitude a decreases as the length of the path (z) increases. This dependence obeys Lambert's law but now in relation to the amplitude of the light wave. This modification of the law is achieved when the roots are extracted from the left-hand and right-hand parts of Eq. (1.21c), i.e.,

$$a = a_0 e^{-(2\pi/\lambda)\varkappa nz}, \tag{1.26}$$

where $a = \sqrt{I}$, $a_0 = \sqrt{I_0}$, and $e^{-(2\pi/\lambda)\varkappa nz}$ is equal, of course, $\sqrt{e^{-(4\pi/\lambda)\varkappa nz}}$. Inserting (1.26) into (1.25) and doing some simple transformations, the following expression can be written:

$$\mathscr{E} = a_0 e^{i\omega[t-(\tilde{n}/c)z]}, \tag{1.27}$$

where

$$\tilde{n} = n(1-i\varkappa) \tag{1.28a}$$

is called the *complex refractive index*. Sometimes this quantity is also defined as

$$n = n - i\mathscr{K}, \tag{1.28b}$$

where $\mathscr{K} = n\varkappa = K\lambda/4\pi$.

As can be seen, the motion of a plane light wave in an absorbing medium (which obeys Lambert's law) and that in a transparent medium resemble each other; Equations (1.25) and (1.27) are similar. The latter only contains the complex refractive index $\tilde{n}$ instead of n, but the real part of $\tilde{n}$ has a physical significance for the refraction of light, while the imaginary part represents the effect of light absorption.

1.2.3. Propagation of light in transparent anisotropic media

When a medium is anisotropic, its dielectric permittivity ε and magnetic permeability μ are no longer scalars but vectors, and therefore the velocity of light in such a medium depends on the direction of light propagation. Consequently, an analytical theory for wave propagation in anisotropic media is much more complicated than that for wave motion in isotropic media, and is beyond the scope of this book.[13] Hence only a short description of some basic phenomena of optical anisotropy will be given, concentrating in first place on transparent *birefringent crystals*.

Although the emphasis is given to homogeneous, non-conducting ($\sigma = 0$), electrically anisotropic but magnetically isotropic crystals, nevertheless media with magnetic anisotropy also exist. The effects of this anisotropy are, however, small and need not be considered. A basic difference, from the point of view of light propagation, between an isotropic and an anisotropic medium is shown in Fig. 1.4, which illustrates the directions of the electromagnetic field vectors

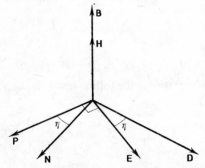

Fig. 1.4. Directions of the electromagnetic vectors (**E, D, H, B**), of the Poynting vector (**P**) and of the wave normal (**N**) in an electrically anisotropic medium.

E, D, H, and **B,** as well as the vectors of the wave energy flow (the Poynting vector) **P** and of the normal **N** of the wavefront. As can be seen, the vectors **E** and **D** are no longer parallel to each other in the anisotropic medium, and form an angle η. The same angle is to be found between **N** and **P**. Consequently, **D, H,** and **N,** on the one hand, as well as **E, H,** and **P,** on the other, form two systems of orthogonal vectors. Both systems are rotated relatively to one another by angle η around their common vector **H**. The latter coincides, in both media, with **B**.

<hr>

[13] The reader interested in this field is referred to more advanced books on optics, e.g., [1.1] and [1.2], or to monotopical treatises [1.5] and [1.6].

There are two basic groups of electrically anisotropic crystals. The first includes those which have two directions, known as *optic axes*, along which light waves polarized in any azimuth travel with the same single velocity, whereas for any other direction there are two velocities. The other group contains those crystals which have a single optic axis, and are therefore known as *uniaxial crystals*, whereas the former are called *biaxial crystals*. Although of lesser interest to crystallographers, uniaxial crystals are more frequently used in microscopes and other optical instruments. The best known are quartz and calcite (Iceland spar). Their basic optical properties, as well as those of some other birefringent crystals, are listed in Table 1.2.

At the beginning of our discussion of light propagation in electrically aniso-

TABLE 1.2

Basic optical properties of some characteristic birefringent crystals (at 20°C)

Name and chemical symbol of crystal	Principal refractive indices (for a spectral line marked in brackets)	Mean dispersion $n_F - n_C$	Abbe number V	Wavelength range of transmittance [μm]
Quartz SiO_2	$n_o = 1.54424$ (D) $n_e = 1.55335$ (D)	0.0078 0.0080	70 69	0.18–4
Calcite $CaCO_3$	$n_o = 1.65835$ (D) $n_e = 1.48640$ (D)	0.0134 0.0072	50 68	0.2–2
ADP* $NH_4H_2PO_4$	$n_o = 1.527$ (e) $n_e = 1.481$ (e)	0.010 0.007		0.2–1.1
KDP** KH_2PO_4	$n_o = 1.5099$ (d) $n_e = 1.4711$ (d)	0.0111 0.0073	56 65	0.19–1.16
Sodium nitrate $NaNO_3$	$n_o = 1.5874$ (D) $n_e = 1.3361$ (D)			
Rutile TiO_2	$n_o = 2.649$ (e) $n_e = 2.946$ (e)			
Mica (muscovite) $KAl_2(OH, F)_2[AlSi_3O_{10}]$	$n_\gamma = 1.5977$ (D) $n_\beta = 1.5936$ (D) $n_\alpha = 1.5601$ (D)			0.4–5.5
Gypsum $CaSO_4 \cdot 2H_2O$	$n_\gamma = 1.5296$ (D) $n_\beta = 1.5226$ (D) $n_\alpha = 1.5205$ (D)			

 * Ammonium dihydrogen phosphate
** Potassium dihydrogen phosphate

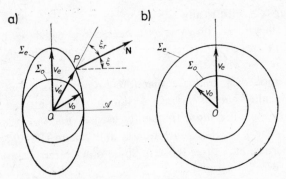

Fig. 1.5. Wave surface for a negative uniaxial crystal: a) section in a plane parallel to the optic axis ($\mathscr{A}$), b) section in a plane perpendicular to the optic axis.

tropic media, let any point O be chosen as a hypothetical light source inside a uniaxial crystal (Fig. 1.5). The source radiates a monochromatic wave which spreads out, in general, as two slightly split wavefronts Σ_o and Σ_e. The first one (Σ_o) is spherical and is called the *ordinary wave*, while the other (Σ_e) has the form of a biaxial ellipsoid and is called the *extraordinary wave*. The ellipsoid touches the sphere Σ_o along the optic axis of crystal,[14] and can be both inscribed into the sphere and described on it. Both waves, ordinary Σ_o and extraordinary Σ_e, are polarized linearly and the directions of their vibrations are perpendicular to each other.

One of the axes of the biaxial ellipsoid coincides with the direction of the optic axis of crystal ($\mathscr{A}$). Only in this direction both waves spread with the same speed equal to the velocity v_o of the ordinary wave Σ_o, but along any other direction the velocity v'_e of the extraordinary wave Σ_e is greater (Fig. 1.5) or smaller (Fig. 1.6) than v_o. An extreme value v_e of v'_e is along directions perpendicular to the optic axis $\mathscr{A}$. The velocity v'_e can be readily derived from the relation

$$v'^2_e = v^2_o\cos^2\xi + v^2_e\sin^2\xi, \tag{1.29}$$

where ξ is the angle between the optic axis $\mathscr{A}$ and the wavefront normal **N**. If $v_e > v_o$, one is said to have a *negative uniaxial crystal* (Fig. 1.5). On the other hand, if $v_e < v_o$, one has a *positive uniaxial crystal* (Fig. 1.6). The velocities v_o and v_e are two *principal velocities*. The corresponding *principal refractive indices*, known as the *ordinary index* n_o and the *extraordinary index* n_e, are therefore given by

---

[14] The sphere and the ellipsoid never quite touch along the optic axis if a crystal displays what is known as *optical activity*. Quartz, for example, is such an optically active crystal.

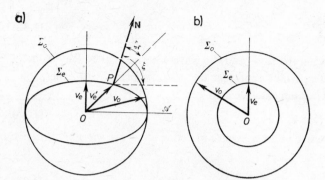

Fig. 1.6. Wave surface for a positive uniaxial crystal: a) section in a plane parallel to the optic axis ($\mathscr{A}$), b) section in a plane perpendicular to the optic axis.

$$n_0 = \frac{c}{v_o}, \qquad\qquad (1.30\text{a})$$

and

$$n_e = \frac{c}{v_e}, \qquad\qquad (1.30\text{b})$$

where c is the velocity of light in vacuum. For intermediary velocities v'_e of the extraordinary wave, i.e., for $v_o < v'_e < v'_e$ or $v_o > v'_e > v_e$, one has

$$n'_e = \frac{c}{v'_e}, \qquad\qquad (1.30\text{c})$$

and consequently to Eq. (1.29) the refractive index n'_e can be derived from the relation

$$\frac{1}{n'^2_e} = \frac{\cos^2\xi}{n^2_o} + \frac{\sin^2\xi}{n^2_e}. \qquad\qquad (1.31)$$

Now, the concept of the light ray defined at the beginning of this chapter must be redefined for the extraordinary wave, because extraordinary rays are perpendicular to the wavefront Σ_e only along the optic axis of crystal and in a particular section as shown in Figs. 1.5b and 1.6b. Moreover, these rays are never perpendicular to their wave surface. In general, the ray is represented by the direction of energy flow, i.e., by the Poynting vector, such as OP in Fig. 1.5 (or Fig. 1.6). When the light wave expands, the point P moves with a speed v'_{er}, which can be called *ray velocity* (or *energy velocity*). This quantity is intermediate between v_o and v_e, and may be determined from the relation

$$\frac{1}{v_{er}'^2} = \frac{\sin^2\xi_r}{v_e^2} + \frac{\cos^2\xi_r}{v_o^2}, \tag{1.32}$$

where ξ_r is the angle between the ray e and the wavefront normal $\mathbf{N}$. Ray velocity v_{er}' differs from the speed v_e' which represent the *phase velocity*.[15] In general, we may predicate the following dependence between v_{er} and v_e':

$$v_e' = v_{er}'\cos\xi_r. \tag{1.33}$$

By analogy with the refractive index n_e', one can define a *ray* (or *energy*) *refractive index*:

$$n_{er}' = \frac{c}{v_{er}'} = n_e'\cos\xi_r. \tag{1.34}$$

According to the above explanation, the surfaces Σ_o and Σ_e in Figs. 1.5 and 1.6 represent two particular sections of the *ellipsoid of the wavefront* (or *wave*) *normals*.[16] These are two shells, one of which is a sphere of radius v_o, while the other is ovaloid (the surface of revolution). The sphere Σ_o represent the ordinary wave with a velocity v_o independent of the direction of light propagation, whereas the ovaloid Σ_e represents the extraordinary wave with velocity v_e' depending on the angle ξ between the direction of the wavefront normal $\mathbf{N}$ and the optic axis $\mathscr{A}$. Both velocities are equal to each other when $\xi = 0$, i.e., when $\mathbf{N}$ and $\mathscr{A}$ are parallel. In this particular case, ray velocity is also equal to phase velocity.

By analogy with the wave-normal surface, one may also speak about the *ray surface* determined by the end points of the position vectors of length v_{er}' plotted from a fixed origin in all directions of a crystal.

These surfaces are useful for qualitative discussions of birefringence phenomena, but for this same purpose, as well as for a quantitative analysis, other surfaces known as *refractive index surfaces* (or, in short, *index surfaces*) will be found more convenient (Figs. 1.7 and 1.8). According to Eqs. (1.30), they have an inverse relationship to the wave-normal surfaces, and consist of a sphere of radius n_o and a biaxial ellipsoid. The ellipsoid touches the sphere along the optic axis $\mathscr{A}$. One of the semi-axes of the ellipsoid is equal to the ordinary refractive index n_o, and the other is equal to the extraordinary refractive index n_e.

---

[15] Born and Wolf stated in their book [1.1] on p. 609 that ray velocity as derived from the Poynting vector shares with it a certain degree of arbitrariness. However, it is a useful notion although, like phase velocity (also called *wave normal velocity*), it has "no directly verifiable physical significance".

[16] These surfaces are also called the *index ellipsoid* or the *reciprocal ellipsoid*. However, the rather vague term *optical indicatrix* is widely used. Born and Wolf [1.1] suggest that the best name is *ellipsoid of wave normals* or, in short, *normal surface*.

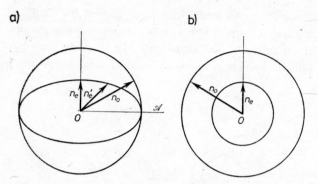

Fig. 1.7. Index surfaces for a negative uniaxial crystal ($n_e < n_o$).

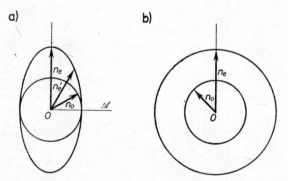

Fig. 1.8. Index surfaces for a positive uniaxial crystal ($n_e > n_o$).

In all these diagrams the differences between n_o and n_e are very much greater than those encountered in real crystals.

Biaxial crystals have three refractive indices: n'_α, n_β, and n'_γ. The first one (n'_α) varies from a minimum value n_α to n_β, whereas the third one (n'_γ) varies from a maximum value n_α to n_β. Thus these crystals possess three principal refractive indices: the *low index* n_α, *medium index* n_β, and *high index* n_γ,[17] which fulfil the relation: $n_\alpha < n_\beta < n_\gamma$. Consequently, the index surfaces of biaxial crystals are more complicated than those of uniaxial crystals. They consist of two interpenetrating sheets intersecting each other at the optic axes. Figure 1.9 shows three characteristic sections of these surfaces. Each of these sections

[17] There is no general agreement as regards the denotation of these indices. The symbols n_x, n_y, n_z or n_p, n_m, n_g are also in use. The latter originate from the French nomenclature (the suffixes p, m, and g denote *petit*, *moyen*, and *grand*).

consist of a circle and ellipse. The most interesting is the section with a plane
containing the optic axes $\mathscr{A}_1$ and $\mathscr{A}_2$ (Fig. 1.9b). This section shows an additional
basic parameter of biaxial crystals, viz. the acute angle 2χ between the optic
axes. If $2\chi < 90°$, the crystal is known as a *positive biaxial crystal*, but when

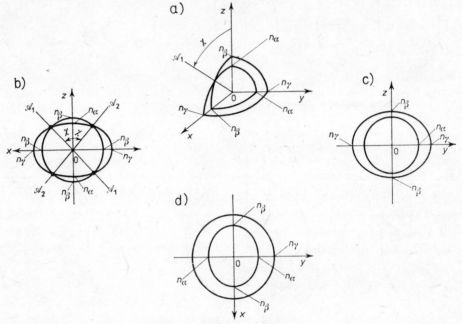

Fig. 1.9. Sections of the index surfaces for a biaxial crystal with $n_\alpha < n_\beta < n_\gamma$.

$2\chi > 90°$, the biaxial crystal is *negative*. The angle 2χ relates to the principal
refractive indices by the following formula:

$$\frac{1}{n_\beta^2} = \frac{\sin^2\chi}{n_\gamma^2} + \frac{\cos^2\chi}{n_\alpha^2} = \cos^2\chi \left(\frac{1}{n_\alpha^2} - \frac{1}{n_\gamma^2} \right) + \frac{1}{n_\gamma^2}. \tag{1.35}$$

When 2χ is equal to zero or $180°$, the two optic axes, $\mathscr{A}_1$ and $\mathscr{A}_2$, coincide, and
the crystal becomes a positive or negative uniaxial crystal, respectively. In the
first case ($\chi = 0$), $\cos\chi = 1$, and $n_\beta = n_\alpha = n_o$, while in the second case ($\chi = 90°$),
$\cos\chi = 0$, and $n_\beta = n_\gamma = n_e$.

A very important feature of anisotropic crystals is their *birefringence* or
double refraction. For uniaxial crystals, the birefringence (B) is defined as the
difference between the principal refractive indices n_e and n_o,

$$B = n_e - n_o. \tag{1.36}$$

This difference is the *principal birefringence*. It characterizes the crystal as an anisotropic material and occurs when a light ray traverses the crystal perpendicularly to its optic axis. For any other direction of light propagation, the birefringence is smaller than $n_e - n_o$ because it is equal to $n_e' - n_o$.

For biaxial crystals, one distinguishes three principal values of birefringence: $n_\gamma - n_\alpha$, $n_\gamma - n_\beta$, and $n_\beta - n_\alpha$. In practice, the first value is the most important, as it enables the maximum birefringence of a biaxial crystal to be shown. This value occurs when a light ray traverses the crystal normally to its plane containing the optic axes. For other directions of light propagation, the birefringence of the biaxial crystal is equal to $n_\gamma' - n_\alpha'$. In particular cases, $n_\gamma' - n_\alpha' = n_\gamma - n_\beta$, or $n_\gamma' - n_\alpha' = n_\beta - n_\alpha$.

1.3. Monochromacy and coherence of light, and related notions

Monochromacy and coherence of light are among the basic concepts of physical optics, and are frequently dealt with in both theoretical and practical light microscopy. There is a close relationship between both concepts, as well as between them and some quantities which characterize the propagation of real light waves in real media. Among these quantities are, above all, the *group velocity of light* and *spectral dispersion of refractive index*.

1.3.1. Monochromatic and heterochromatic light

Equations (1.6) and (1.9) or (1.15) represent a *simple harmonic wave* motion travelling in the positive z-direction. At any given value of $z = z_0$ the wave disturbance continues in time from $t = -\infty$ to $t = +\infty$, and at any instant $t = t_0$ the wave extends in space from $z = -\infty$ to $z = +\infty$. That infinity in time and space is symbolically marked by the pecked ends of sine curves in Fig. 1.3. A wave of this type has only a single wavelength λ_0 and a single frequency ν_0, and is said to be *monochromatic*. However, such an idealization can never be realized in practice; perfectly monochromatic light does not exist. Even light waves emitted by individual atoms of the same kind differ slightly from each other in wavelength and frequency. Besides, each emission of light does not repeat itself periodically to infinity, but takes the form of a train of sine waves more likely to start and stop gradually as shown in Fig. 1.10. The disturbance represented in this figure is no longer a simple harmonic wave motion since it follows a sine curve over a finite range only, gradually starting from and stopping to zero beyond that range. Strictly speaking, such a sine curve represents a non-periodic wave motion, because it does not repeat itself to infinity. However, it can be shown, by using the Fourier transform (see, e.g., Ref. [1.4], that the form of waves as presented in Fig. 1.10 can be treated as a summation of a continuous

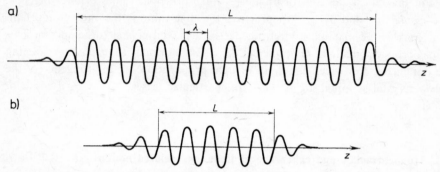

Fig. 1.10. Wave trains of finite length: a) nearly pure sine wave, b) very impure sine wave.

series of simple harmonic waves of different wavelengths and frequencies the spectral distribution of which has the form of a Gaussian curve (Fig. 1.11). A wave may be said to be *almost monochromatic (quasi-monochromatic)* if wavelengths λ (or frequencies ν) differ appreciably from zero only within a narrow range around a predominant wavelength λ_0 or predominant frequency ν_0. In this case one usually speaks of a wave group or a wave packet. Its monochromacy is defined by the spectral half-width $\Delta\lambda$ or $\Delta\nu$. The greater the length L of a wave packet

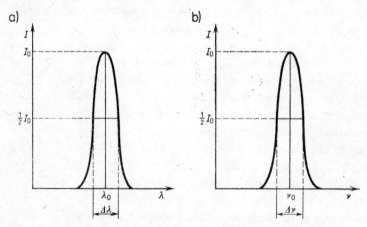

Fig. 1.11. Wavelength (a) and frequency (b) distribution of a wave train of finite length.

(Fig. 1.10), the smaller is the spectral half-width, thus the light is more monochromatic.

Sometimes it is useful to express the monochromacy of light by a normalized quantity

$$M = \frac{\Delta\lambda}{\lambda_0},$$ (1.37a)

or

$$M = \frac{\Delta\nu}{\nu_0}.$$ (1.37b)

Some authors call M the degree of light monochromacy, but a more consistent name for this quantity would rather be the degree of heterochromacy (non-achromacy) because $M \to 0$ as the spectral half-width $\Delta\lambda \to 0$.

In general, radiation from any real light source is *heterochromatic* or at most quasi-monochromatic. When analysed by means of a spectrograph (or spectroscope), it produces either a *continuous*, a *band*, or a *line spectrum*. The latter is a characteristic of elements and their isotopes, and is emitted by atoms in a gaseous state. When the gas is excited, by a glow discharge, for example, it radiates light quanta of discrete energy $W = h\nu$,[18] which are displayed by the spectrograph as bright lines known as *emission spectral lines*. Three of the simplest atomic line spectra are shown in Fig. I,[19] whereas Table 1.3 specifies the most important spectral lines which are used as monochromatic or quasi-monochromatic light in much optical and microscopical research. However, neither these or other spectral lines represent a single wavelength λ_0 or a single frequency ν_0. In reality, the intensity distribution $I(\lambda)$ or $I(\nu)$ of any spectral line is described more or less by Gaussian curves as shown in Fig. 1.11. In this case $\Delta\lambda$ or $\Delta\nu$ are called the *half-width of spectral line* or, for short, the *linewidth*. That and some other parameters of several spectral lines are given in Table 1.4.

If an unexcited gaseous substance is in the path of the white light, which produces a continuous spectrum (see Fig. I, top), dark spectral lines appear instead of bright ones. These constitute a *line absorption spectrum*, whose structure is, in general, simpler than that of an emission spectrum.

Band spectra are produced by compounds in a gaseous state and characterize molecules, whereas continuous spectra are a feature of incandescent solids, liquids, and high pressure gases. The former have a structure resembling line spectra, in

[18] The quantity h is *Planck's constant* equal to $6.626176 \times 10^{-34} \ \mathrm{J \cdot Hz^{-1}}$.
[19] Figures numbered with Roman numerals are assembled at the end of the book.

TABLE 1.3

The most useful spectral lines

Denotation	Wavelength [nm]	Colour	Element
A	768.2*	Red	Potassium K
r	706.5	Red	Helium He
C	656.3	Red	Hydrogen H
C′	643.9	Red	Cadmium Cd
D	589.3**	Yellow	Sodium Na
d	587.6	Yellow	Helium He
e	546.1	Green	Mercury Hg
F	486.1	Blue	Hydrogen H
F′	480.0	Blue	Cadmium Cd
g	435.8	Violet	Mercury Hg
G	434.0	Violet	Hydrogen H
h	404.7	Violet	Mercury Hg

* Two close lines with wavelength $\lambda_1 = 769.9$ nm and $\lambda_2 = 766.4$ nm.
** Double line with wavelength $\lambda_1 = 589.6$ nm and $\lambda_2 = 589.0$ nm.

TABLE 1.4

Monochromacy and coherence of some light sources and spectral lines

Light source or spectral line	Peak (mean) wavelength λ_0 [nm]	Linewidth $\Delta\lambda$ [nm]	Linewidth $\Delta\nu$ [s^{-1}]	Coherence length l_{coh}	$N = \dfrac{l_{coh}}{\lambda_0}$
Visible sun light	550	200	2×10^{14}	1.5 μm	3
Double line of sodium	589.3	0.6	5×10^{11}	0.6 mm	10^3
Green line of mercury Hg^{198}	546.1	0.5×10^{-3}	6×10^8	55 cm	10^6
Orange line of krypton Kr^{85}	606	0.4×10^{-3}	3.6×10^8	85 cm	1.4×10^6
He-Ne laser with stabilized frequency	632.8	0.3×10^{-5}	5×10^6	60 m	10^8
He-Ne laser, special performance	1132	0.3×10^{-8}	10^3	300 km	3×10^{11}
Ruby laser, free-running	694.3	0.7	4.3×10^{11}	0.7 mm	10^3
Ruby laser, single mode	694.3	2.3×10^{-4}	1.3×10^8	2 m	3.2×10^6

which the lines are concentrated in groups. When the compound becomes dissoci-ated, the line spectra of its constituent elements can be revealed. The continuous spectrum, on the other hand, has no detectable structure and contains radiation of all possible wavelengths. Such radiation is emitted, for instance, by the well-known halogen (quartz-iodine) lamps (Fig. 1.12).

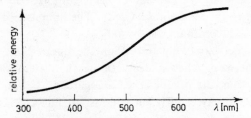

Fig. 1.12. Emission spectrum of a tungsten–halogen lamp.

Microscopy makes wide use of some light sources whose spectrum shows several peaks against a continuous background. Each of these peaks can easily be extracted by means of a suitable optical filter, allowing a quasi-monochromatic light to be obtained. Such sources include, for instance, the high-pressure mercury lamp HBO. Its emission spectrum is shown in Fig. 1.13.

Both the emission and absorption spectra constitute a basis for the emission and absorption *spectral analysis* of elements and compounds. These two tech-

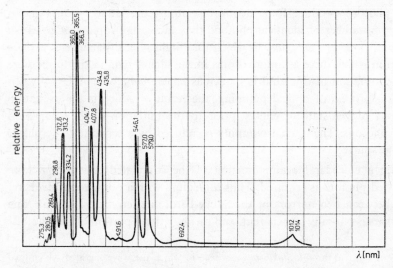

Fig. 1.13. Emission spectrum of a high pressure mercury lamp (HBO).

niques form an important field of quantitative microscopy, although absorption spectral microanalysis is predominant. The latter is covered by Chapter 14.

A particular kind of radiation provides *luminescence*. This phenomenon is a result of a specific interaction between light and matter (solids, liquids, and gases). The interaction depends on light absorption followed by an emission of photons, the wavelength of which is longer than that of the light originally absorbed. That wavelength sequence is known as *Stokes' law*.[20] The luminescent light produces either a line, a band, or a continuous spectrum.

If the emission only occurs immediately after light absorption, the luminescence is specified as the *fluorescence*. Otherwise, when a delay (more than 10^{-8} s) occurs between emission and absorption, the luminescence is regarded as *phosphorescence*. Phosphorescent emission decays slowly, in contrast to fluorescent emission, which vanishes suddenly when the excitation is interrupted. The phenomenon of fluorescence is the basis of fluorescence microscopy, which is covered in Chapter 9.

The ordinary process of light absorption and emission is as follows: photons first excite atoms from a lower to a higher state of energy, and are then emitted spontaneously and without any phase correlation when the atoms fall from the higher to the lower state of energy. The situation is, however, quite different when an excited atom is struck by an identical photon such as may be emitted by the atom. In this case two photons are emitted, both having the same frequency and the same phase. If there is a strong accumulation of identically excited atoms, these two photons produce four identical photons, the latter produce more, and so on, the number of identical photons thus increasing rapidly. A result of this process is *stimulated emission*, in which total light intensity increases if there is a medium in which there are more atoms in the higher than in the lower state of energy. A situation of this kind is known as *population inversion*. Usually it is obtained by a process of optical pumping, the result of which is the excitation of atoms to a higher state of energy prior to the initiation of the stimulated emission of photons. The optical pumping is induced by a separate (external or internal) light source.

To obtain a considerable amplification of the stimulated emission, a long path of emitted photons is required. Thus a medium with a population inversion is enclosed between two mirrors of the Fabry-Pérot interferometer. At least one of these mirrors is partially transparent, and through this a portion of the photons

[20] It is also known as the *anti-Stokes luminescence*, in which the wavelength of the emitted light is smaller than that of the absorbed light.

from the stimulated emission passes outside. This is in fact the principle on which the *laser*[21] is based.

Actually there are many kinds and types of lasers. The principle on which two of them are constructed is shown in Figs. 1.14 and 1.15. The first one represents a *continuous-wave gas laser*, whereas the other is a *pulsed solid-state laser*. Among

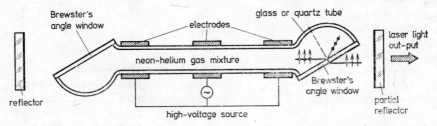

Fig. 1.14. Continuous-wave gas laser (helium–neon laser) with Brewster's angle windows.

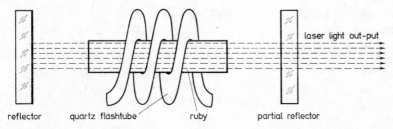

Fig. 1.15. Pulsed solid-state laser (ruby laser).

gas lasers the best known is the *helium–neon laser*. This is largely used in microscopy, especially in holographic microscopy. When appropriately prepared, it is a source of the most monochromatic and most coherent light. Table 1.5 specifies the main spectral lines of several lasers commercially available. For the interested microscopist, more details regarding lasers may be found in Ref. [1.11].

1.3.2. *Passage of heterochromatic light through a spectrally dispersive medium*

So far it has been assumed that the velocity of the passage of light waves through a medium is independent of the wave frequency. Such as assumption is true only if the light propagates in a vacuum, while in a real material, like glass, the velocity of light is different for various wave frequencies. Such a material is called *dispersive*.

---

[21] "Laser" is an abbreviation of the phrase: "light amplification by the stimulated emission of radiation".

TABLE 1.5

Main wavelengths (λ) emitted by different lasers

Lasing (active) material	Kind of emission*	λ [nm]		
Helium-neon (He–Ne)	cw	632.8		
Ruby ($Cr^{3+}: Al_2O_3$)	p	694.3		
Neodymium or neodymium-doped glass (Nd)	p	1058.0		
Neodymium-doped yttrium aluminium garnet (Nd: YAG)	p or cw	532.0** 1064.0		
Argon (Ar)	cw (or p)	334.0	351.1	363.8
		457.9	465.8	472.7
		476.5	488.0	496.5
		501.7	514.5	528.7
Helium–cadmium (He–Cd)	cw	325.0	441.6	
Carbon dioxide (CO_2 or CO_2+N_2+He)	cw (or p)	10600		
Nitrogen (N_2)	p	337.1		
Xenon (Xe)	cw	2026.1	3507.0	
Krypton (Kr)	cw	406.7	413.1	415.4
		568.2	647.1	676.4
Neon (Ne)	cw	332.4		
Copper vapour (Cu)	cw or p	510.5		
Dye	cw or p	300–900		
Semiconductor (GaAs)	cw or p	780–940		

   * cw = continuous-wave, p = pulsed,
  ** second harmonic.

Thus a wave packet constituting a disturbance of finite length moves in a dispersive medium with a velocity which differs from that of any simple harmonic wave component of the packet.

To explain this phenomenon, let us consider two superposed plane waves with only slightly different frequencies and velocities propagated in the positive z-direction. According to (1.9), these waves can be expressed as

$$E_1 = a\sin(\omega_1 t - k_1 z), \tag{1.38a}$$
$$E_2 = a\sin(\omega_2 t - k_2 z). \tag{1.38b}$$

By summation (1.38a) and (1.38b), a resultant wave arises which, after some mathematical transformations, may be expressed as

$$E = a'\sin(\omega t - kz), \tag{1.39}$$

where

$$a' = 2a\cos\tfrac{1}{2}(t\,d\omega - z\,dk) \tag{1.40}$$

is its amplitude, while ω and k denote the mean angular frequency and mean wave number, respectively, i.e., $\omega = (\omega_1 + \omega_2)/2$ and $k = (k_1 + k_2)/2$, and

$$d\omega = 2(\omega - \omega_1) = 2(\omega_2 - \omega),$$
$$dk = 2(k - k_1) = 2(k_2 - k).$$

For a given value of time $t = t_0$, a graphical representation of Eq. (1.39) is shown in Fig. 1.16. As can be seen, the resultant wave E, unlike the component waves

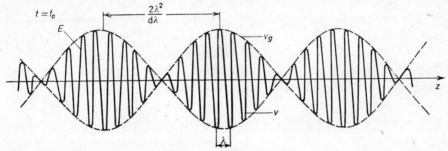

Fig. 1.16. Phase velocity (v) and group velocity (v_g) for two superposed sine waves of slightly different frequencies.

E_1 and E_2, no longer has a constant but a periodically varying amplitude. In Fig. 1.16 this variation is marked by the pecked cosine lines which constitute an envelope with repeated minimum and maximum values equal to $a' = 0$ and $a' = 2a$, respectively. The space interval between two successive peaks of the phase component of the resultant wave E is equal to $2\pi/k$ or λ, whereas such an interval between two adjacent peaks of the amplitude component (a') is equal to $4\pi/dk$ or $2\lambda^2/d\lambda$. Thus the resultant wave is divided into groups (packets) the amplitude maxima of which move with a velocity known as *group velocity*

$$v_g = \frac{d\omega}{dk}, \tag{1.41}$$

whereas the phase of this wave propagates, according to (1.8), with the *phase velocity*[22] $v = \omega/k = \lambda/T$. In other words, the wave shown as a continuous line

---

[22] The phase velocity is also called *wave velocity* [1.4, 1.7]. This quantity is used here so far, and when later in this work mention is made of the speed of light without any additional explanation what is meant is phase velocity.

in Fig. 1.16 propagates along the z-axis with velocity v, while the amplitude envelope moves with velocity v_g.

According to (1.7) and (1.8), $d\omega = d(2\pi v/\lambda) = 2\pi(\lambda dv - vd\lambda)/\lambda^2$, and $dk = d(2\pi/\lambda) = -(2\pi/\lambda^2)d\lambda$. Then Eq. (1.41) can be written

$$v_g = v - \lambda\frac{dv}{d\lambda}, \tag{1.42}$$

where $dv/d\lambda$ expresses a differential variation of wave velocity caused by a small variation of wavelength. This formula shows that v_g is smaller than v if v increases with λ, i.e., if the derivative $dv/d\lambda$ is positive, and vice versa, v_g is greater than v if $dv/d\lambda < 0$. In a particular case, when $dv/d\lambda = 0$, i.e., when there is no dispersion of a medium in which the light propagates, the group velocity is equal to phase velocity.

It is sometimes convenient to have another expression for v_g, i.e., one related to the refractive index n of a medium in which the light propagates. Taking into account Eq. (1.2a), the velocity $v = c/n$, and $dv = -(c/n^2)dn$. Then

$$v_g = v\left(1 + \frac{\lambda}{n}\frac{dn}{d\lambda}\right), \tag{1.43}$$

where $dn/d\lambda$ expresses the *spectral dispersion of refractive index*.[23] The above formula shows that v_g is smaller than v if the derivative $dn/d\lambda$ is negative, i.e., if the refractive index n decreases as λ increases. For transparent materials, such a dependence is typical, and thus described as *normal dispersion*, to distinguish it from *anomalous dispersion*. The latter is observed in spectral regions for which absorption bands occur. When λ is continuously varied and passes through an absorption band, the refractive index n increases with increasing λ, and the derivative $dn/d\lambda$ becomes positive. In this case the group velocity v_g is greater than the phase velocity v. Both velocities are the same only in free space because the refractive index of the latter is constant for all wavelengths and $dn/d\lambda = 0$. However, certain dielectrics exhibit nearly constant n over a considerable range of wavelengths, especially in the infrared region. But most media are dispersive, and pure monochromatic light is rare in nature, thus group velocity presents the major features of what we see.

Figure 1.16 represents a train of wave packets along the z-axis at a fixed

---

[23] In short, $dn/d\lambda$ is generally known as *dispersion*. In contrast to other refractive index dispersion quantities (see Subsection 1.5.3), the term *differential dispersion* is also in use. In the author's opinion the name "dispersion" relates rather to the variation of the refractive index n with wavelength λ. In particular, this variation expresses the derivative $dn/d\lambda$.

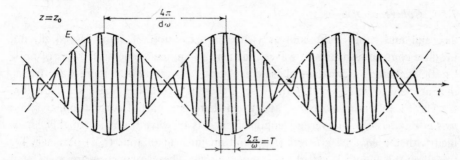

Fig. 1.17. The same wave train as in Fig. 1.16 but in course of time at a fixed point of space.

value of time $t = t_0$, whereas Fig. 1.17 shows the same train but in course of time at a given point $z = z_0$ of the wave path. In the latter case the time interval between two successive peaks of the phase component of the resultant wave E is equal to $2\pi/\omega$, while such an interval between two adjacent peaks of the amplitude envelope is equal to $4\pi/d\omega$. The squared amplitude gives the light intensity whose peaks re-occur at time intervals equal to $2\pi/d\omega$. These intervals cannot be observed as resolved peaks of intensity because they are extremaly short (equal to the order of nano- or even picoseconds).

Although the relations (1.42) and (1.43) are true in general, nevertheless the wave train considered above is not a fully adequate example because every real

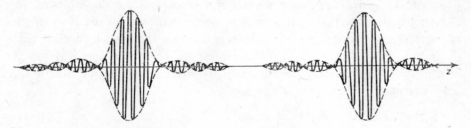

Fig. 1.18. Train of wave packets.

wave train consists of a great number of monochromatic waves whose angular frequencies (ω) and wave numbers (k) lie within intervals $\Delta\omega$ and Δk, respectively. Thus the real train is similar rather to that shown in Fig. 1.18. Such a train consists of wave packets whose resultant amplitude is non-zero only in certain intervals along the path of wave propagation (z-axis). The non-zero interval is called the *length of the wave packet*.

1.3.3. Coherence of light

Thermal and other light sources (with the exception of some lasers) do not produce continuous trains of waves, but tend to emit sequences of trains of finite length. If there is a constant phase correlation between two or more wave trains, they are said to be *coherent*. Otherwise, if there is a complete phase decorrelation, the wave trains are *incoherent*, and when there is a partial correlation, they are *partially coherent*. An average length of packets of wave trains is roughly identified with the *coherence length* (l_{coh}), and an interval of time (t_{coh}) necessary for travelling this length is called the *coherence time*. These quantities are associated with each other because

$$l_{coh} = vt_{coh}, \tag{1.44}$$

where v is the velocity of light.

The coherence of light is closely related to light monochromacy. One can prove that

$$\Delta \nu = \frac{1}{t_{coh}} = \frac{v}{l_{coh}}, \tag{1.45a}$$

and

$$\Delta \lambda = \frac{\lambda_0^2}{l_{coh}}, \tag{1.45b}$$

where $\Delta \nu$, $\Delta \lambda$, and λ_0 are the parameters of monochromacy earlier defined. The above relations show that light is more monochromatic the greater is its length of coherence, and vice versa, the coherence is better the more monochromatic the light is. Sometimes it is useful to express the coherence length by means of a number which defines how much l_{coh} is greater than the wavelength λ_0. If N is that number, then

$$l_{coh} = N\lambda_0, \tag{1.46a}$$

$$\Delta \nu = \frac{\nu_0}{N}. \tag{1.46b}$$

and

$$\Delta \lambda = \frac{\lambda_0}{N}. \tag{1.46c}$$

For some light sources and spectral lines the above defined parameters are given in Table 1.5.

The coherence length l_{coh} and coherence time t_{coh} relate to the so-called *temporal coherence*, which may be considered as a characteristic of a light field along the direction of its propagation. However, there is another type of coherence, known as *spatial coherence*, which may be regarded as a characteristic of the light field in a transverse direction to the direction of light propagation [1.8]. What is the difference between these two notions? To explain the distinction, let $E_1(P_1; t)$ and $E_2(P_2; t)$ denote the wave vibration of a light field at two arbitrary points P_1 and P_2 of a space at the observation time t. If a phase difference between E_1 and E_2 is independent of t for all points of the space, the light field is spatially coherent. Next, let $E_1'(P; t_1)$ and $E_2'(P; t_2)$ denote the same vibration of the light at an arbitrary point P of the space at two different values t_1 and t_2 of the time. If now phase difference between E_1' and E_2' depends only on the time difference $t_2 - t_1$, and not on t_1 and t_2 separately, the light field is temporally coherent.

Perfectly monochromatic light waves are both spatially and temporally coherent, but such waves do not exist, and conversely, no real light field is perfectly incoherent. Between these two extreme situations, real light waves exhibit infinite degrees of partial coherence.

Light coherence is closely related to the interference of light waves because only coherent (or nearly coherent) waves produce observable interference patterns, from which the coherence length, coherence time, and *degree of partial coherence* can be determined. However, it is worth mentioning that an exact theory of coherence is highly complicated. For the interested reader, more details regarding this problem may be found in Refs. [1.1], [1.4], and especially in Refs. [1.9], [1.10].

1.4. Polarized light

The orientational characteristics of wave vibrations in time and space are termed the *polarization* of the wave. As was stated in Section 1.2, the electric vector **E** of the electromagnetic wave (Fig. 1.2) plays a dominant role in optical phenomena. Thus one would expect a plane containing this vector and the direction of wave propagation (*z*-axis) to be distinguished as the *plane of light polarization*. For certain historical reasons [1.1, 1.12], however, it is the plane containing the magnetic vector **H** what has come to be known as the plane of light polarization. This nomenclature is not accepted by all writers, and some of them, including the author of the present book, define the direction of polarization and plane of polarization with respect to the electric vector. To avoid confusion, it is better, in accordance with more recent practice, to omit the terms "direction of polariz-ation" and "plane of polarization" and to use "direction of vibration" or "plane of vibration" instead.

In the type of light waves represented in Figs. 1.3, 1.10, and 1.16–1.18, the wave vibrations take place in a single plane, i.e., the plane of the paper. Such a light wave is said to be *plane* or *linearly polarized*. In this case the tip of the electric vector describes a single line in the plane perpendicular to the direction of light propagation. By contrast, if the orientation of this vector in space and time is random, the light is said to be *unpolarized*. Between perfectly polarized and completely unpolarized light beams one distinguishes different states of *elliptical polarization* in which the tip of the electric vector describes an ellipse in any plane perpendicular to the direction of light propagation. According to the sense of rotation of the electric vector as "viewed" by an observer receiving the light energy, one speaks of *right-handed* and *left-handed elliptical polarization*. This property is called the *handedness* of the polarization. If the observer "sees" a clock-wise rotation of the electric vector, the elliptical polarization is said to be right-handed, and when he "sees" a counterclockwise rotation, the polarization is said to be left-handed. In a particular case, when the minor axis of the polarization ellipse is equal to the major axis, the light is *circularly polarized* (consequently right-handed or left-handed). In another extreme case, when the minor axis tends to zero, the light becomes linearly polarized. These various states of po-larization can be achieved in a pure form by using certain optical devices, but in nature they tend to occur rather in the presence of some unpolarized light. Therefore, natural light[24] is a mixture which can be regarded as *partially ellipti-cally, partially linearly,* or *partially circularly polarized light*.

There are several procedures and mathematical methods (the Poincaré sphere, Stokes vector, Jones calculus, Mueller matrix) for quantitative descriptions of the state and *degree of polarization* of light beams. They are reviewed in Refs. [1.6], [1.12]–[1.15], and in this section only the basic properties of the polaroids and some other means for the production and analysis of polarized light will be given.

1.4.1. Dichroic polarizers for use in microscopy

As was stated in Subsection 1.2.3, birefringent crystals, in general, split one inci-dent light beam into two beams—ordinary and extraordinary—polarized at

---

[24] Some authors use the term "natural" instead of "unpolarized" light. This terminology is not accurate as most light found in nature is *partially polarized*.

right angles to each other. There are some crystals (e.g., tourmaline) which transmit one of the split beams much better than the other. This phenomenon is termed *dichroism*. It was discovered by Biot in 1815, and more than 100 years later was utilized by Land to produce the most useful *linear polarizers*, namely *polaroids*. Instead of using a single dichroic crystal, the effect of linearly polarized light was initially achieved by using an array of tiny dichroic microcrystals (herapathyt), which were embedded and aligned in a plastic sheet. Today polaroids are largely manufactured by bringing organic polymers with long-chain molecules (polyvinyl alcohol) into alignment, and so into birefringence, by stretching, and then treating them with certain chemicals (iodine) to achieving the dichroism. Various types of polaroids are commercially available from the Polaroid Corporation, Cambridge, Mass., USA [1.12, 1.14]. The main type is the H type described above (stretched polyvinyl alcohol treated with iodine).

A typical polaroid for use in microscopy consists of two circular glass plates between which is cemented a polarizing sheet. Frequently it is mounted in a metallic frame on which the direction of vibration of the light being transmitted is marked. This direction is called the *axis of polaroid*. The basic parameters of any polarizer are its *principal transmittances*, $\tau_{\parallel}$ and $\tau_{\perp}$, which are defined as the intensity transmission coefficients relating to light linearly polarized parallel to and perpendicular to the polarizer axis. Thus $\tau_{\parallel} = I_{\parallel}/I$ and $\tau_{\perp} = I_{\perp}/I$. Here I is the intensity of the incident light linearly polarized, $I_{\parallel}$ and $I_{\perp}$ are, respectively, the intensities of the light emerging from the polarizer orientated in such a manner that its axis is parallel and next perpendicular to the vibration direction of the incident light. The quantity

$$\varepsilon = \frac{\tau_{\parallel}}{\tau_{\perp}} = \frac{I_{\parallel}}{I_{\perp}} \tag{1.47}$$

is said to be *extinction coefficient*.[25]

It is found that the transmittance τ_{β} for the light linearly polarized at an angle β to the polarizer axis is given by

$$\tau_{\beta} = \tau_{\parallel}\cos^2\beta + \tau_{\perp}\sin^2\beta. \tag{1.48a}$$

For an ideal polarizer, $\tau_{\parallel} = 1$, $\tau_{\perp} = 0$ and

$$\tau_{\beta} = \cos^2\beta. \tag{1.48b}$$

The last relation is generally known as *Malus law*, while Eq. (1.48a) can be considered as a generalized form of this law.

Another value of transmittance arise when unpolarized light strikes a linear

[25] Some authors [1.15] define this quantity as $\varepsilon = \tau_{\perp}/\tau_{\parallel}$.

polarizer. In general, unpolarized light of intensity I can be regarded as consisting of two orthogonally polarized components, each of intensity $I/2$, the vibration directions of which are parallel and perpendicular to the polarizer axis. In this case the principal transmittances of the polarizer are formally defined as

$$\tau_{\parallel} = \frac{I_{\parallel}}{0.5I}, \quad \text{and} \quad \tau_{\perp} = \frac{I_{\perp}}{0.5I}, \tag{1.49}$$

although a common transmittance defined as $\tau = I'/I = (I_{\parallel} + I_{\perp})/I$ may be more useful. Taking into account relations (1.49) we have

$$\tau = \tfrac{1}{2}(\tau_{\parallel} + \tau_{\perp}). \tag{1.50a}$$

Usually $\tau_{\perp}$ is very small, and for most work with polarized light one can accept

$$\tau \approx \tfrac{1}{2}\tau_{\parallel}. \tag{1.50b}$$

The transmittance τ of typical polaroids is equal to 0.2–0.4.

Another very important parameter of any polarizer is the *degree of polarization* (or the *polarizance*) defined as

$$P = \frac{\tau_{\parallel} - \tau_{\perp}}{\tau_{\parallel} + \tau_{\perp}}. \tag{1.51}$$

This parameter for a good polaroid may be higher than 0.9999 or 99.99%. Such a polarizer transforms an unpolarized light beam into a beam with nearly perfect linear polarization. However, the greater the degree of polarization P the smaller is transmittance τ. Besides, the values of these parameters depend upon the wavelength (see Table 1.6).

It is also useful to know the transmittance of two polarizers arranged one behind the other, when the first is struck by an unpolarized light beam. The answer may be obtained immediatley if the rule of calculation of the resultant transmittance (τ_r) of a series of light absorbing filters is kept in mind. Let τ_1, τ_2, τ_3, ... will be the transmittances of individual filters. Then

$$\tau_r = \tau_1 \tau_2 \tag{1.52a}$$

for two filters,

$$\tau_r = \tau_1 \tau_2 \tau_3 \tag{1.52b}$$

for three filters, and so on. Now, let $\tau'_{\parallel}$, $\tau'_{\perp}$ and $\tau''_{\parallel}$, $\tau''_{\perp}$ denote the principal transmittances of two polarizers, and consider two extreme situations: (1) where the axes of both polarizers are parallel, (2) where the axes are perpendicular to each other, i.e., the polarizers are *crossed*. In the first case $\tau'_{\parallel}$ coincides with $\tau''_{\parallel}$ and $\tau'_{\perp}$ coincides with $\tau''_{\perp}$, so the resultant principal transmittances of these

TABLE 1.6

Transmittance (τ) and degree of polarization (P) of several types of polaroids for use in the visible spectrum (H and K types) and for the infrared (HR type)

Wavelength λ [nm]	HN-22		HN-38		KN-36		λ [μm]	HR	
	τ [%]	P [%]	τ [%]	P [%]	τ [%]	P [%]		τ [%]	P [%]
375	5.5	99.9909	27.0	92.857	21.0	99.052	0.9	35.5	99.72
400	10.5	99.9905	33.5	88.732	25.5	99.217	1.0	36.0	87.01
450	22.5	99.9987	40.5	95.181	32.5	99.908	1.25	35.5	98.32
500	27.5	99.9993	43.0	98.844	35.5	99.986	1.5	36.5	99.99
550	24.0	99.9992	41.0	99.829	37.0	99.989	1.75	36.5	99.99
600	21.5	99.9991	39.9	99.924	39.5	99.992	2.0	36.5	99.73
650	23.5	99.9991	41.0	99.927	41.5	99.981	2.25	34.0	98.83
700	29.5	99.9990	43.0	99.837	44.0	95.55	2.5	12.0	71
750	34.5	99.9971	45.0	99.115	46.0	30			

two polarizers are $\tau_{||} = \tau'_{||}\tau''_{||}$ and $\tau_\perp = \tau'_\perp \tau''_\perp$, and the common transmittance, by analogy to (1.50a), is

$$\tau_0 = \tfrac{1}{2}(\tau'_{||}\tau''_{||} + \tau'_\perp \tau''_\perp). \tag{1.53a}$$

If the polarizers are identical, $\tau'_{||} = \tau''_{||} = \tau_{||}$ and $\tau'_\perp = \tau''_\perp = \tau_\perp$, and Eq. (1.53a) reduces simply to

$$\tau_0 = \tfrac{1}{2}(\tau^2_{||} + \tau^2_\perp). \tag{1.53b}$$

For typical polarizers $\tau^2_\perp$ is extremely small in comparison with $\tau^2_{||}$, and

$$\tau_0 = \tfrac{1}{2}\tau^2_{||} \tag{1.53c}$$

is generally accepted.

Another situation occurs when the polarizers are crossed. Now, $\tau'_{||}$ coincides with $\tau''_\perp$ and $\tau'_\perp$ coincides with $\tau''_{||}$, so $\tau_{||} = \tau'_{||}\tau''_\perp$ and $\tau_\perp = \tau'_\perp \tau''_{||}$. Hence, the common transmittance of two crossed polarizers is given by

$$\tau_{90} = \tfrac{1}{2}(\tau'_{||}\tau''_\perp + \tau'_\perp \tau''_{||}), \tag{1.54a}$$

or

$$\tau_{90} = \tau_{||}\tau_\perp, \tag{1.54b}$$

if the polarizers are identical ($\tau'_{||} = \tau''_{||} = \tau_{||}$; $\tau'_\perp = \tau''_\perp = \tau_\perp$).

Next, let the angle α (*azimuth*) between the axes of the polarizers under consideration differ from $0°$ and $90°$ (Fig. 1.19). It is found that in this case the transmittance τ_α takes the form

$$\tau_\alpha = \tau_{90} + (\tau_0 - \tau_{90})\cos^2\alpha, \tag{1.55a}$$

which may be also written as

$$\tau_\alpha = \tau_0 \cos^2\alpha \approx \tfrac{1}{2}\tau^2_{||}\cos^2\alpha, \tag{1.55b}$$

because the quantities τ_{90} and $\tau^2_\perp$ are usually very small. The last equation simply expresses the Malus law for a series of two polarizers when one of them is struck by an unpolarized light beam. Consequently, Eq. (1.55a) can be regarded in this case as a generalized form of this law.

It is also worth mentioning that the ratio

$$\varepsilon = \frac{\tau_0}{\tau_{90}} \tag{1.56a}$$

is called the *extinction coefficient*, which characterizes the degree of light extinguishing by two crossed polarizers.[26] Taking into account Eqs. (1.53c) and (1.54b) we have

---

[26] This quantity is also defined as $\varepsilon = \tau_{90}/\tau_0$ [1.15].

$$\varepsilon = \frac{\tau_{||}}{2\tau_{\perp}},$$

(1.56b)

when unpolarized light is incident on a pair of polarizers of identical performance.

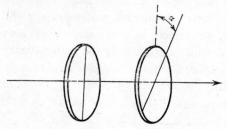

Fig. 1.19. Two polarizers whose azimuth α differs from 0° and 90°.

1.4.2. *Birefringent polarizers as components of microscope systems*

If one of two light beams, ordinary or extraordinary, polarized linearly by a double refraction crystal is not absorbed dichroically but suppressed by total reflection, a linear polarizer is achieved. Among such *birefringent polarizers*, the best known, historically, is the *nicol*, invented by W. Nicol in 1828. It is composed of two prisms of calcite, suitably cut along cleavage planes and cemented together with Canada balsam, which permit only the extraordinary beam to pass, while the ordinary beam is eliminated by total reflection at the interface (Fig. 1.20). This polarizer has, however, some disadvantages. First of all, its

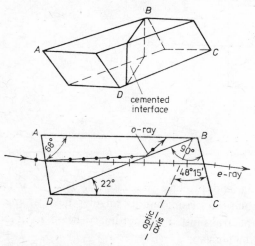

Fig. 1.20. Nicol polarizer and its cross section through the corners A, B, C and D.

entrance and exit faces are oblique to the incident light beam which means that the transmitted extraordinary beam is displaced laterally and not perfectly linearly polarized but is slightly elliptical, so that an astigmatism occurs. Moreover, the *acceptance angle* (*angular field*) of the nicol is rather small (equal to about 25°). Therefore, some other birefringent polarizers, and especially the *Glan–Thompson* or *Glan-Foucault prisms*, are largely in use today. They also consist of two calcite (or quartz) prisms, which permit only the extraordinary beam to pass, but their entrance and exit faces are normal to the incident light beam. The Glan–Thompson prism provides a large angular field (up to 40°), and is at present the most common birefringent polarizer (Fig. 1.21). Its apex angle

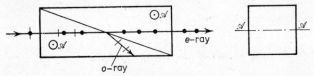

Fig. 1.21. Side elevation and end elevation of a typical Glan–Thomson polarizer (the optic axis is indicated by $\mathscr{A}$).

varies according to specific use, but is usually equal to about 20°. The Glan–Foucault polarizer is similar except that there is an air gap between the component prisms instead of a layer of cement (Fig. 1.22). As a result of this gap it can be used in ultraviolet light, whereas Canada balsam and other optical cements

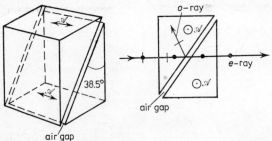

Fig. 1.22. Glan–Foucault polarizer and its cross section in a plane perpendicular to the optic axis ($\mathscr{A}$).

absorb light in this spectral region and produce fluorescence. This polarizer is also shorter than the Glan–Thompson prism, but its angular field is small (about 8°) and the intensity of the transmitted extraordinary beam is much reduced by multiple reflections in the air gap.

Another group of birefringent prisms is formed by double refracting prisms and plates, which transmit both ordinary and extraordinary rays. These elements are usually employed as *beam-splitters*, *beam-recombiners* and *phase compensators* in some polarizing and interference-polarizing instruments, especially in double refracting interference microscopes (see Section 1.6 and Chapters 7 and 16).

1.4.3. Polarization by reflection

Let a ray (*S*) strike obliquely an interface between two dielectric media of different refractive indices (Fig. 1.23). It is well known from experience that a portion of ray energy from one medium enters the second medium as a *refracted ray T*

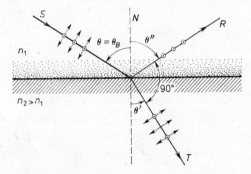

Fig. 1.23. Illustrating Brewster's angle (θ_B).

and the rest of the incident light energy returns, as a *reflected ray R*, into the first medium. The rays *S*, *T*, and *R* are all in the plane which contains the interface normal *N* and the *incident ray S*. This plane is called the *plane of incidence*, whereas angles θ, θ', and θ'' between *N* and *S*, *N* and *T*, *N* and *R* are called the *angle of incidence*, the *angle of refraction*, and the *angle of reflection*, respectively.

If the incident ray *S* is unpolarized, it can be treated as a sum of two components, the vibration directions of which are parallel (∥) and perpendicular (⊥) to the plane of incidence. In Fig. 1.23 the parallel component is marked by arrows and the perpendicular component is distinguished by circles with dots.

It can be shown (see Subsection 1.5.4) that for a particular angle of incidence $\theta = \theta_B$, the reflected ray contains only the perpendicular component of the incident light, so that it is completely linearly polarized in a plane perpendicular to the plane of incidence. The *polarization angle* θ_B is known as *Brewster's angle*. It is defined by the relation

$$\tan \theta_B = \frac{n_2}{n_1}, \tag{1.57}$$

where n_1 is the refractive index of the medium from which the light strikes the interface and n_2 is the refractive index of the medium from which the light is reflected. It can also be shown that for an angle of incidence $\theta = \theta_B$, the sum $\theta_B + \theta' = 90°$, and that the angle between the transmitted ray T and reflected ray R is also equal to $90°$.

On the other hand, the refracted ray T contains both the perpendicular and parallel component, although the latter predominates. The degree of polarization of this ray is given by

$$P = \frac{\tau_{||} - \tau_{\perp}}{\tau_{||} + \tau_{\perp}} = \frac{1}{1 + 2\cot^2(\theta - \theta')}. \tag{1.58a}$$

This quantity is rather small, but can be greatly increased by using a series of dielectric plates set at Brewster's angle (θ_B) to the incident ray (S), as shown in Fig. 1.24. If m is the number of plates, the polarizance of such a device is [1.12]

$$P = \frac{1 - \left(\dfrac{2n}{n^2 + 1}\right)^{4m}}{1 + \left(\dfrac{2n}{n^2 + 1}\right)^{4m}}, \tag{1.58b}$$

where n is the refractive index of the plates. The above formula is true when the plates are in an air medium, and if all the reflected rays R_3, R_4, R_5, ... are rejected on the outside or absorbed before they strike another (previous) plate of the system. If, for instance, $n = 1.52$ and $m = 5$, the degree of polarization of the transmitted ray T is equal to 70%, but when $m = 10$, then $P = 99\%$. As can be seen, this system becomes effective if it consists of a great number of plates.

Such a polarizing device (Fig. 1.24) is known as the *pile-of-plates polarizer*, and although it is not used in microscopical instruments nevertheless the principle on which it is based is widely applied in lasers with what is known as *Brewster windows* (Fig. 1.14). In this case m in the formula (1.58b) is the number of times the laser light passes back and forth through the Brewster window. Here this number is very high, and the laser beam becomes completely linearly polarized.

Various forms of polarizers of this type have also been made by using both the polarizing and interference effects in multiple thin films deposited on glass substrates. An example of a such polarizer is shown in Fig. 1.25. This is a compact device consisting of a pile of dielectric thin films (*TF*) of alternately high (*H*)

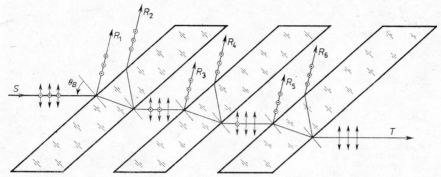

Fig. 1.24. Pile-of-plates polarizer.

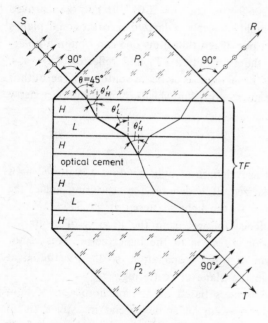

Fig. 1.25. Thin film interference polarizer.

and low (L) refractive indices n_H and n_L. By suitably adjusting these indices and the thicknesses t_H and t_L of the films, one can use the right-angle prisms (P_1 and P_2) to produce the reflected beam R at 90° to the incident (S) and transmitted (T) beams. To obtain this result as well as the efficient linear polarization of the transmitted and reflected beams, the following relations must be fulfilled:

$$\tan \theta'_H = \frac{n_L}{n_H}, \tag{1.59a}$$

$$\theta'_H + \theta'_L = 90°, \tag{1.59b}$$

$$2nt_H \cos \theta'_H = \frac{\lambda}{2}, \tag{1.59c}$$

$$2nt_L \cos \theta'_L = \frac{\lambda}{2}, \tag{1.59d}$$

$$\frac{\sin \theta}{\sin \theta'_H} = \frac{n_H}{n}, \tag{1.59e}$$

where n is the refractive index of the prisms P_1 and P_2. If, for instance, $n_H = 2.3$ (zinc sulphide ZnS) and $n_L = 1.32$ (cryolite Na_3AlF_6), then $\theta'_H = 29°50'$, $\theta'_L = 60°10'$, $t_H = 68.7$ nm, $t_L = 206.9$ nm, and $n = 1.617$. It has been proved that the beams T and R may be highly linearly polarized in orthogonal planes as indicated in Fig. 1.25. By using a pile of ten thin films one can obtain a polarizance better than 99% for both the transmitted (T) and reflected (R) rays. Such an interference polarizer has been used as a beam-splitter by the author of this book for phase-interference microscopy and shearing microinterferometry (see Subsections 5.8.2 and 16.2.3).

1.4.4. Polarization by scattering

Particles of a size comparable to a wavelength of light produce a scattered light which is, in general, partially polarized. If σ denotes the scattering angle, the polarizance is proportional to $\cos \sigma$, so the light scattered at 90° is completely linearly polarized. Such a dependence follows from the process of scattering which does not change the vibration direction of the light vector. This vector has no component in the direction of wave propagation, so light scattered at 90° has no vibration component in the plane of scattering.

In microscopy there are no polarizers based on the phenomenon of light scattering, but this phenomenon can occur in some special investigations of objects which scatter the light. In nature, however, the atmosphere causes sunlight to be partially polarized by scattering, reaching a maximum polarizance ca. 30% at a point 90° from the sun [1.17].

1.4.5. Retarders used in microscopy

Any light vibration (electric vector **E**) can be treated as composed of two orthogonal components, $\mathbf{E}_{\parallel}$ and $\mathbf{E}_{\perp}$, one of which is parallel (∥) and the other perpendicular ($\perp$) to a reference axis (Fig. 1.26). The amplitudes $E_{||} = |\mathbf{E}_{||}|$ and $E_\perp = |\mathbf{E}_\perp|$ of these components are given by

$$E_{||} = E\cos\alpha, \tag{1.60a}$$

$$E_\perp = E\sin\alpha, \tag{1.60b}$$

where E is the amplitude of the original light vector $\mathbf{E}$ and α is the angle (azimuth) between the vector $\mathbf{E}$ and the reference axis.

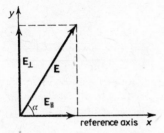

Fig. 1.26. Resolution of the electric vector ($\mathbf{E}$) of unpolarized light into two mutually incoherent orthogonal components ($\mathbf{E}_{||}$ and $\mathbf{E}_\perp$).

The use of polarized light frequently needs some phase alteration between the orthogonal components $\mathbf{E}_{||}$ and $\mathbf{E}_\perp$. An optical device which is able to retard one of these components relative to the other is called a *retarder*. The amount of phase change produced by this device is known as its *retardance*. Each retarder has a particular direction with respect to which the retardance is defined. This direction is said to be the *reference axis* (for short, axis) of the retarder.

We distinguish between pure and impure retarders. The effect of the former is solely that of phase alteration. Frequently, the mechanism of the phase retardation cannot be completely separated from other effects, such as a change of reflectance, and in this case the retarder is said to be impure. This problem is discussed in greater detail in Ref. [1.12].

The most common retarders are made of birefringent crystals (quartz, mica, gypsum) or birefringent synthetic foils (cellophane, stretched polyvinyl alcohol). As was previously mentioned in Subsection 1.2.3, if a plate is cut parallel to the optic axis of a uniaxial crystal, and is normally struck by a parallel beam of light, this will in general produce a longitudinal splitting of the incident wave Σ into the ordinary and extraordinary wavefronts, Σ_o and Σ_e (Fig. 1.27). Between these wavefronts a phase difference φ will occur, which depends upon the thickness t of the plate (BP) and its birefringence $n_e - n_o$. However, no wavefront splitting will occur if the plate BP is struck by a beam of linearly polarized light whose

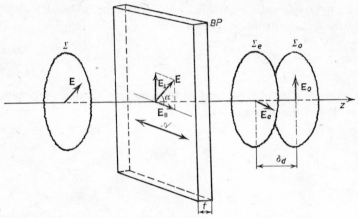

Fig. 1.27. Wavefront (Σ) axially split into two components (Σ_o and Σ_e) by a birefringent plate. Vibration directions (E_o and E_e) of these components are perpendicular and parallel to the optic axis of the plate.

direction of vibration is parallel or perpendicular to the optic axis $\mathscr{A}$ of the plate BP.

Usually the phase difference φ is expressed by

$$\varphi_d = \frac{2\pi \delta_d}{\lambda} = \frac{2\pi}{\lambda}(n_e - n_o)t, \tag{1.61a}$$

where

$$\delta_d = (n_e - n_0)t \tag{1.61b}$$

is the optical path difference caused by the birefringence. A desired retardance may be simply obtained by making a plate of the proper thickness. It is, however, interesting to note that the expression (1.61a) is approximative, as it does not include some phase alterations caused by different reflections of light at each interface for the o-component and e-component. Strictly speaking, the phase difference produced by a birefringent plate is [1.12]:

$$\varphi = \varphi_e - \varphi_o, \tag{1.62}$$

where

$$\varphi_e = \arctan\left[\left(\frac{n_e^2 + 1}{2n_e}\right)\tan\left(\frac{n_e}{n_e - n_o}\right)\varphi_d\right],$$

$$\varphi_o = \arctan\left[\left(\frac{n_o^2 + 1}{2n_o}\right)\tan\left(\frac{n_0}{n_e - n_o}\right)\varphi_d\right].$$

In general, the difference $\varphi - \varphi_d$ is small, and the approximative value φ_d can be accepted as the retardance for many practical purposes, but the difference is nevertheless very important in accurate measurements using polarized light.

In practice, it is better to produce birefringent retarders by aiming to obtain the desired retardance rather than the thickness t resulting from Eq. (1.61b). Otherwise, the injurious effects of reflection can be greatly reduced by coating the birefringent plate with antireflecting coatings (see Subsection 1.6.3).

If the electric vector $\mathbf{E}$ of linearly polarized light forms an angle α with respect to the optic axis ($\mathscr{A}$) of a uniaxial birefringent plate (Fig. 1.27), it may be considered to consist of two coherent components, $\mathbf{E}_{||}$ and $\mathbf{E}_{\perp}$, whose vibration directions are parallel and perpendicular to the optic axis $\mathscr{A}$. The transmitted light will have the same directional components $\mathbf{E}_e$ and $\mathbf{E}_o$, parallel to $\mathbf{E}_{||}$ and $\mathbf{E}_{\perp}$, respectively, but with their phases displaced. Depending on the phase retardation φ_d of one component relative to the other, the transmitted light will, in general,

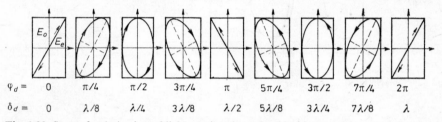

$\varphi_d =$	0	$\pi/4$	$\pi/2$	$3\pi/4$	π	$5\pi/4$	$3\pi/2$	$7\pi/4$	2π
$\delta_d =$	0	$\lambda/8$	$\lambda/4$	$3\lambda/8$	$\lambda/2$	$5\lambda/8$	$3\lambda/4$	$7\lambda/8$	λ

Fig. 1.28. State of polarization of light passing through a birefringent plate.

be polarized elliptically with different ellipticity (Fig. 1.28). For some particular cases the transmitted light may, however, be polarized linearly and circularly. The same effect is obtained if the plate is cut parallel to the axes of a biaxial crystal.

If, however, the incident light is unpolarized, the orthogonal components $\mathbf{E}_{||}$ and $\mathbf{E}_{\perp}$ must be regarded as temporally incoherent. In this case the azimuth α (Fig. 1.27) changes rapidly in function of time, so that a rapid variation of the amplitudes of these components occurs and it is not possible to obtain a stable state of the polarization of light passing through the birefringent plate.

The direction in which orientated light vibrations pass slowest through a birefringent material is held to be the reference axis of the retarder and is known as the *slow axis*. Similarly, the direction in which orientated light vibrations pass faster is called the *fast axis*. The latter could also be treated as the reference axis, but the former has been accepted as a standard (especially in polarized light microscopy). The optic axis of a positive uniaxial crystal (e.g., quartz) is the slow axis, while that of a negative uniaxial crystal is the fast axis. For biaxial crystals the slow axis is their acute bisectrix.

In microscopy the most useful retarders are the *quarter-wave plate* (its retardance $\varphi = \pi/2$), *half-wave plate* ($\varphi = \pi$), and *full-wave plate* ($\varphi = 2\pi$).

If a beam of linearly polarized light falls normally upon a $\lambda/4$-plate, the transmitted light is, in general, polarized elliptically. The ellipticity is determined by the azimuth α (Fig. 1.27). If $\alpha = 45°$ and $E_e = E_o$, the emergent light becomes circularly polarized. On the other hand, if a beam of elliptically or circularly polarized light falls upon the $\lambda/4$-plate, the emergent beam can be polarized linearly (for a concrete azimuthal orientation of the $\lambda/4$-plate). In combination

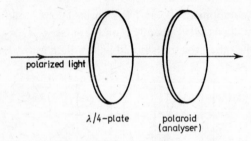

polarized light

$\lambda/4$-plate   polaroid
                    (analyser)

Fig. 1.29. Sénarmont's compensator.

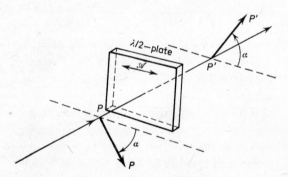

Fig. 1.30. Alteration of the direction of vibration of linearly polarized light by using a half-wave plate.

with a polarizer (analyser), such as nicol or polaroid, the $\lambda/4$-plate forms the *Sénarmont compensator* (Fig. 1.29), which is largely used in variable phase contrast microscopy and interference-polarizing microinterferometry for the variation and measurements of optical path differences (see e.g., Sections 5.7 and 16.5).

If this plate is cemented together with a polaroid so that the axes of these elements form an angle of 45°, this combination constitutes a *circular polarizer* when the polaroid faces the incident beam. Such a combination is very helpful in suppressing injurious glare in optical instruments, and especially in incident light microscopes (see e.g. Chapter 8).

The main property of the $\lambda/2$-plate is its effect of rotating the plane of vibration ($P'P'$) of the emergent light through an angle 2α (Fig. 1.30), where α is the angle between the reference axis ($\mathscr{A}$) of the $\lambda/2$-plate and the plane of vibration (PP) of the incident light. In general, this retarder is less useful in microscopy than the previous ($\lambda/4$-plate) or λ-plate.

When combined with a polarizer, the λ-plate constitutes a very sensitive device for the analysis of polarized white light. Its main characteristic is that it produces the retardance 2π for only a small central region of the visual spectrum. When the analysing polarizer (analyser) is crossed with the plane of vibration of linearly polarized light which falls upon the λ-plate, and this plate is rotated about the axis of the incident light beam, a high variation of colours of the emergent beam is obtained. In a particular orientation of this plate, when the azimuth $\alpha = 45°$ (Fig. 1.31), green and yellow light are extinguished and the blue

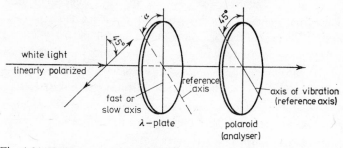

Fig. 1.31. Full-wave retarder used as sensitive tint plate.

and red regions of the spectrum are transmitted, which together give a very sensitive purple colour. If the λ-plate is then slightly rotated, the purple colour changes rapidly into red or violet. For that reason the λ-plate is known as the *sensitive tint plate* (it is also called the *first order red compensator* or *Red I compensator*).

The most suitable material for the manufacture of birefringent retarders is quartz. But the use of this crystal involves some difficulties relating to thickness. For example, a $\lambda/4$-plate of quartz for $\lambda = 550$ nm would only be about 15 μm thick. Such a thin plate is difficult to make. The difficulty is overcome by making two thicker plates (Fig. 1.32), then polishing one plate until a difference in thicknesses $t_2 - t_1$ of the desired amount is obtained, and finally cementing them together with their axes ($\mathscr{A}_1$ and $\mathscr{A}_2$) crossed. Such a double plate is said to be a *subtractive retarder*. Neglecting reflection on surfaces, its retardance is given by

$$\varphi_d = \frac{2\pi}{\lambda} [(n_e - n_o)(t_2 - t_1)]. \tag{1.63}$$

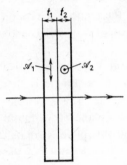

Fig. 1.32. Subtractive retarder.

Birefringent retarders are less efficient over a wide spectral range than birefringent polarizers. Therefore, the retarders described above have a great *spectral dispersion of retardance*. This dispersion is sometimes a useful property of the λ-plate, but detracts from the usefulness of the $\lambda/4$-plate and $\lambda/2$-plate. This gave rise to the need for producing *achromatic retarders*, which were obtained in several ways, for instance by using the phenomenon of total internal reflection. A well known quarter-wave retarder based on this phenomenon is the *Fresnel rhomb*, which is almost achromatic [1.12]. Achromatic retarders, however, less important for microscopy than for polarimetry.

When utilized without any additional device, the retarders described above have a constant retardance. However, there is another group of retarders, i.e., those with variable or controllable retardance. Among these are the *quartz wedge*, *Berek compensator*, and *Ehringhaus compensator*. These form typical attachments to more advanced polarizing microscopes (see Chapter 15), and usually have retardance values varying from zero to a few wavelengths (sometimes to several dozen wavelengths). In applied optics the best known variable retarders are, however, the *Babinet compensator* and the *Soleil compensator*. These are only occasionally used in microscopy.

The *Pockels cell* and *Kerr cell* are the best known representatives of electrically controllable retarders. The former is based on the phenomenon of the *induced birefringence* of some crystals in the direction of an applied electric field, and its retardance can be suitably controlled by adjusting the voltage applied to cell electrodes made of a mesh of transparent material. Pockels cells are standardly made of ADP and KDP crystals (see Table 1.2). The Kerr cell, on the other hand, is based on the phenomenon of induced birefringence in a direction perpendicular to an electric field applied to a liquid with highly dipolar molecules. A liquid of this kind is nitrobenzene which is typically employed for manufacturing Kerr cells. Both kinds of cell are largely used in laser technique.

For the interested reader, more details regarding different retarders and compensators may be found in Refs. [1.12] and [1.18].

1.5. Refraction, reflection, and transmission of light

When the optical properties of a medium abruptly change, only part of the travelling light passes through, while the remainder reflects at the discontinuity. The proportions in which the division of light occurs depends on the character of the discontinuity, as well as on the angle of incidence and state of polarization

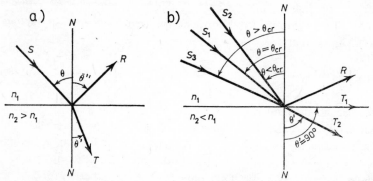

Fig. 1.33. Reflection and refraction at a plane surface: a) the first medium is optically less dense than the second medium, b) the first medium is optically denser than the second one.

of the incident light beam. In this section we shall consider the behaviour of light when the discontinuity is a plane interface between two media as shown in Fig. 1.33.

1.5.1. Basic laws of light reflection and refraction

When considering Fig. 1.33a, it is worth calling to mind the basic laws of reflection and refraction. To begin with then, the incident ray *S*, the refracted ray *T*, and the reflected ray R lie in the same plane as the normal N at the point of incidence. Secondly, the angle of reflection θ'' is equal to the angle of incidence with opposite sign,[27] i.e.,

$$\theta = -\theta''. \tag{1.64}$$

Thirdly, the ratio of the sines of the angle of incidence and of the angle of refraction is a constant (n_{21}), i.e.,

$$\frac{\sin\theta}{\sin\theta'} = n_{21} = \frac{n_2}{n_1}. \tag{1.65}$$

The constant n_{21} is said to be the *relative index of refraction*, i.e., the refractive index of the second medium with respect to that of the first medium.[28] If n_1 and n_2 are the absolute refractive indices (related to the vacuum) of the first and second medium, respectively, then $n_{21} = n_2/n_1$, and, in accordance with Eq. (1.2a), $n_{21} = v_1/v_2$, where v_1 and v_2 are the light velocities in the first and second medium, respectively. Equation (1.65) is known as *Snell's law of refraction*.

When the first medium is *optically less dense*[29] than the second ($n_1 < n_2$), the angle of refraction is smaller than the angle of incidence and takes a maximum value for $\theta = 90°$. This value (θ'_{cr}) is known as the *critical angle of refraction*. Equation (1.65) shows that

$$\theta'_{cr} = \arcsin\left(\frac{n_1}{n_2}\right). \tag{1.66}$$

If, however, the second medium is less dense than the first one ($n_1 > n_2$), the angle of refraction θ' is greater than the angle of incidence (Fig. 1.33b), and the well known phenomenon of *total internal reflection* occurs for angles of incidence θ greater from a critical value θ_{cr} relating to the angle of refraction $\theta' = 90°$. In this case there are refracted rays (T_2) only for $\theta < \theta_{cr}$, whereas for $\theta > \theta_{cr}$ there are only reflected rays (R), and for $\theta = \theta_{cr}$ there is a particular refracted

---

[27] In Fig. 1.33 the angles are measured from the normal (N), so θ and θ' are positive, and θ'' is negative. This usage agrees with the accepted convention that an anti-clockwise rotation from the normal (or from any other reference axis) is positive, and a clockwise rotation is negative.

[28] For short, the medium from which a ray goes out is called the *first medium*, and that which the ray enters is called the *second medium*.

[29] A medium is optically less dense (or rarer) if its refractive index is smaller than that of another medium, and vice versa, the first medium is optically denser (or more dense) when its refractive index is greater than that of the second medium.

ray (T_1), which propagates along the interface. By substituting $\theta' = 90°$ into Eq. (1.65), the *critical angle of incidence* is found to be

$$\theta_{cr} = \arcsin\left(\frac{n_2}{n_1}\right). \tag{1.67}$$

It will be readily seen that for any two media and reversed rays, and in accordance with the *reversibility (reciprocity) theorem* of optical rays [1.19], the critical angle of incidence is equal to the critical angle of refraction, i.e., $\theta_{cr} = \theta'_{cr}$. Therefore, some authors do not differentiate between these two notions and call both the *critical angle*.

The phenomenon of total internal reflection is made wide use of in different microscope systems containing right-angle and other light reflecting prisms. The critical angle for glass surrounded by air is equal to $\arcsin(1/n)$, where n is the refractive index of glass. For different types of optical glass, this index varies from about 1.46 to 1.92, so the corresponding values of the critical angle are

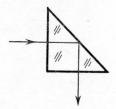

Fig. 1.34. Right-angle prism.

from 43° to 32°. Thus, the hypotenuse face of an isosceles right-angle prism (Fig. 1.34) reflects totally light striking normally one of the cathetus faces of a prism made of any optical glass.

1.5.2. Some lesser known phenomena related to total internal reflection

In contrast to the situation shown in Fig. 1.33, total internal reflection exhibits quite different features if a thin film of refractive index n_2 is placed between two media whose refractive indices, n_1 and n_3, are greater than n_2 (Fig. 1.35). This allows an incident ray S to be propagated across the thin film TF, even if the angle of incidence θ is greater than the critical angle $\theta_{cr} = \arcsin(n_2/n_1)$. This phenomenon was discovered by Hall in 1902 and is known as *frustrated total reflection* (FTR). The intensity of the transmitted ray T is reduced in proportion to the thickness of thin film; when the film thickness t exceeds a few wavelengths, the transmitted ray vanishes. If t is suitably selected, one can obtain transmitted and reflected beams, T and R, of any desired relative intensity (Fig. 1.36). Since the thin film TF is non-absorbing, the result is an efficient beam-splitter. The state of polarization of both beams does, however, differ slightly. The FTR phenomenon has also been used in the production of interference filters, high

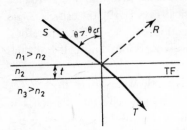

Fig. 1.35. Frustrated total reflection (FTR).

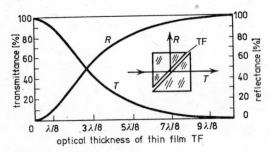

Fig. 1.36. FTR beam-splitter.

frequency light modulators for the infrared region, and some devices for laser technology [1.19, 1.21].

In connection with total internal reflection and the FTR it is necessary to mention *evanescent waves* and the *Goos–Hänchen effect*. Strictly speaking, the ray tracing in Figs. 1.33–1.35 is not entirely correct. In reality, the reflected ray and the incident ray do not have a common point at the interface of two media. In 1943, Goos and Hänchen discovered that a light beam totally internally reflected by a plane dielectric interface appears to undergo a lateral shift (s) as shown in Fig. 1.37. This shift is very small (of the order of one micrometre) and slightly different for light vector components parallel ($\parallel$) and perpendicular ($\perp$) to the plane of incidence. If the angle of incidence θ is greater than the critical angle θ_{cr}, the shifts $s_{\parallel}$ and $s_{\perp}$ are given by the expressions

$$s_{||} = \frac{\lambda n_1}{\pi n_2^2} \frac{\tan\theta}{(\sin^2\theta - \sin^2\theta_{cr})^{1/2}}, \qquad (1.68a)$$

$$s_{\perp} = \frac{\lambda}{\pi n_1} \frac{\tan\theta}{(\sin^2\theta - \sin^2\theta_{cr})^{1/2}}. \qquad (1.68b)$$

These formulae show that $s_{||} > s_{\perp}$ and the shifts $s_{||}$ and $s_{\perp}$ tend towards infinity when $\theta \to \theta_{cr}$. If, e.g., $n_1 = 1.52$ (glass slide) and $n_2 = 1.33$ (water), then $\theta_{cr} = \arcsin(1.33/1.52) = 61°$, and $s_{||} = 1.36 \,\mu m$, $s_{\perp} = 1.04 \,\mu m$ for $\lambda = 550$ nm and $\theta = 65°$.

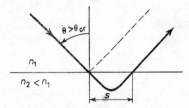

Fig. 1.37. Goos–Hänchen shift.

The Goos–Hänchen effect cannot be explained on the basis of geometrical optics. It proceeds from the electromagnetic theory of light [1.22], and is a consequence of evanescent waves. For an angle of incidence $\theta > \theta_{cr}$, the evanescent wave occurs in the second (less dense) medium, propagates along the interface,

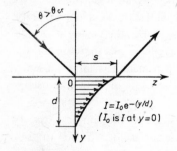

Fig. 1.38. Evanescent wave.

and its intensity I vanishes (evanesces) exponentially with the distance y perpendicular to the interface, as shown in Fig. 1.38. Mathematically, I may be expressed as $I = I_0 e^{-(y/d)}$, where I_0 is the intensity incident on the interface and d is the decay depth defined as a distance y at which $I = I_0/e$. The depth d is given by

$$d = \frac{\lambda}{4\pi} (n_1^2 \sin^2\theta - n_2^2)^{-1/2} \qquad\qquad\qquad (1.69a)$$

and is a measure of the depth to which the evanescent wave penetrates into the less dense medium.

Equation (1.69a) can also be written as

$$d = \frac{\lambda}{4\pi n_1} \cdot \frac{1}{(\sin^2\theta - \sin^2\theta_{cr})^{1/2}} . \qquad\qquad\qquad (1.69b)$$

As can be seen, the decay depth d decreases with increasing θ and for $\theta > \theta_{cr}$ is of the order of λ or smaller. For the above example of two media (glass and water), $d = 0.12$ μm if $\theta = 65°$ and $\lambda = 550$ nm. When $\theta \to \theta_{cr}$, the decay depth $d \to \infty$.

Both the Goos–Hänchen effect and evanescent waves are of great importance in many areas of contemporary optics, especially in integrated optics [1.23] and internal reflection spectroscopy [1.24]. Some attempts to apply evanescent waves to holography [1.25, 1.26] and also to microscopy [1.27] have been undertaken with success.

A useful modification of the phenomenon of total internal reflection occurs if the second medium absorbs the light in some degree. Frequently such a medium is qualified as "lossy". Its refractive index is complex, i.e., consists of a real part and of an imaginary part (see Subsection 1.2.2). The exponential penetration of light into the second medium is now accompanied by an absorption loss of light energy propagated as an evanescent wave (Fig. 1.38). As a consequence, the reflected energy is smaller than the incident energy, i.e., the totally internally reflected light beam is attenuated. This type of light reflection is therefore known as *attenuated total reflection* (ATR). Today this phenomenon is largely used in absorption spectrophotometry [1.24]. Some potential uses for it in microscopy have also been suggested [1.27].

Absorption losses caused by ATR can be measured in the simplest case by using a right–angle prism (Fig. 1.34), the hypotenuse surface of which is in good optical contact with the substance to be investigated. The prism must be highly transparent in the desired spectral range. It is only used as a tool to study a medium with a complex refractive index. In order to amplify the effects of a small loss of reflected light, *multiple reflection ATR technique* is used (for more details see, e.g., [1.2]). If ϱ denotes the reflectance of a medium under examination, then $1-\varrho$ is the absorption loss. If $1-\varrho \ll 1$, then after m reflections the measured loss is m-fold greater.

1.5.3. Refractive index dispersion formulae

In Subsection 1.3.2 it has been stated that the refractive index depends on the wavelength, i.e., $n = n(\lambda)$. The analytical form of this dependence becomes highly complicated if a precision of 10^{-5} in the refractive index has to be taken into account. Different dispersion formulae were therefore suggested, but none is generally valid.

The *Cauchy formula*

$$n = A + \frac{B}{\lambda^2} + \frac{C}{\lambda^4} + \ldots \tag{1.70}$$

is the best known in practice. It contains three (or even more) constants, A, B, and C, characterizing the medium concerned. These constants are usually determined experimentally by measuring the refractive indices n for three different values of λ. For many practical purposes, it is sufficient to take into account

TABLE 1.7

Refractive indices of several liquids and optical glasses for different wavelengths at 20°C

Substances	Wavelengths [nm]				
	435.8	486.1	546.1	587.6	656.3
	Refractive indices				
Water	1.3440	1.3377	1.3345	1.3330	1.3312
Ethyl alcohol	1.3696	1.3662	1.3630	1.3617	1.3599
Glycerine	1.4795	1.4750	1.4710	1.4695	1.4672
Benzene	1.5230	1.5134	1.5050	1.5014	1.4966
Immersion oil*	1.5320	1.5260	1.5202	1.5167	1.5127
HD liquid**	1.5420	1.5324	1.5236	1.5171	1.5107
Carbon disulphide	1.6750	1.6523	1.6370	1.6277	1.6181
Optical glass***					
FK3	1.4732	1.4694	1.4662	1.4645	1.4623
BK7	1.5267	1.5224	1.5187	1.5168	1.5143
KF9	1.5362	1.5305	1.5258	1.5234	1.5204
BAK2	1.5512	1.5463	1.5421	1.5400	1.5332
LF7	1.5927	1.5848	1.5783	1.5750	1.5709
F5	1.6238	1.6146	1.6072	1.6034	1.5987
SF8	1.7177	1.7046	1.6942	1.6889	1.6825

* Immersion oil, type A, from R. P. Cargille Laboratories, Cedar Grove, N. J. 07009, USA.
** One of the high dispersion liquids from R. P. Cargille Laboratories.
*** After a catalogue of Schott Mainz.

only the first two terms of Eq. (1.70). Cargille refraction liquids, for instance, comply well with the Cauchy formula in the red to blue spectral regions.

Equation (1.70) expresses the fact that the refractive index n decreases as λ increases, so that the slope $dn/d\lambda$ of the dispersion curve $n = n(\lambda)$ is negative since the constants A, B, and C are positive for transparent media. This is a *normal dispersion*, which is a feature of the majority of materials used in microscopy and instrumental optics (Table 1.7 and Fig. 1.39).

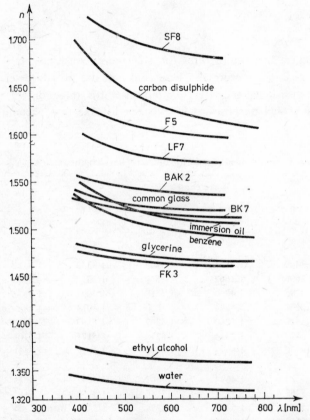

Fig. 1.39. Dispersion curves of several transparent materials (FK3, BK7, BAK2, LF7, F5, SF8—some optical glasses after a catalogue of Schott Mainz).

An alternative dispersion formula was put forward by Hartmann. It has the form

$$n = n_0 + \frac{D}{(\lambda - \lambda_0)^{1.2}}, \qquad (1.71a)$$

or

$$n = n_0 + \frac{D}{\lambda - \lambda_0}, \tag{1.71b}$$

where D is a constant and n_0 is the refractive index for a particular wavelength λ_0. In the view of some authors the Hartmann formula expresses the dispersion of glasses more precisely than the Cauchy formula. It is also more widely used by researchers of polymers. However, none of these formulae expresses the variation of the refractive index over the full spectral range from UV to IR with a better precision than 10^{-5}. For optical glasses, the best result is obtained by using a formula given in Schott's current catalogue:

$$n^2 = A_0 + A_1 \lambda^2 + A_2 \lambda^{-2} + A_3 \lambda^{-4} + A_4 \lambda^{-6} + A_5 \lambda^{-8}, \tag{1.72}$$

where A_0 to A_5 are constants listed (with an accuracy of $\pm 3 \times 10^{-6}$) for each glass type, resulting from its components. Sometimes a shorter version:

$$n = A_0' - A_1' \lambda^2 + A_2' \lambda^{-2} \tag{1.73}$$

of the expression (1.72) is used (constants A_0', A_1', and A_2' are, of course, others than A_0, A_1, and A_2 in Eq. (1.72)). The last version expresses a typical S-shaped dispersion curve with an inflection point between the UV and IR regions. The second term in Eq. (1.73) represents a decrease of n in the infrared region, while the third term expresses an increase of n in the ultraviolet region.

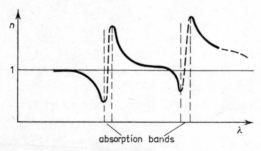

absorption bands

Fig. 1.40. Anomalous dispersion.

Formulae (1.71) to (1.73) relate to materials without abrupt absorption bands. Near and within such bands the dispersion is anomalous (Fig. 1.40), and the dependence $n(\lambda)$ is usually described by the *Sellmeier formula*

$$n^2 = 1 + \sum_m \frac{C_m \lambda^2}{\lambda^2 - \lambda_m^2}, \tag{1.74a}$$

where λ_m denote the wavelengths of successive absorption maxima and C_m are constants. In principle, Eqs. (1.71) to (1.73) can be regarded as approximations of the Sellmeier dispersion curve for a spectral region between absorption bands. In particular, Eq. (1.74a) takes the form

$$n^2 = 1 + \frac{C_1 \lambda^2}{\lambda^2 - \lambda_1^2} \tag{1.74b}$$

for a medium with only one absorption band.

For many practical purposes some standard dispersion quantities are in use. Among them the most important are the *mean dispersion*

$$D = n_F - n_C \tag{1.75}$$

and the *Abbe number* (also known as the *dispersion index, V-number*, or *constringence* [1.4]) usually defined as

$$V_d = \frac{n_d - 1}{n_F - n_C} \quad \text{or} \quad V_D = \frac{n_D - 1}{n_F - n_C}, \tag{1.76}$$

where n_F and n_C are the refractive indices at the hydrogen F and C line, respectively, n_d is the refractive index at the helium d line, and n_D is the refractive index at the sodium D line (see Table 1.3). To correct the chromatic aberrations of microscope objectives (and other optical systems), some other V-numbers are in use which can be generally defined as

$$V_{\lambda_2} = \frac{n_{\lambda_2} - 1}{n_{\lambda_1} - n_{\lambda_3}}, \tag{1.77}$$

where n_{λ_1}, n_{λ_2}, and n_{λ_3} are the refractive indices for wavelengths $\lambda_1 < \lambda_2 < \lambda_3$, respectively. The reciprocal of the V-number is called the *dispersive power* of the medium between wavelengths λ_1 and λ_3 [1.4].

To obtain a more detailed specification of the ability of optical glasses to reduce the secondary spectrum of achromatic objectives, use is also made of *relative partial dispersions* defined as

$$P_{\lambda_1, \lambda_2} = \frac{n_{\lambda_1} - n_{\lambda_2}}{n_F - n_C}, \tag{1.78}$$

where n_{λ_1} and n_{λ_2} are the refractive indices for two different wavelengths selected as required within the visual spectral range. In particular, the following partial dispersions are specified: $P_{F,d}$ at the F and d spectral lines, $P_{F,e}$ at the F and e lines, and $P_{g,F}$ at the g and F lines. A good insight into the dispersion properties of different types of optical glasses is obtained by plotting the above mentioned partial dispersions against the Abbe number. A characteristic feature of such plots is the straight line along which the majority of optical glass types are to be found. These are known as *normal glasses*, but only types not found on the straight line but as far away as possible can be used together with normal glasses to correct the secondary spectrum. This is what is known as an *apochromatic correction* of microscope objectives.

1.5.4. Reflectance and transmittance at interfaces

For the design of microscopes, as well as for microscopical research, it is necessary to be familiar with all aspects of the behaviour of light at the interfaces of different media. A boundary between two media not only refracts and reflects light in accordance with the basic laws of refraction and reflection, but also changes the amplitude, phase, and polarization properties of reflected and transmitted rays with respect to the corresponding features of incident rays.

Light behaviour at a single dielectric interface. The reflectance (ϱ) and transmittance[30] (τ) for an interface depend upon the state of polarization of the incident light beam. These quantities for a single dielectric interface are described by the *Fresnel formulae*:

$$\varrho_{\parallel} = \frac{I_{R\parallel}}{I_{S\parallel}} = \frac{\tan^2(\theta - \theta')}{\tan^2(\theta + \theta')}, \tag{1.79a}$$

$$\varrho_{\perp} = \frac{I_{R\perp}}{I_{S\perp}} = \frac{\sin^2(\theta - \theta')}{\sin^2(\theta + \theta')}; \tag{1.79b}$$

$$\tau_{\parallel} = \frac{I_{T\parallel}}{I_{S\parallel}} = \frac{4\sin^2\theta'\cos^2\theta}{\sin^2(\theta + \theta')\cos^2(\theta - \theta')}, \tag{1.80a}$$

$$\tau_{\perp} = \frac{I_{T\perp}}{I_{S\perp}} = \frac{4\sin^2\theta'\cos^2\theta}{\sin^2(\theta + \theta')}, \tag{1.80b}$$

where θ is the angle of incidence (Fig. 1.33), θ' is the angle of refraction, I denotes the intensity, subscripts S, R, and T represent the incident, reflected, and transmitted (refracted) beams, respectively, and suffixes $\parallel$ and $\perp$ denote the light vector components,[31] parallel and perpendicular to the plane of incidence (see Fig. 1.23).

---

[30] Some authors, e.g., Born and Wolf [1.1], and Klein [1.2], use the terms "reflectivity" and "transmissivity" instead of "reflectance" and "transmittance".

[31] Frequently the German subscripts p (*parallel*) and s (*senkrecht*) are in use instead of the suffixes $\parallel$ and $\perp$, respectively.

The vibration directions of unpolarized light vary rapidly and at random, in an irregular manner. For light of this kind, the reflectance and transmittance are given by average values:

$$\varrho = \tfrac{1}{2}(\varrho_{\|}+\varrho_{\perp}), \tag{1.81}$$

$$\tau = \tfrac{1}{2}(\tau_{\|}+\tau_{\perp}). \tag{1.82}$$

Equations (1.79) are more interesting than (1.80). Their graphical representation is given in Fig. 1.41 for two opposing situations: (a) the first medium

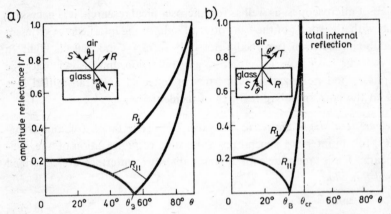

Fig. 1.41. Graphical representations of the Fresnel formulae for reflection at dielectric interfaces: a) the first medium is air and the second medium is glass, b) the first medium is glass and the second one is air.

is air and the second is glass, (b) the first medium is glass and the second is air. In order to show the lower parts of the curves in Fig. 1.41 more clearly, the absolute value $|r|$ of the *amplitude reflectance* instead of the *intensity reflectance* ϱ is plotted as function of the angle of incidence ($|r| = \sqrt{\varrho}$). As is shown by Eqs. (1.79) and the graphs, the reflectance $\varrho_{\|}$ vanishes for $\theta+\theta' = 90°$, and the reflected beam R consists of a perpendicular component only; hence this beam is completely linearly polarized. The corresponding angle of incidence is the Brewster angle θ_B (compare Subsection 1.4.3). For angles of incidence $\theta \neq \theta_B$, the reflected beam becomes partially linearly polarized. The degree of polarization is given by

$$P_{\varrho} = \frac{\varrho_{\perp}-\varrho_{\|}}{\varrho_{\perp}+\varrho_{\|}}. \tag{1.83a}$$

Two examples of variation of P_{ϱ} in function of θ are shown in Fig. 1.42.

In contrast to the reflected beam R (Fig. 1.33), the transparent beam T is always partially polarized. From Eqs. (1.80) we obtain

$$\tau_{||} = \frac{\tau_{\perp}}{\cos^2(\theta - \theta')},$$

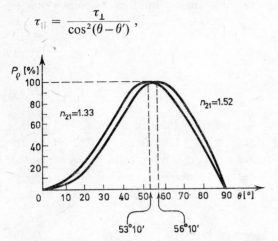

Fig. 1.42. Variation with angle of incidence (θ) of the degree of polarization (P_{ϱ}) of the light reflected from a dielectric interface. Drawn for $n_{21} = n_2/n_1 = 1.33$ (e.g., air–water interface), and for $n_{21} = 1.52$ (e.g., air–glass interface).

and because $\cos^2(\theta - \theta') < 1$, the transmittance $\tau_{||} > \tau_{\perp}$, hence the parallel component prevails. The degree of polarization of the transmitted beam,

$$P_{\tau} = \frac{\tau_{||} - \tau_{\perp}}{\tau_{||} + \tau_{\perp}}, \qquad (1.83b)$$

is given by Eq. (1.58a). Two exemplary variations of P_{τ} in function of θ are shown in Fig. 1.43.

Reflectance and transmittance for normally incident light cannot be obtained from Eqs. (1.79) and (1.80) because these equations become indeterminate at

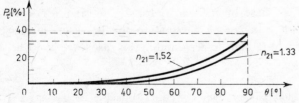

Fig. 1.43. Variation with angle of incidence (θ) of the degree of polarization (P_{τ}) of the light passing through dielectric interfaces. Drawn for $n_{21} = 1.33$ (e.g., air–water interface) and $n_{21} = 1.52$ (e.g., air–glass interface).

$\theta = 0$ and $\theta' = 0$. Moreover, since the plane of incidence is undefined, one cannot talk of parallel or perpendicular components. However, taking into account Snell's law of refraction, in this case the reflectance ϱ and transmittance τ may be found to have the forms:

$$\varrho = \left(\frac{n_{21}-1}{n_{21}+1} \right)^2, \tag{1.84}$$

$$\tau = \frac{4n_{21}}{(n_{21}+1)^2}, \tag{1.85}$$

where $n_{21} = n_2/n_1$, n_1 is the refractive index of the first medium, and n_2 that of the second medium (Fig. 1.33). If, e.g., $n_1 = 1$ (air) and $n_2 = 1.52$ (glass), the reflectance $\varrho = 0.043$ and transmittance $\tau = 0.957$. When these values are added, the sum is unity. This result is, of course, in agreement with the *law of the conservation of energy*. In general, for the boundary of two transparent media[32] this law can be written as

$$\varrho_{\parallel} + \tau_{\parallel} = 1, \quad \varrho_{\perp} + \tau_{\perp} = 1, \quad \varrho + \tau = 1. \tag{1.86}$$

For many microscopical purposes it is useful to know the phase jumps or phase variations of the reflected beam with respect to the phase of the incident beam. This problem is discussed in detail by several authors (see, e.g., Refs. [1.1], [1.4] and [1.12]), although they are not always in agreement. In the present writer's opinion the most reliable and comprehensive interpretation is given in Ref. [1.12], and this will be summarized here.

In general, the phase of each component, $T_{\parallel}$ and $T_{\perp}$, of the transmitted light beam is equal to the phase of the corresponding component of the incident beam, but the phase of the reflected components, $R_{\parallel}$ and $R_{\perp}$, depends on the relative values of the angle of incidence and the angle of refraction, θ and θ', respectively. If the first medium is less dense than the second, then $\theta' < \theta$, and the phase of the component $R_{\perp}$ differs by π from the phase of the corresponding component $S_{\perp}$ of the incident beam for all the angles of incidence θ. The parallel component $R_{\parallel}$ also differs in phase by π from $S_{\parallel}$ for $\theta < \theta_B$ and for $\theta > \theta_B$ does not.[33] If, however, $\theta + \theta' = \pi/2$, or θ is equal to the Brewster angle θ_B, the amplitude of the component $R_{\parallel}$ falls to zero, and in this case there is no need to talk about its phase.

---

[32] It is worth noting that another interpretation of Eq. (1.85) and the relation $\varrho + \tau = 1$ is given by Clarke and Grainger [1.12] for the boundary of two transparent media.

[33] The authors of the books [1.1] and [1.4] state that the phase of $R_{\parallel}$ is the same as that of $S_{\parallel}$ for $\theta < \theta_B$ and differs by π for $\theta > \theta_B$.

There are, however, more complicated phase relations between the reflected and incident beams when the first medium is denser than the second medium. For angles of incidence θ less than θ_B, both the parallel component $R_{\parallel}$ and the perpendicular component $R_{\perp}$ are in phase with the corresponding components $S_{\parallel}$ and $S_{\perp}$ of the incident beam. Whereas for θ greater than the critical angle $\theta_{\rm cr}$, the phase variations, $\varphi_{\parallel}$ and $\varphi_{\perp}$, of the components $R_{\parallel}$ and $R_{\perp}$ are expressed by the following formulae:[34]

$$\varphi_{\parallel} = 2\arctan\left(\frac{\sqrt{\sin^2\theta - n_{21}^2}}{n_{21}^2\cos\theta}\right) - \pi, \tag{1.87a}$$

$$\varphi_{\perp} = 2\arctan\left(\frac{\sqrt{\sin^2\theta - n_{21}^2}}{\cos\theta}\right). \tag{1.87b}$$

The quantity $-\pi$ in Eq. (1.87a) shows that the component $R_{\perp}$ is ahead of the component $R_{\parallel}$, and is advanced relative to the incident beam. In the remaining range of θ, i.e., for $\theta_B < \theta < \theta_{\rm cr}$, the component $R_{\perp}$ is in phase with $S_{\perp}$ and the component $R_{\parallel}$ is out of phase by $-\pi$ with respect to $S_{\parallel}$. All these phase variations are graphically presented in Fig. 1.44 for the relative refractive index $n_{21} = 0.66$ (e.g., $n_1 = 1.52$ and $n_2 = 1$). In this figure we also see plotted the phase difference $\varphi = \varphi_{\perp} - \varphi_{\parallel}$ between the reflected components. This difference vanishes for $\theta = 90°$ (grazing incidence of light) and for $\theta = \theta_{\rm cr}$ (i.e., $\sin\theta_{\rm cr} = n_{21}$). Between these two values there is a minimum of φ; it occurs at θ given by

$$\cos\theta = \sqrt{\frac{1 - n_{21}^2}{1 + n_{21}^2}}, \tag{1.88}$$

and its value

$$\varphi_{\min} = 4\arctan n_{21}. \tag{1.89}$$

The above discussion is important for several microscopical techniques, especially for interference microscopy and polarized light microscopy. For the latter it is also useful to know the change of the azimuth α, i.e., of the angle between the plane of light vibration and the plane of incidence when incident light is linearly polarized. Light of this kind can also be resolved into two components, parallel and perpendicular to the plane of incidence, but now these components are coherent. It is assumed that azimuth α is in the range from $-\pi/2$ to $+\pi/2$, and is positive when the plane of vibration turns anti-clockwise for an observer who receives the light rays. If $s_{\parallel}$ and $s_{\perp}$, $r_{\parallel}$ and $r_{\perp}$, $t_{\parallel}$ and $t_{\perp}$

---

[34] In Refs. [1.1] and [1.4] the equation (1.87a) is given without $-\pi$.

are the amplitude components of the incident, reflected, and transmitted beams, respectively, then

$$\tan \alpha_S = \frac{s_\perp}{s_{||}}, \quad \tan \alpha_R = \frac{r_\perp}{r_{||}}, \quad \tan \alpha_T = \frac{t_\perp}{t_{||}}, \tag{1.90}$$

where α_S, α_R, and α_T are the azimuthal angles for the incident, reflected, and transmitted beams, respectively. It can be found [1.1] that for $n_2 > n_1$,

$$\tan \alpha_R = - \frac{\cos(\theta - \theta')}{\cos(\theta + \theta')} \tan \alpha_S, \tag{1.91a}$$

$$\tan \alpha_T = \cos(\theta - \theta') \tan \alpha_S. \tag{1.91b}$$

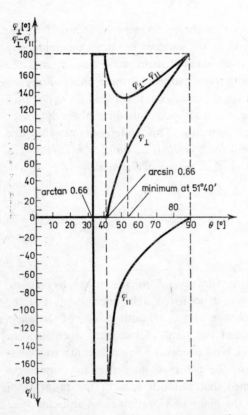

Fig. 1.44. Variation with angle of incidence (θ) of the phase change ($\varphi_{||}$ and $\varphi_\perp$) and phase difference ($\varphi = \varphi_\perp - \varphi_{||}$) at total internal reflection. Drawn for $n_{21} = 0.66$ (e.g., glass–air interface).

From these relations it is easy to see that $\alpha_R = \alpha_S$ only for $\theta = \theta' = 0$ and for $\theta = \pi/2$, whereas $\alpha_T = \alpha_S$ only for $\theta = 0$. Otherwise,

$$|\tan \alpha_R| > |\tan \alpha_S|, \tag{1.92a}$$

and

$$|\tan \alpha_T| < |\tan \alpha_S|. \tag{1.92b}$$

These two inequalities show that the plane of vibration of the reflected beam R is rotated away from the plane of incidence, whereas the plane of vibration of the refracted beam T is turned towards the plane of incidence. For a better illustration of this phenomenon, the complete behaviour of α_R and α_T for n_{21} = 1.52 and $\alpha_S = 45°$ is shown in Fig. 1.45.

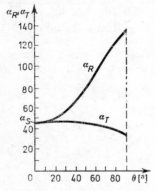

Fig. 1.45. Variation with angle of incidence (θ) of the azimuthal angles α_R and α_T for $n_{21} = 1.52$ and $\alpha_S = 45°$.

Another situation occurs for $n_1 > n_2$ when linearly polarized light is internally reflected. Since the reflected components undergo phase jumps of different amounts (see Eqs. (1.87)), the reflected beam R_3 (Fig. 1.33b) will in consequence become elliptically polarized; hence it is only possible to describe the azimuth α_R of the major axis of the ellipse. For $\theta > \theta_{cr}$, both the azimuth and the ellipticity may be changed, whereas in the range $\theta_B < \theta < \theta_{cr}$ only the azimuth is turned, and for $\theta < \theta_B$ a change in the azimuth and handedness occur.

Reflection by metallic surfaces. Though highly interesting and very important for metallographic microscopy, the interaction of light with metals cannot be widely discussed in this book (for more details see, e.g., Refs. [1.1] or [1.28]), and only some basic facts relating to the reflection of light by metallic surfaces will be given. First of all, it is worth noting that a description of surface optical phenomena is more complicated for metals than for dielectrics since the angle of reflection, refractive index, and some other quantities are complex. If the first medium is a dielectric and the second is metal, Snell's law takes the form

$$n_1 \sin\theta = \tilde{n}_2 \sin\tilde{\theta}', \tag{1.93}$$

where $\tilde{n}_2$ is the complex refractive index of the second medium (see Eq. (1.28)) and $\tilde{\theta}'$ is the complex angle of refraction. Whereas the reflectance at normal incidence is given by

$$\varrho = \left| \frac{\tilde{n}_{21}-1}{\tilde{n}_{21}+1} \right|^2 = \frac{(n_{21}-1)^2+n_{21}^2 \varkappa^2}{(n_{21}+1)^2+n_{21}^2 \varkappa^2}, \tag{1.94}$$

where $\tilde{n}_{21} = \tilde{n}_2/n_1$, $n_{21} = n_2/n_1$, n_2 is the real part of the complex refractive index $\tilde{n}_2$, and $\varkappa$ is the extinction coefficient defined by Eq. (1.22). When $(n+1)^2 \ll n^2\varkappa^2$, the reflectance $\varrho \approx 1$, and if $\varkappa \to 0$, the relation (1.94) takes the form of Eq. (1.84).

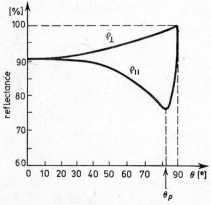

Fig. 1.46. Variation with angle of incidence (θ) of the reflectances $\varrho_{||}$ and $\varrho_\perp$ of a metallic surface (e.g., aluminium surface).

At oblique incidence an exact description of light reflection of metals is complicated. The reflectances $\varrho_{||}$ and $\varrho_\perp$ are, however, enough well expressed by the following formulae [1.12]:

$$\varrho_{||} = \frac{\left(n_{21} - \dfrac{1}{\cos\theta}\right)^2 + n_{21}^2 \varkappa^2}{\left(n_{21} + \dfrac{1}{\cos\theta}\right)^2 + n_{21}^2 \varkappa^2}, \tag{1.95a}$$

$$\varrho_{\perp} = \frac{(n_{21}-\cos\theta)^2 + n_{21}^2\varkappa^2}{(n_{21}+\cos\theta)^2 + n_{21}^2\varkappa^2}.$$ (1.95b)

If the angle of incidence $\theta \to 0$, these expressions reduce to Eqs. (1.94).

A graphical representation of Eqs. (1.95) for a typical metal, e.g., for polished massive aluminium, is shown in Fig. 1.46. As can be seen, $\varrho_{\parallel}$ and $\varrho_{\perp}$ are now much greater than previously for a dielectric interface (Fig. 1.41), and at a certain angle of incidence θ_p, called the *principal angle of incidence*, the reflectance $\varrho_{\parallel}'$ has a minimum which, however, does not fall to zero. Therefore, the reflected light is not polarized linearly at this or at any other angle of incidence.

In general, light reflected by a metallic surface is polarized elliptically. The parallel and perpendicular components, $R_{\parallel}$ and $R_{\perp}$, reflect with a phase difference $\varphi_{\parallel}-\varphi_{\perp}$, which continuously changes from 0 to π as θ varies from 0 to 90° (Fig. 1.47). When linearly polarized light is incident, it also becomes elliptically polarized after reflection, provided its azimuth α_S is other than zero or 90°.

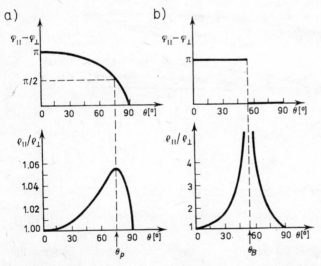

Fig. 1.47. Variation with angle of incidence (θ) of the phase difference $\varphi_{\parallel}-\varphi_{\perp}$ and of the reflectance ratio $\varrho_{\parallel}/\varrho_{\perp}$ on light reflection from a typical metal (a) and from a transparent dielectric (b). Drawn for air–silver and air–glass interfaces.

A matter of considerable interest, especially for metallographers, is the set of relations between some directly measurable quantities used to determine n and $\varkappa$ of metals. These quantities include, e.g., the principal angle of incidence θ_p, the phase difference $\varphi_{\parallel}-\varphi_{\perp}$, and the ratio $\varrho_{\parallel}/\varrho_{\perp}$ when linearly polarized light falls upon the metallic surface at $\theta = \theta_p$ and with the azimuth $\alpha_S = 45°$. The desired relations are found to be [1.12]:

$$[\sin\theta_p \tan\theta_p]^4 = n^4(1+\varkappa^2)^2 - 2n^2(1-\varkappa^2)\sin^2\theta_p - \sin^4\theta_p, \tag{1.96}$$

$$\tan\alpha_{Rp} = \sqrt{\frac{\varrho_{\parallel} \text{ at } \theta_p}{\varrho_{\perp} \text{ at } \theta_p}}, \tag{1.97}$$

$$\alpha_{Rp} = \arctan\varkappa, \tag{1.98}$$

where α_{Rp} is the *principal azimuth* of the reflected beam. Measuring α_{Rp} (or $\tan\alpha_{Rp}$) and θ_p yields $\varkappa$ from Eq. (1.98) and next n from Eq. (1.96). The azimuthal angle is measured with respect to the normal to the plane of incidence and its value is obtained in the simplest case by adjusting the Sénarmont compensator at extinction. It is interesting to note that for $\theta = \theta_p$ the phase difference $\varphi_{\parallel} - \varphi_{\perp} = \pi/2$, and, if $n^2(1+\varkappa^2) \gg 1$, $\tan\alpha_R$ has a minimum value (Fig. 1.47). Some other useful formulae relating to n and $\varkappa$ can be found in Ref. [1.1].

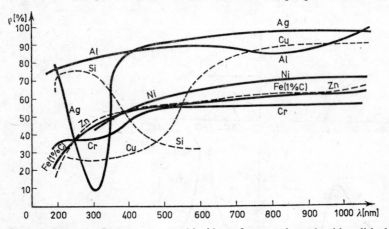

Fig. 1.48. Spectral reflectance at normal incidence for several metals with polished surfaces

The reflectance of metals strongly depends on the wavelength of incident light (Fig. 1.48). Thus the above discussion should be related to monochromatic light.

Reflectance and transmittance of a dielectric plate. In microscopy it is often of great interest to know not only the reflectance and transmittance of a single interface but also the total light intensity reflected and transmitted by a dielectric, e.g., glass plate. The situation sketched in Fig. 1.49 is one where $S1$, $R1$ and $T1$ represent rays incident at, reflected from and transmitted through the first (entering) surface of the dielectric plate, respectively; similarly S2, R2, and T2 denote rays incident at, reflected from, and transmitted through the second (emerging) surface of this plate (geometrically, but not with respect to their intensities, the rays T1 and S2 represent one and the same ray). The corresponding intensities

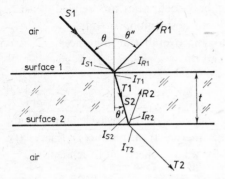

Fig. 1.49. Reflectance and transmittance of a transparent plane parallel plate. Illustrating the notations relating to Eqs. (1.99).

of all these rays at the first and second surfaces are denoted by I_{S1}, I_{R1}, I_{T1} and I_{S2}, I_{R2}, I_{T2}. They relate to each other as follows:

$$I_{T1} = I_{S1} - I_{R1}, \tag{1.99a}$$

$$I_{S2} = I_{T1} e^{-Kt/\cos\theta'}, \tag{1.99b}$$

$$I_{T2} = I_{S2} - I_{R2}, \tag{1.99c}$$

where the exponential term in Eq. (1.99b) describes a possible light absorption by the plate according to Lambert's law (see Eq. (1.21a)). If there is no absorption of light, the absorption coefficient $K = 0$ and the exponential term is unity, and in this case $I_{S2} = I_{T1}$.

Let us assume that the plate is surrounded by air and its thickness t is so great as to not produce interference effects typical for thin films (see Subsection 1.6.3). In that situation the overall transmission and reflection of the plate is a combination of the effect of reflection on the plate surfaces and the effect of optical absorption. For light incident other than normally, the overall reflectance resulting from the combined effect of both glass surfaces of the plate and for a particular wavelength λ is given by [1.29]

$$\varrho = \frac{\varrho_{\parallel}}{2}\left[1 + \frac{(1-\varrho_{\parallel})^2 \tau_i^2}{1 - \varrho_{\parallel}^2 \tau_i^2}\right] + \frac{\varrho_{\perp}}{2}\left[1 + \frac{(1-\varrho_{\perp})^2 \tau_i^2}{1 - \varrho_{\perp}^2 \tau_i^2}\right], \tag{1.100}$$

where $\varrho_{||}$ and $\varrho_{\perp}$ are defined as previously (Eqs. (1.79)), and τ_i is the *internal transmittance*. The last quantity results from Eq. (1.99b) because

$$\tau_i = \frac{I_{S2}}{I_{T1}} = \mathrm{e}^{-Kt/\cos\theta'}. \tag{1.101}$$

For light incident normally, $\varrho_{||} = \varrho_{\perp}$ and both quantities are defined by Eq. (1.84), whereas $\tau_i = \exp[-Kt]$. If there is no absorption, $\tau_i = 1$.

Correspondingly, the general relation for the overall transmittance incorporating both the reflection on plate surfaces and the absorption effect is given by

$$\tau = \frac{\tau_i}{2} \frac{(1-\varrho_{||})^2}{1-\varrho_{||}^2 \tau_i^2} + \frac{(1-\varrho_{\perp})^2}{1-\varrho_{\perp}^2 \tau_i^2}. \tag{1.102}$$

The absorption coefficient K normally depends on the wavelength of light and produces a spectral dependence of the internal transmittance τ_i. Thus the light transmission (T) of a dielectric plate for a large spectrum is defined by

$$T = \frac{\int\limits_{\lambda_1}^{\lambda_2} S(\lambda)\tau(\lambda)V(\lambda)\mathrm{d}\lambda}{\int\limits_{\lambda_1}^{\lambda_2} S(\lambda)V(\lambda)\mathrm{d}\lambda}, \tag{1.103}$$

where $S(\lambda)$ is the spectral energy distribution of the light source, $\tau(\lambda)$ is defined by Eq. (1.102), $V(\lambda)$ is the spectral sensitivity of the eye (or of another photoreceptor). From the whole visual spectrum a wavelength range from $\lambda_1 = 380$ nm to $\lambda_2 = 770$ nm is normally accepted by the eye.

The interested reader may find more details regarding light reflection from and transmission through dielectric plates in Refs. [1.29] and [1.30].

1.6. Interference of light

Interference of light is one of the most important phenomena found in microscopy. By this we mean not only interference microscopy and polarized light microscopy, where interference effects are directly observed by the microscopist, but also the phase contrast method, holographic microscopy, and even the typical bright-field technique, whose working principles are based on the diffraction and interference of light.

1.6.1. Interference between two monochromatic waves

Let us consider two linearly polarized plane light waves with superimposed directions of vibration and identical frequencies ω. Their amplitudes a_1 and

a_2, as well as their phases ψ_1 and ψ_2, are, however, different. Both waves propagate in an isotropic medium along the same direction z. Thus they are superposed and it is interesting to know the light intensity resulting from the superposition effect of both waves at any point $z = z_0$. In these circumstances, the waves under consideration may be represented by

$$E_1 = a_1 \sin(\omega t - \psi_1), \quad E_2 = a_2 \sin(\omega t - \psi_2). \tag{1.104}$$

According to the principle of superposition of wave disturbances, the resultant wave E is simply the sum of E_1 and E_2, then

$$E = a_1 \sin(\omega t - \psi_1) + a_2 \sin(\omega t - \psi_2)$$
$$= \sin \omega t (a_1 \cos \psi_1 + a_2 \cos \psi_2) - \cos \omega t (a_1 \sin \psi_1 + a_2 \sin \psi_2). \tag{1.105a}$$

It is convenient to put

$$a_1 \sin \psi_1 + a_2 \sin \psi_2 = A \sin \Psi, \quad a_1 \cos \psi_1 + a_2 \cos \psi_2 = A \cos \Psi.$$

Then Eq. (1.105a) reduces to

$$E = A \sin(\omega t - \Psi), \tag{1.105b}$$

where A and Ψ are the amplitude and phase of the resultant wave, respectively. They are defined by

$$A^2 = a_1^2 + a_2^2 + 2 a_1 a_2 \cos(\psi_2 - \psi_1) \tag{1.106}$$

and

$$\tan \Psi = \frac{a_1 \sin \psi_1 + a_2 \sin \psi_2}{a_1 \cos \psi_1 + a_2 \cos \psi_2}. \tag{1.107}$$

The first of the last two equations expresses the resultant intensity $I = A^2$, and may be written as

$$I = I_1 + I_2 + 2\sqrt{I_1 I_2} \cos \psi, \tag{1.108a}$$

where I_1 and I_2 are the intensities due to the separate waves ($I_1 = a_1^2$ and $I_2 = a_2^2$), whereas ψ is the phase difference of these waves ($\psi = \psi_2 - \psi_1$). As can be seen, the resultant intensity I is not simply the sum of the separate intensities I_1 and I_2, and depends on wave phase difference ψ. In such a situation the two waves are said to *interfere*.

Equation (1.108a) is frequently written as

$$I = I_1 + I_2 + I_{12}, \tag{1.108b}$$

where

$$I_{12} = 2\sqrt{I_1 I_2} \cos \psi \tag{1.109}$$

is the interference term. Depending on ψ this term varies from $+2\sqrt{I_1 I_2}$ to $-2\sqrt{I_1 I_2}$, and takes zero values if the phase difference is an odd multiple of $\pi/2$ ($\psi = \pm\pi/2,\ \pm 3\pi/2,\ \pm 5\pi/2,\ ...$). In this connexion, one can distinguish three particular situations in the interference phenomenon (Fig. 1.50):

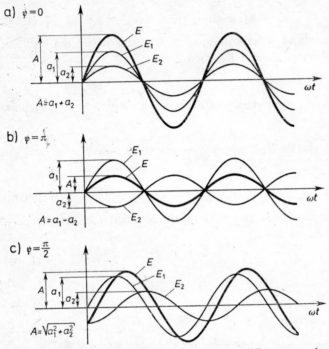

a) $\psi = 0$

$A = a_1 + a_2$

b) $\psi = \pi$

$A = a_1 - a_2$

c) $\psi = \dfrac{\pi}{2}$

$A = \sqrt{a_1^2 + a_2^2}$

Fig. 1.50. Two-wave interference. Drawn for phase difference equal to 0 (a), π (b) and $\pi/2$ (c).

(a) The phase difference is an even multiple of π ($\psi = 0,\ \pm 2\pi,\ \pm 4\pi,...$). Two waves which fulfil this condition are said to be *in phase*. They give a maximum resultant intensity equal to

$$I = I_1 + I_2 + 2\sqrt{I_1 I_2} = \left(\sqrt{I_1} + \sqrt{I_2}\right)^2 = (a_1 + a_2)^2, \tag{1.110a}$$

and one has a *constructive interference*.

(b) The phase difference is an odd multiple of π ($\psi = \pm\pi,\ \pm 3\pi,\ ...$). In this case the waves are said to be in *antiphase* and they give a minimum resultant intensity equal to

$$I = I_1 + I_2 - 2\sqrt{I_1 I_2} = \left(\sqrt{I_1} - \sqrt{I_2}\right)^2 = (a_1 - a_2)^2. \tag{1.110b}$$

Hence there is a *destructive interference*.

(c) The phase difference is an odd multiple of $\pi/2$ ($\psi = \pm\pi/2, \pm3\pi/2, \ldots$). Now, the interference term I_{12} is equal to zero, and

$$I = I_1 + I_2 = a_1^2 + a_2^2, \tag{1.110c}$$

i.e., the resultant intensity is equal to the sum of intensities due to both waves. In this case the principle of superposition is valid not only in respect to the wave amplitudes but also to the intensities.

According to Eq. (1.11), the phase difference ψ is related to the difference Δ between the optical paths traversed by the two waves by the relation

$$\psi = \frac{2\pi}{\lambda} \Delta, \tag{1.111}$$

where λ is the wavelength. Taking into account this formula, constructive interference (Fig. 1.50a) occurs if Δ is a multiple of λ and destructive interference if Δ is an odd multiple of $\lambda/2$.

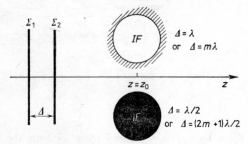

Fig. 1.51. Homogeneous (fringe-less) two-wave interference.

If the amplitudes a_1 and a_2 of the waves under consideration are the same ($a_1 = a_2 = a$), then Eq. (1.106) reduces simply to

$$I = 2a^2(1+\cos\psi) = 4a^2\cos^2\tfrac{1}{2}\psi, \tag{1.112}$$

and the resultant intensities are expressed for the above three cases by

$$
\begin{array}{llll}
I = 4a^2 & \text{for} \quad \psi = \pm 2m\pi & \text{or} \quad \Delta = \pm m\lambda, & (1.113\text{a}) \\
I = 0 & \text{for} \quad \psi = \pm(2m+1)\pi & \text{or} \quad \Delta = \pm(2m+1)\lambda/2, & (1.113\text{b}) \\
I = 2a^2 & \text{for} \quad \psi = \pm(2m+1)\pi/2 & \text{or} \quad \Delta = \pm(2m+1)\lambda/4, & (1.113\text{c})
\end{array}
$$

where $m = 0, 1, 2, 3, \ldots$

It has been assumed that two perfectly monochromatic waves have plane wavefronts Σ_1 and Σ_2 (Fig. 1.51), which propagate along the same direction z. The wavefronts are parallel to each other, and in such a situation the phenomenon of interference can be observed as a uniform field whose brightness varies according to the variations of the optical path difference Δ of the wavefronts. If, e.g., $\Delta = 0$ (or $\Delta = \pm m\lambda$), the interference field (*IF*) is maximally bright, and when $\Delta = \lambda/2$ (or $\Delta = \pm (2m+1)\ \lambda/2$), the interference field is maximally dark. This kind of interference is called *uniform* or *homogeneous interference*.

Let us now take two monochromatic waves propagating along different but intersecting directions z_1 and z_2 (Fig. 1.52). The wavefronts Σ_1 and Σ_2 are

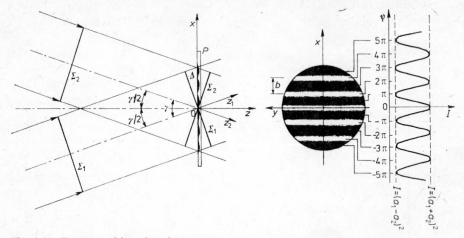

Fig. 1.52. Two-wave fringe interference.

no longer parallel to each other, but form an angle γ. At any value of time the optical path difference Δ between the wavefronts is no longer constant but continuously variable. Now the interference field occurring in the region where the wavefronts intersect is no longer homogeneous, but displays straight bright and dark fringes. What we have here is *fringe interference*. The centres of the bright fringes occur at those points of the interference field where $\Delta = 0, \pm \lambda, \pm 2\lambda, \dots$ (or phase difference $\psi = 0, \pm 2\pi, \pm 4\pi, \dots$), whereas for the centres of the dark fringes $\Delta = \pm \lambda/2, \pm 3\lambda/2, \dots$ (or $\psi = \pm \pi, \pm 3\pi, \dots$). Interfringe spacing (*b*), i.e., the distance between two neighbouring bright or dark fringes in a plane (*P*) perpendicular to the bisectrix of the angle γ is defined by the expression

$$b = \frac{\lambda}{2\sin\frac{1}{2}\gamma}, \tag{1.114a}$$

where λ is the wavelength. If, however, the interference pattern is observed or recorded in a plane P, the normal N of which makes an angle β with respect to the bisectrix z (Fig. 1.53), then Eq. (1.114a) takes the form

$$b = \frac{\lambda}{2\sin\frac{1}{2}\gamma\cos\beta} \cdot \tag{1.114b}$$

Fringe interference also occurs in the situation illustrated in Fig. 1.51, when at least one of the wavefronts is curved. If the wavefront Σ_1 is spherical, an interference field with ring fringes will appear.

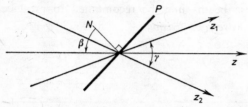

Fig. 1.53. Illustrating the notation relating to Eqs. (1.114).

Let $I_{\max}$ and $I_{\min}$ denote the maximum and minimum intensities of the bright and dark fringes, respectively. The *visibility* (or *contrast*) of the interference fringes may be expressed by a simple relation

$$V = \frac{I_{\max}-I_{\min}}{I_{\max}+I_{\min}} \cdot \tag{1.115}$$

This is a very important parameter because in certain cases it is a measure of the degree of coherence of the interfering light waves. The waves are perfectly coherent if $V = 1$, and completely incoherent when $V = 0$. If $0 < V < 1$, the waves are partially coherent.

Two light waves are coherent only if their phase differences are constant in function of time. A common light source emits a succession of finite wave trains whose phase relation changes randomly at the rate of 10^8 or more per second. When such trains from two independent sources are superposed, they also interfere and their instantaneous resultant intensity is still defined by Eq. (1.106) or (1.108), but $\cos\psi$ will change rapidly in the range between $+1$ and -1. Neither the eye nor other photoreceptors record such rapid intenisty variations, but can appreciate only an average value of intensity over a period of time which includes many variations of $\cos\psi$ between $+1$ and -1. Because these variations are random, i.e., incoherent, the average value of intensity is equal to $I = a_1^2 + + a_2^2 = I_1 + I_2$. Hence no interference effects are recorded.

In order to observe the interference of two light waves, their phase difference ψ should be constant over a sufficiently long period of time. In practice, such constant-phase waves are generated by splitting a single initial wave into two waves. There are two fundamentally different ways of performing this operation. The first one depends on a division of a single wavefront into two components in such a manner that each of them retains a fraction (preferably a half) of the orginal amplitude. This operation is known as *amplitude division*. It is usually accomplished by means of a semitransparent mirror. The second way depends on the separation, by means of a double slit for instance, of two light beams from a single wavefront. Next, these beams may be recombined to produce interference. This operation is called *wavefront division*.

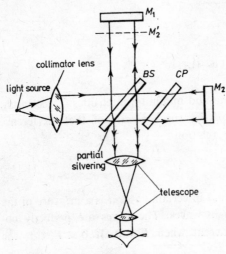

Fig. 1.54. Michelson interferometer.

A classical example of an interference system based on amplitude division of the wavefront is the *Michelson interferometer* (Fig. 1.54). Its basic elements are a beam-splitter *BS* and two plane mirrors M_1 and M_2 adjusted at right angles to each other. Light falls upon the beam-splitter (a half-silvered plate), and is split into two beams which are reflected in the mirrors, and then recombined by the *BS*. A compensating plate *CP* is usually inserted into one of the beams to equalize the optical paths of both beams. The principle of this instrument can be easily understood if the image M_2' of the mirror M_2 is regarded as being produced by the plate *BS* in the space of the mirror M_1. If there are no relevent phase shifts upon reflections, the optical path difference between the two partial wavefronts is given by

$$\Delta = 2nt\cos\theta,$$

(1.116a)

where n is the refractive index of the surrounding medium (air), t is the distance between M_1 and M'_2, and θ is the angle of incidence at the mirrors. Normally $\theta = 0$, and

$$\Delta = 2nt. \tag{1.116b}$$

The optical path difference Δ can be varied at will by the axial displacement of one of the mirrors which changes t. Starting from $t = 0$, one can determine the coherence length of used light by moving the mirror until the interference disappears. When M_1 and M'_2 are parallel to each other, the uniform interference is observed in the centre of the field of view. On the other hand, if M_1 and M'_2 are not parallel, fringe interference occurs.

One of the simplest interference systems with wavefront division was developed by Young in 1801 (Fig. 1.55). Monochromatic light from a small source

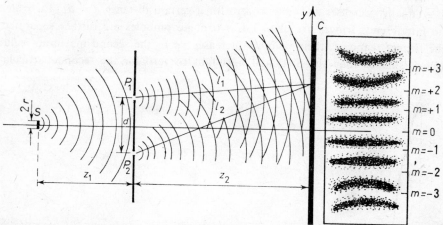

Fig. 1.55. Young's interference.

S passes through two pinholes P_1 and P_2 situated equidistant from the source. Thus in P_1 and P_2 the original light beam is divided by splitting its wavefront. When the portions of the wavefront at P_1 and P_2 are mutually coherent, interference occurs in the region of space where there is the superposition of waves spreading from the pinholes. If a screen C is located at a sufficiently long distance z_2 from the pinholes, fringe interference can be observed on it. The maxima of light intensity (centres of bright fringes) will occur for

$$y = \frac{z_2}{d} m\lambda, \tag{1.117a}$$

and intensity minima (centres of dark fringes) for

$$y = \frac{z_2}{d}(2m+1)\frac{\lambda}{2},$$ (1.117b)

where y is the distance from the optical axis passing through the centre of the light source S symmetrically with respect to the pinholes P_1 and P_2, whereas $m = 0, \pm 1, \pm 2, \ldots$ Equations (1.117) give the interfringe spacing

$$b = \frac{z_2}{d}\lambda.$$ (1.118)

The Young interference system is a very useful device for the study of the effects of the spatial coherence of light. If the distance d between pinholes P_1 and P_2 is varied, while the other distances are fixed, the fringe visibility V (see Eq. (1.115)) changes. This quantity has a high maximum for $d \to 0$. As d increases, V gradually decreases and falls to zero for a certain distance $d = d_1$, for which no interference fringes are observed. When the pinholes are further separated, the fringes reappear, their visibility increases up to the second maximum value (much smaller than the first) and next falls to zero for the second particular

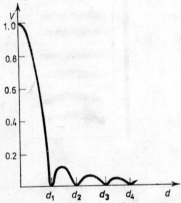

Fig. 1.56. Visibility of interference fringes as a function of pinhole separation d in Young's set-up (Fig. 1.55).

distance $d = d_2$. If we separate the pinholes still further we again obtain the effect shown in Fig. 1.56 for a circular light source. A similar effect is obtained if d is constant, but we vary the light source diameter $2r$ (Fig. 1.55).

There is, however, a siginificant difference between the successive interference patterns. If for $0 < d < d_1$ we have the centre of a bright interference fringe at a given y, then for $d_1 < d < d_2$ the centre of a dark fringe appears at y; for $d_2 < d < d_3$ the same holds good as for $0 < d < d_1$, and for $d_3 < d < d_4$ as for $d_1 < d < d_2$, and so on.

Strictly regarding the problem by using the notation of light coherency, the fringe visibility V can be more generally expressed by

$$V = 2|\gamma_{12}| \frac{\sqrt{I_1 I_2}}{I_1 + I_2}, \tag{1.119}$$

where γ_{12} is the *complex degree of coherence*, $|\gamma_{12}|$ is the modulus of γ_{12}, whereas I_1 and I_2 are the total intensities at the pinholes P_1 and P_2, respectively. When $I_1 = I_2$, the visibility V is equal simply to $|\gamma_{12}|$. Specifically, for a uniformly luminous circular source of radius r, the complex degree of coherence for two pinholes, P_1 and P_2 (Fig. 1.55), separated by a small distance d, is given by

$$\gamma_{12}(D) = \frac{2J_1(D)}{D}, \tag{1.120}$$

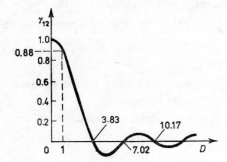

Fig. 1.57. Plot of Eq. (1.120).

where $J_1(D)$ is the first-order Bessel function, the argument of which

$$D = \frac{2\pi}{\lambda} d \sin \sigma \approx \frac{2\pi}{\lambda} d \frac{r}{z_1}, \tag{1.121}$$

where σ is the angular radius of the source seen from the middle point of the distance d between the pinholes P_1 and P_2, whereas z_1 is the distance between the source and the pinholes. The plot of Eq. (1.120) and several values of D for which γ_{12} takes extreme and zero values are shown in Fig. 1.57. When $D = 1$, the degree of coherence $\gamma_{12} = 0.88$. This value is usually taken as a measure for almost coherent light, while the light is practically incoherent if $|\gamma_{12}| \leqslant 0.05$. Otherwise, light is regarded as partially coherent ($0.05 < |\gamma_{12}| < 0.88$).

To obtain a good visibility of interference fringes, the modulus of γ_{12} should be not smaller than 0.88 when $D = 1$, i.e., for

$$d = d_{coh} = 0.16 \frac{\lambda}{\sin \sigma} \approx 0.16\lambda \frac{z_1}{r}. \tag{1.122}$$

This relation results simply from Eq. (1.121) by substituting $D = 1$. The distance d_{coh} between two pinholes, P_1 and P_2 (Fig. 1.55), is the diameter of the *coherence patch*, over which light waves are coherent for many practical purposes.

On the contrary, the interference fringes vanish completely if $\gamma_{12} = 0$. For the first time this situation occurs for $D = 3.83$, i.e., for

$$d = d_1 = 0.61 \frac{\lambda}{\sin \sigma} \approx 0.61\lambda \frac{z_1}{r}. \tag{1.123}$$

This relation, like Eq. (1.122), results from Eq. (1.121) by substituting $D = 3.83$.

Frequently Eq. (1.122) is used to determine the maximum size of a light source to be used in optical instruments, especially in microscopes, where a given area is to be coherently illuminated. If the diameter of this area is d_{coh}, the radius of the light source should be equal to

$$r = 0.16\lambda \frac{z_1}{d_{coh}}. \tag{1.124}$$

As can be seen, the source may increase in direct proportion to the distance z_1 between the source and the area being coherently illuminated.

These and other spatial coherence effects can be easily demonstrated by means of an ordinary microscope using a condenser with a slit diaphragm and an object

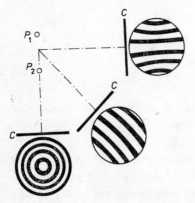

Fig. 1.58. Shape of Young's interference fringes depending on the direction of observation.

slide with two slits instead of the circular light source and two pinholes (see Subsection 3.2.2).

Fringes produced by means of Young's interference system do not occur in a fixed region of space. They may appear at any distance z_2 from the pinholes P_1 and P_2 (Fig. 1.55) where the two interfering waves overlap. Such interference fringes are called *non-localized*. Their shape depends on the direction of observation or on the relative orientation of the pinholes and the screen C (Fig. 1.58), and results simply from the geometry of Young's system. As can be seen in Fig. 1.55, for a given interference order the optical path difference is constant, i.e.,

$$l_2 - l_1 = \text{const.} \tag{1.125}$$

This relation is the equation of a hyperboloid of revolution the foci of which are at the pinholes P_1 and P_2. Different sections (C, Figs. 1.58 and 1.59) through a group of such hyperboloids of various interference orders determine the shape of the interference fringes.

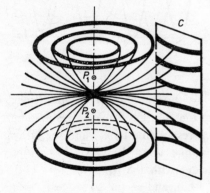

Fig. 1.59. Interference hyperboloids.

1.6.2. Interference between two heterochromatic waves

Up to now it has been assumed that interfering light waves are perfectly monochromatic. Now let us put aside this assumption and induce heterochromatic light to produce interference fringes, e.g., by using Young's system (Fig. 1.55). Firstly, for the sake of simplicity, let a wave train contain only two slightly different wavelengths λ_1 and λ_2. Only waves of the same wavelength are able to interfere constructively and destructively. Thus the two split waves of wavelength λ_1 produce one fringe pattern (see Fig. IIa), and the two waves of wavelength λ_2 produce another (see Fig. IIb). Both patterns are mutually incoherent and superpose only incoherently, i.e., by intensities (see Fig. IIc). The interfringe spacings in the corresponding patterns are, according to Eq. (1.118), equal to

$$b_1 = \frac{z_2}{d}\,\lambda_1, \tag{1.126a}$$

$$b_2 = \frac{z_2}{d}\,\lambda_2. \tag{1.126b}$$

If $\lambda_2 > \lambda_1$, then also $b_2 > b_1$. At the centre of the interference patterns, i.e., at $y = 0$ (Figs. 1.55 and 1.60), the zero-order fringes coincide. But as the distance y from these fringes increases the fringes of higher orders corresponding to λ_1

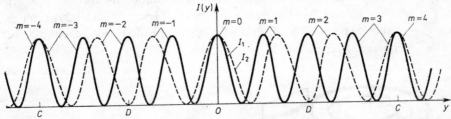

Fig. 1.60. Two-wave interference of bichromatic light.

occur behind the fringes corresponding to λ_2. For a number of interfringe spacings the decoincidence of fringes is so great that a bright fringe corresponding to λ_1 falls on the neighbouring dark fringe corresponding to λ_2, and the latter disappears (points D in Fig. 1.60). If we continue along the y-axis, the fringe of the m-th order corresponding to λ_2 coincides with the fringe of $(m+1)$-th order corresponding to λ_1 (points C in Fig. 1.60), and so on as we go further along the y-axis.

Let q denotes the difference between the orders of the interference fringes belonging to λ_1 and λ_2. Thus the same optical path difference Δ can be expressed by

$$\Delta = m\lambda_1 = (m-q)\,\lambda_2. \tag{1.127}$$

This equation shows that

$$m = q\,\frac{\lambda_2}{\lambda_2 - \lambda_1}. \tag{1.128}$$

If q is equal to $0, 1, 2, \ldots$, Eq. (1.128) gives the interference orders m, corresponding to λ_1, for which the bright fringes of the two interference patterns coincide.

The coincidence reoccurs after a number of interfringe spacings b_1. This number results simply from Eq. (1.128), and is given by

$$M = (q+1)\frac{\lambda_2}{\lambda_2 - \lambda_1} - q\,\frac{\lambda_2}{\lambda_2 - \lambda_1} = \frac{\lambda_2}{\lambda_2 - \lambda_1}. \tag{1.129}$$

Let us now consider a light source S (Fig. 1.55) generating three different wavelengths λ_1, λ_2, and λ_3. Each of them produces an individual interference pattern on a screen C. Figure 1.61, like Fig. 1.60, shows the intensity distributions I_1, I_2, and I_3 corresponding to each pattern. The resulting interference pattern is a summation of the individual intensity distributions. As can be seen, the situation is now much more complicated then for two wavelengths (Fig. 1.60). The interference fringes corresponding to λ_1, λ_2, and λ_3 coincide in the centre of the screen at $y = 0$, but when advancing along the y-axis, it is difficult to find a point where there is another coincidence.

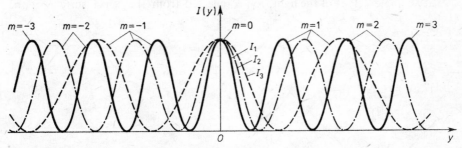

Fig. 1.61. Two-wave interference of polichromatic light.

If in Young's system (Fig. 1.55) the pinholes P_1 and P_2 are illuminated by white light, then the screen C is covered by many interference patterns of different colours since the white light is composed of an infinite number of wavelengths. In this case a coincidence of bright fringes occurs only at the centre of the screen, where the combination of all the wavelengths produces a white central fringe (see Fig. IId). As the distance from this fringe increases, the decoincidence of the individual interference patterns is more and more noticeable, and after passing a few coloured bands the interference fringes disappear and a bright white field occurs. The interference of white light can often be of practical use, especially in interference microscopy, as this kind of interference enables the zero-order fringe to be easily identified. Apart from that, it is largely used for qualitative and semiquantitative studies of thin films, for differential interference contrast, phase-interference contrast, and other microscopical techniques.

1.6.3. *Thin film interference: basic phenomena and applications in microscopy*

Interference of light is easily produced by a thin film the two surfaces of which divide the amplitude of an incident light wave. Thin films are familiar to all microscopists. In particular, microtom sections of tissues, numerous biological membranes, and oxide layers on metallographic samples belong to this class of objects. Thus, thin film interference plays a very important part in many micro-scopical methods, and especially in reflected contrast microscopy (see Chapter 8). There are two main types of interference: that produced by parallel-sided films and that produced by thin films of varying thickness. In the first case interference fringes of *equal inclination* (*Haidinger fringes*) occur, whereas in the second case interference fringes of *equal thickness* (*Fizeau fringes*) are observed.

Two-beam interference fringes of equal inclination. The principle of this type of interference is shown in Fig. 1.62. A light beam, *S*, strikes a transparent film, *TF*, at an angle θ. Part of the light, R_1, is reflected from the upper surface of the

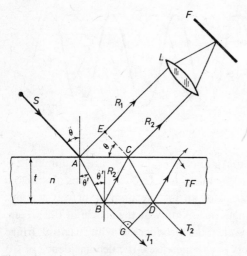

Fig. 1.62. Principle of formation of interference fringes of equal inclination on a thin film.

film, and another part, R_2, from the bottom surface. These two parts can interfere with each other. A portion of the light, T_2, reflected internally from the upper surface of the film can also interfere with the light T_1, which traverses the film directly. The interference can be observed either with the naked eye when its accommodation is relaxed or in the focal plane *F* of a lens *L*.

The optical path difference Δ between the superposed rays R_2 and R_1, or T_2 and T_1, is given by

$$\Delta = (AB+BC)n - AE = (BC+CD)n - BG = 2nt\cos\theta', \tag{1.130}$$

where θ' is the angle of refraction as shown in Fig. 1.62, whereas n and t are the refractive index and the thickness of the film, respectively. This optical path difference produces a phase difference

$$\psi = \frac{2\pi}{\lambda}\Delta = \frac{2\pi}{\lambda}2nt\cos\theta'. \tag{1.131}$$

Between the reflected rays, R_1 and R_2, there is a difference: the first one is reflected by an optically denser medium, while the second ray is reflected by a less dense medium (air). Hence in the first case there is a jump of phase by π (see Subsection 1.5.4). No phase jump occurs between the transmitted rays T_1 and T_2. Thus the resultant phase difference ψ for the reflected rays, R_2 and R_1, is given by

$$\psi = \frac{2\pi}{\lambda}\Delta - \pi = \frac{2\pi}{\lambda}2nt\cos\theta' - \pi, \tag{1.132}$$

whereas for the transmitted rays, T_2 and T_1, the phase shift ψ is defined by Eq. (1.131).

The interference is constructive when $\psi = 2m\pi$, and it is destructive if $\psi = (2m \mp 1)\pi$. Thus for the reflected rays, R_1 and R_2, constructive interference occurs when $\Delta = (m+\frac{1}{2})\lambda$, i.e., if

$$2nt\cos\theta' = (2m+1)\tfrac{1}{2}\lambda, \tag{1.133a}$$

and destructive interference results when $\Delta = m\lambda$, i.e.,

$$2nt\cos\theta' = m\lambda, \tag{1.133b}$$

where $m = 0, 1, 2, \ldots$ The opposite holds good for the transmitted rays, T_1 and T_2. They interfere constructively if

$$2nt\cos\theta' = m\lambda, \tag{1.134a}$$

and destructively when

$$2nt\cos\theta' = (2m+1)\tfrac{1}{2}\lambda. \tag{1.134b}$$

An arrangement for observing the interference fringes of equal inclination in both the reflected rays from and the transmitted light through a transparent plane parallel film is shown in Fig. 1.63. It will be seen that the fringes by transmission are complementary to those observed by reflection. The former, however, show less contrast than the other because the transmitted rays which have undergone two reflections (ray T_2 in Fig. 1.62) are much less intense than the others. For a particular bright or dark fringe the optical path difference Δ, defined by Eq. (1.130), is constant so that for $nt = $ const the same angle of refraction θ', or

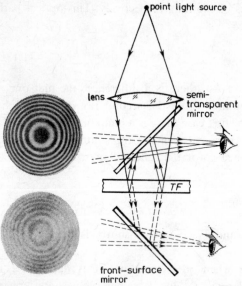

Fig. 1.63. Arrangement for observation of interference fringes of equal inclination in both reflected and transmitted light.

the same angle of incidence θ, is represented by the fringe of a given interference order. Such interference fringes are therefore known as fringes of equal inclination.

Interference fringes of equal thickness. These are much more useful in microscopy than fringes of equal inclination. They are produced by a transparent film with two nonparallel surfaces (Fig. 1.64), and can be either observed with the naked eye directly focused at the film or displayed on a screen by means of a lens. A pure pattern of interference fringes of equal thickness, i.e., without fringes of equal inclination, occurs only when a parallel light beam strikes a film with unparallel surfaces. Then a local optical path difference between two reflected rays R_1 and R_2, as well as between the transmitted rays T_1 and T_2, is defined by the same expressions for constructive and destructive interference as previously for fringes of equal inclination (Eqs. (1.130)–(1.134)). But now the optical thickness nt

is variable along the thin film *TF*, while the angle of incidence θ is constant, and a particular fringe of a given interference order represents the same optical thickness; the fringes are therefore known as *fringes of equal thickness*. They are easily observed in both reflected and transmitted light by using an arrangement as shown in Fig. 1.63, in which a small light source is placed exactly in the focal point of the lens. In that case a parallel light beam falls normally onto the film

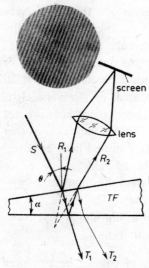

Fig. 1.64. Principle of formation of interference fringes of equal thickness in a thin film.

$(\theta = 0)$. In reflected light the bright fringes are then observed along lines where the optical thickness

$$nt = (2m+1)\tfrac{1}{4}\lambda,\tag{1.135a}$$

and the dark fringes along the lines where

$$nt = m\tfrac{1}{2}\lambda.\tag{1.135b}$$

In transmitted light the opposite holds good: Eq. (1.135a) is valid for dark fringes and Eq. (1.135b) for bright fringes.

A single interfringe spacing *b* corresponds to a variation of the optical thickness *nt* by $\lambda/2$. If the angle α between the plane surfaces of the film is small, then

$$b = \frac{\lambda}{2n\alpha}.\tag{1.136}$$

This formula shows that as the angle α becomes smaller the interfringe spacing b increases; in the extreme case that $\alpha = 0$, uniform interference occurs.

Inteference colours of thin films. Interference colours produced by thin films are not rare in microscopy and are frequently used to estimate the thickness of thin specimens, microtom sections for instance. These colours are, in general, very large or infinitely wide interference fringes of equal thickness occurring in white light when the angle α (Fig. 1.64) between the two surfaces of a thin film is very small or equal to zero. In this case for a given optical thickness nt of the film, a particular wavelength λ_0 produces constructive interference, while other particular wavelengths $\lambda_1 < \lambda_0$ and $\lambda_2 > \lambda_0$ produce destructive interference, or vice versa. Thus the intensity of the light of wavelength λ_0 becomes amplified, while that of the light of wavelengths λ_1 and λ_2 is reduced, or vice versa. It is a characteristic feature of this phenomenon that some interference hues are not spectrum colours. Purple, for instance, is a composition of violet and red. Another characteristic is the fact that interference colours for transmitted light are complementary with respect to the colours observed in reflected light.

Table 1.8 shows the interference colours of transparent thin films as a function of their optical path difference Δ or optical thickness. This table may be used for an approximate estimation of the thickness of a thin film if its refractive index is known (for an air film this index is simply unity).

Newton's interference rings. Interference fringes may, of course, be produced by an air film which is formed, for example, by two glass plates P_1 and P_2 (Fig. 1.65). In this instance constructive and destructive interference are defined by the same relations as were previously valid for a solid film (Eqs. (1.130)–(1.136)) although there is a small difference in phase jumps. The reflected ray R_1 does

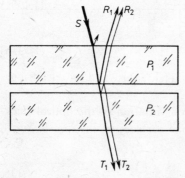

Fig. 1.65. Illustrating the formation of interference fringes on an air film.

TABLE 1.8

Scale of interference colours [1.31]

Optical path difference Δ [μm]	Interference colours of thin films	
	In reflected light	In transmitted light
	Interference colours of polarized light	
	Between crossed polarizers	Between parallel polarizers
0.000	Black	Bright white
0.040	Iron-grey	White
0.097	Lavender-grey	Yellowish white
0.158	Greyish blue	Brownish white
0.218	Clearer grey	Brownish yellow
0.234	Greenish white	Brown
0.259	White	Light red
0.267	Yellowish white	Carmine
0.275	Pale straw-yellow	Dark reddish brown
0.281	Straw-yellow	Deep violet
0.306	Light yellow	Indigo
0.332	Bright yellow	Blue
0.430	Brownish yellow	Greyish blue
0.505	Reddish orange	Bluish grey
0.536	Red	Pale green
0.551	Deep red	Yellowish green
0.565	Purple	Lighter green
0.575	Violet	Greenish yellow
0.589	Indigo	Golden yellow
0.664	Sky-blue	Orange
0.728	Greenish blue	Brownish orange
0.747	Green	Light carmine
0.826	Lighter green	Purplish red
0.843	Yellowish green	Violet-purple
0.866	Greenish yellow	Violet
0.910	Pure yellow	Indigo
0.948	Orange	Dark blue
0.998	Bright orange-red	Greenish blue
1.101	Dark violet-red	Green
1.128	Light bluish violet	Yellowish green
1.151	Indigo	Dirty yellow
1.258	Greenish blue	Flesh colour
1.334	Sea-green	Brownish red
1.376	Diamond green	Violet

not change its phase by π when it reflects at the glass–air interface, but a phase jump by π occurs for rays R_2, as well as for the transmitted rays T_1 and T_2, because they reflect from a denser medium (glass). Consequently, an additional phase difference, equal to π, appears between the rays R_1 and R_2, and one equal to 2π between the rays T_1 and T_2. But a phase jump by 2π does not alter the character of interference fringes.

If an air film is formed by a spherical glass surface and a plane surface, circular interference fringes of equal thickness, known as *Newton's rings*, are

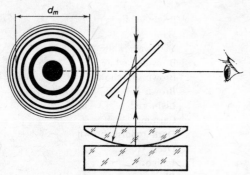

Fig. 1.66. Observation of Newton's interference rings.

observed under normal incidence of light (Fig. 1.66). This interference pattern enables the radius r of the spherical surface to be determined from the relation

$$d_m^2 = 4rm\lambda, \tag{1.137a}$$

by measuring the diameter d_m of the m-th dark ring, where $m = 1, 2, 3, \dots$ If bright fringes are taken into account, the radius r results from

$$d_m^2 = 4r(m + \tfrac{1}{2})\lambda. \tag{1.137b}$$

The above equations are valid for reflected light, whereas for dark fringes in transmitted light Eq. (1.137b) should be used, and vice versa, Eq. (1.137a) for bright fringes.

Antireflecting interference films. Up to now it has been assumed that there is either an optically less dense (Fig. 1.62 and 1.64) or denser (Fig. 1.65 and 1.66) medium on each side of a thin film. Now, let us consider a transparent film (*TF*) surrounded by an optically less dense medium (M_1), e.g., air, on one side and by a denser medium (M_2), e.g., glass, on the other (Fig. 1.67). The light (S) traverses the medium M_1 and strikes the film TF at an angle θ. As before, part of the light, R_1, is reflected from the upper surface of the film, and another part, R_2, from the bottom surface. In this case, however, both the ray R_1 and the ray R_2 are reflected by optically denser media. Hence both rays take a phase jump equal to π, and the resulting phase difference ψ, or optical path difference Δ, is defined by Eq. (1.131). The case differs for transmitted rays T_1 and T_2 because the first one passes directly through the film and the second is twice reflected inside it. The first reflection, at point B, is from a denser medium, and a jump of phase by π arises. As a result, an overall phase difference between the trans-mitted rays T_1 and T_2 is defined by Eq. (1.132). Thus the effect is the opposite to that where a thin film is surrounded by either a less dense medium (Figs. 1.62 and 1.64) or a denser medium (Figs. 1.65 and 1.66).

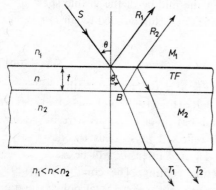

Fig. 1.67. Interference on a thin film surrounded by an optically less dense medium on one side and by a denser medium on the other.

Thin film interference as shown in Fig. 1.67 is largely used to reduce of light reflection from the surfaces of lenses, prisms, plates, and other optical elements. For an interface between two transparent media of refractive indices n_1 and n_2, the percentage of light that is reflected at normal incidence is given by Eq. (1.84), which may be written as

$$\varrho = \left(\frac{n_2 - n_1}{n_2 + n_1} \right)^2 100\%. \tag{1.138}$$

For glass, whose refractive index n_2 is equal to 1.52, Eq. (1.138) yields a reflectance of 4% at each air–glass interface. Within a microscope there are many such air–glass interfaces, and if some precautions are not taken, these individual 4% reflectances are capable of adding up to a reduction of 50% of the initial light intensity. Moreover, the light reflected within the microscope greatly reduces the contrast and overall quality of the microscopic image. This defect is especially critical for the observation of non-metallic or dielectric specimens of low reflectance using an incident-light microscope.

For this reason, the optical elements of modern microscopes are coated with *antireflecting films* to decrease interferentially the reflections of light at air–glass interfaces. This antireflecting coating is based on a simple principle: the film *TF* (Fig. 1.67) is chosen so that its optical thickness nt and refractive index n satisfy the conditions

$$nt = \tfrac{1}{4}\lambda_0, \tag{1.139}$$

$$n = \sqrt{n_1 n_2}, \tag{1.140a}$$

where λ_0 is a selected wavelength, usually in the middle of the visible spectrum. If the surface of the glass with a refractive index n_2 is surrounded by air ($n_1 = 1$), Eq. (1.140a) reduces simply to

$$n = \sqrt{n_2}. \tag{1.140b}$$

Equation (1.139) yields a phase difference $\psi = \pi$ (or optical path difference $\Delta = \lambda_0/2$) between the reflected rays R_1 and R_2 (Fig. 1.67), and these will therefore interfere destructively. Whereas Eq. (1.140a) or (1.140b), as it results from Eq. (1.84), gives equal intensities for both reflected rays. Thus the destructive interference is complete, and the resultant reflected intensity will be zero.

This is the simplest kind of antireflecting coating. The condition (1.139) only obeys the wavelength λ_0 and for neighbouring wavelengths of polichromatic light there will be some reflection. If this condition is fulfilled for the central yellow-green part of the spectrum, some of the blue-violet and red light is reflected. Therefore, when we observe such an antireflecting coating, it shows purple in reflected white light. This colour is typical of most antireflecting coatings of lenses and other optical elements. It should be also noted that the conditions (1.139) and (1.140) only hold at normal incidence of light. At oblique incidence there will be some residual reflection for λ_0 as well. Besides, there are no materials which could be used for the production of antireflecting films which would exactly meet the condition (1.140a). For instance, glass with a refractive index $n_2 = 1.52$ and air ($n_1 = 1$) need a film with a refractive index $n = \sqrt{1.52} = 1.235$. There is, however, no material with such a low refractive index. The most typical materials for antireflecting coatings, using vacuum evaporation, are cryolite Na_3AlF_6 ($n = 1.32$–1.35) and magnesium fluoride MgF_2 ($n = 1.36$–1.38).

In view of the above-mentioned obstacles the spectral reflectivity of a glass–air interface coated with a typical antireflecting film will be represented by graph 2 in Fig. 1.68. The reflectivity of glass may be further decreased by using antireflecting coatings composed of a pile of several different dielectric thin films. The antireflecting effect of such a *multiple-film coating* is shown in Fig. 1.68 by graph 3.

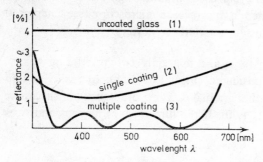

Fig. 1.68. Reduction of light reflections at an air–glass interface by antireflecting coatings.

Interference mirrors. A thin film the optical thickness of which is equal to an odd multiple of $\lambda/4$ and whose refractive index is much higher than that of a substrate (glass) acts as an effective reflector. High-efficiency *reflecting films* are usually made of TiO_2 ($n = 2.45$), ZnS ($n = 2.3$), and Sb_2S_3 ($n = 2.8$).

Glass plates with several dielectric films of an alternating high (H) and low (L) refractive index, each of optical thickness nt equal to $\lambda_0/4$, are known as very effective *interference mirrors*, the reflectance of which is nearly equal to 100%.

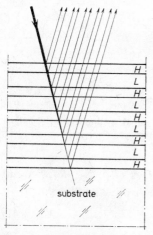

Fig. 1.69. Multilayer interference mirror giving high reflectivity at the desired wavelength λ_0.

In this case the phase difference between the rays reflected from successive thin film interfaces (Fig. 1.69) is equal to 2π or 0, the rays are therefore in phase and interfere constructively, so that a very high reflected intensity is produced.

The mirrors in Fabry–Pérot interferometers and laser resonators are based on this principle. The so-called *cool mirrors* used in some high-power microscope illuminators and *dichroic mirrors* used in fluorescence microscopy (see Chapter 9) also belong to this category of thin film interference devices.

Multiple-beam interference. In the foregoing text only two interfering beams, R_1 and R_2 or T_1 and T_2, were considered (Figs. 1.62–1.68). In reality, a thin film produces multiple reflection with a gradual decrease of intensity of successive reflected and transmitted rays (Fig. 1.70). If the film is homogeneous and does

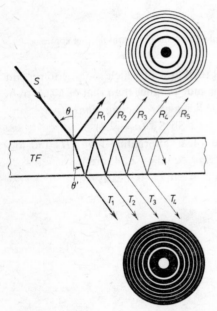

Fig. 1.70. Formation of multiple-beam interference fringes of equal inclination on a plane parallel film *TF*.

not absorb light, the intensities (I) of the reflected rays (R_1, R_2, R_3, ...) and transmitted rays (T_1, T_2, T_3, ...) are described, respectively, as follows:

$$I_{R_1} = \varrho I_S, \quad I_{R_2} = \tau^2 \varrho I_S, \quad I_{R_3} = \tau^2 \varrho^3 I_S, \quad I_{R_4} = \tau^2 \varrho^5 I_S, ..., \tag{1.141}$$

$$I_{T_1} = \tau^2 I_S, \quad I_{T_2} = \tau^2 \varrho^2 I_S, \quad I_{T_3} = \tau^2 \varrho^4 I_S, \quad I_{T_4} = \tau^2 \varrho^6 I_S, ... \tag{1.142}$$

where I_S is the intensity of incident light, ϱ and τ are the reflectance and transmittance of each surface of the film, respectively. The above formulae show that rays generated by reflections more than two in number can only be neglected when the reflectance ϱ of the film is small, as for glass and like dielectrics. Otherwise, interference is produced by a number of waves greater than two and *multiple-wave* (or *multiple-beam*) *interference* occurs. A mathematical description of such interference will be found in several books on physical optics (see, e.g., Refs. [1.32] and [1.33]) and is omitted here; only some concluding formulae will be included.

Appropriate calculation shows that where there is no absorption of light, the intensity distribution is described by

$$I_R = I_S \frac{F\sin^2(\tfrac{1}{2}\psi)}{1 + F\sin^2(\tfrac{1}{2}\psi)} \tag{1.143}$$

for the interference pattern in reflected light, and by

$$I_T = I_S \frac{1}{1 + F\sin^2(\tfrac{1}{2}\psi)} \tag{1.144}$$

for interference in transmitted light, where

$$F = \frac{4\varrho}{(1-\varrho)^2} \tag{1.145}$$

is the *coefficient of finesse* of interference fringes (*Fabry's coefficient*) and ψ is the phase difference as defined earlier (Eq. (1.131) or (1.132)).

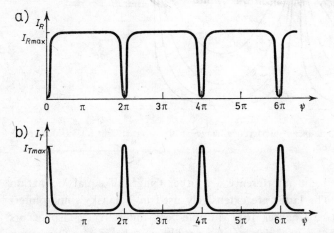

Fig. 1.71. Intensity of multiple-beam interference fringes in reflected light (a) and in transmitted light (b). Drawn for a given (rather high) coefficient of finesse F.

A graphical representation of Eq. (1.143) in function of ψ for a given (rather high) coefficient F is shown in Fig. 1.71a. The intensity maxima (I_{Rmax}) occur for $\sin\frac{1}{2}\psi = +1$ (or -1), and minima (I_{Rmin}) for $\sin\frac{1}{2}\psi = 0$. In this case

$$I_{Rmax} = I_S\frac{F}{1+F} \quad \text{and} \quad I_{Rmin} = 0. \tag{1.146}$$

Similarly, Fig. 1.71b shows an analogical graph of Eq. (1.144). Its maxima and minima of intensity occur for $\sin\frac{1}{2}\psi = 0$ and $\sin\frac{1}{2}\psi = +1$ (or -1), respectively. In this case

$$I_{Tmax} = I_S \quad \text{and} \quad I_{Tmin} = I_S\frac{1}{1+F}. \tag{1.147}$$

As can be seen, the intensity minimum I_{Tmin} is never equal to zero. The graphs show that the intensity minima for reflected light appear where the maxima of intensity occur for transmitted light. The interference pattern in reflected light takes the form of narrow dark fringes and in transmitted light of narrow bright fringes. The width of these fringes largely depends on the coefficient F. An example of this dependence is shown in Fig. 1.72. The greater F (or reflectance ϱ) the narrower are the fringes. It is useful to express their interfringe spacing in radians. This is equal to 2π.

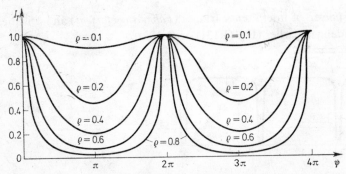

Fig. 1.72. Intensity of multiple-beam interference fringes in transmitted light for various values of reflectance ϱ of a thin film.

Fringes of multiple-beam interference are either fringes of equal inclination or of equal thickness. The latter are extensively used in Tolansky's microinterferometric technique (see Chapter 16). Multiple-beam interference also occurs as uniform field interference on which interference filters are based (see Subsection 1.6.4).

Interference fringes of equal chromatic order. When a thin film with multiple reflection is illuminated by white light, the reflection interference pattern is coloured and narrow dark fringes are normally invisible. If, however, the interference pattern is analyzed by means of a spectroscope or spectrograph, then in the focal plane of this instrument dark lines can be observed on the coloured background of the continuous spectrum of the white light. These dark lines are known as *fringes of equal chromatic order*, usually abbreviated to FECO, or as the *channelled spectrum*. They are sometimes used for the measurement of the thickness of thin films as well as for an evaluation of the roughness of polished surfaces [1.32, 1.34]. These applications of the FECO technique are discussed in Chapter 16.

1.6.4. Monochromatic interference filters for use in microscopy

Interference filters are one of the numerous practical applications of multiple-beam interference in thin films. In its simplest form the interference filter represents a glass plate P_1, on which three layers are vacuum deposited: metallic M_1, dielectric D, and metallic M_2, all protected with another glass plate P_2 (Fig. 1.73).

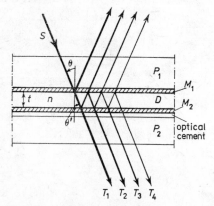

Fig. 1.73. Principle of a simple monochromatic interference filter.

The layers of carefully controlled thickness are arranged in such a way that two partially reflected layers M_1 and M_2 are separated by a transparent dielectric layer sufficiently thin to allow collimated white light to be transmitted only within a narrow spectral band. The dielectric layer is called the *spacer*. Its optical thickness nt determines a *peak wavelength* λ_p of the transmitted band. This wavelenght results from the relation

$$\frac{2\pi}{\lambda_p} 2nt\cos\theta' + \psi_j = m2\pi, \tag{1.148}$$

where ψ_j represents the overall phase jumps produced by the reflecting layers M_1 and M_2, and m is the interference order ($m = 1, 2, 3, \ldots$). This relation is a more general form of Eq. (1.134a).

The interference filter can act in both transmitted and reflected light. Here only the light transmission type will be taken into consideration.

A portion (T_1, Fig. 1.73) of the incident polichromatic light (S) is transmitted throught the filter, whereas another portion (T_2, T_3, T_4, ...) is reflected internally within the spacer before exiting from it. Thus the pathlenght of the second portion is elongated by the spacer so that constructive interference occurs between all the transmitted rays of wavelength λ_p for a given interference order m, whereas the neighbouring wavelenghths are interferentially cancelled. Unfortunately, λ_p is not a single peak wavelength because for another interference order another wavelength is simultaneously reinforced. As can be seen, several peak wavelengths result from Eq. (1.148), which can be rewritten as

$$\lambda_{p(m)} = \frac{2nt\cos\theta'}{m - (\psi_j/2\pi)}. \tag{1.149}$$

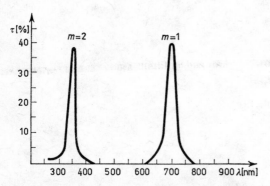

Fig. 1.74. Transmission of a typical metal–dielectric interference filter.

If, e.g., the chosen peak wavelength $\lambda_{p(1)}$ belongs to the first interference order ($m = 1$), then the wavelenght $\lambda_{p(2)}$ belonging to the second interference order ($m = 2$) also appears in the transmitted spectrum (Fig. 1.74). This second peak wavelenght must be blocked by another filter, usually by a coloured glass plate which is cemented with the interference filter instead of the protective glass plate P_2 as shown in Fig. 1.73.

In general, the interference filter is used in collimated light which strikes the filter normally. In this case Eq. (1.149) takes the form

$$\lambda_{p(m)} = \frac{2nt}{m} \qquad (1.150)$$

because the angle of incidence θ is zero (hence the angle of refraction $\theta' = 0$), and phase jumps ψ_j are small and can be neglected.

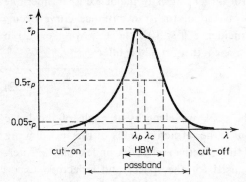

Fig. 1.75. Some graphical definitions and basic parameters of a typical interference filter.

Figure 1.75 shows some basic parameters of a typical interference filter. Those are: *peak transmittance* τ_p (maximum transmittance in the passband), peak wavelength λ_p (the wavelength where the filter has its highest transmittance), and *half bandwidth* (HBW). The latter is defined as the width of the passband at half the peak transmittance τ_p. Sometimes this parameter is also called the *half power bandwidth* or *full width at half maximum* (FWHM). The smaller the half bandwidth the more monochromatic is the filter. Monochromacy is additionally characterized by bandwidths at 10%, 5%, and even 1% of τ_p. Another useful parameter is also the slope (S), which is a measure of the steepness of the transmittance curve. The slope is defined as

$$S = \frac{\lambda_{80} - \lambda_5}{\lambda_{50}} \, 100\%, \qquad (1.151)$$

where λ_{80} is the wavelength at 80%, λ_5 at 5%, and λ_{50} at 50% of peak transmittance. For filters with unsymmetrical passband, the *centre wavelength* λ_c is also specified. This is determined by the centre point of the half bandwidth. For filters with symmetrical transmittance curve, $\lambda_c = \lambda_p$. Each filter has its *cut-on* and *cut-off points*. For bandpass filters, the cut-on point is the wavelength at which the transmittance reaches 5% of τ_p. On the contrary, the cut-off point is the wavelength where the transmittance falls to 5% of τ_p. The distance between both points is called the *passband*.

The peak wavelength and other parameters of interference filters are specified for collimated light at normal incidence. If the angle of incidence θ changes away from normal incidence, the filter transmittance spectrum shifts towards shorter wavelengths. At small angles θ, this shift is very useful in tuning a filter to the required peak wavelength, but for larger angles of incidence a considerable loss of peak transmittance and deformation of the transmittance curve occur. For collimated light and angles of incidence $\theta < 10°$, the following equation can be used to determine the peak wavelength of a tilted filter:

$$\lambda_\theta = \lambda_p \sqrt{1 - \left(\frac{n}{n_{\text{eff}}}\right)^2 \sin^2\theta}, \qquad (1.152)$$

where λ_θ and λ_p are, respectively, the peak wavelengths at angle of incidence (tilt angle) θ and at normal incidence, n is the refractive index of the surrounding medium ($n = 1$ for air), and n_{eff} is the effective refractive index of the filter. This index depends on the materials used for the spacer and reflecting layers, and can be determined by measuring λ_θ at different angles of incidence θ.

Divergent and convergent light are most typical in practice. If the total solid angle of the cone of light is smaller than 20°, an eventual shift of the peak wavelength can be evaluated by using half the full cone angle in Eq. (1.152) and dividing the result by half. As a rule, typical interference filters exhibit insignificant shifts of the peak wavelength for up to 5° solid angle.

The peak wavelength of an interference filter is also a function of the temperature of the filter. The shift is towards longer wavelengths and is equal to about 0.1 nm for each 5.5°C change in temperature. Thus it is important to face interference filters correctly. As a general rule, the highly reflecting (shiny or metallic looking) side of the filter should always face the light source. The thermal load will then be minimized, as the absorbing (blocking) coloured glass is facing away from the light source and most of the unwanted light is reflected back.[35]

Interference filters shown in Fig.1.73 are called *metal–dielectric filters*. Their typical half bandwidth is equal to 10–15 nm, and roughly results from the following equation:

[35] In general, interference filters tend to become increasingly unstable. As a rule, their peak wavelength progressively shifts towards longer wavelengths. This shift can amount to as much as a few nanometres after several years of use. Therefore, the peak wavelength position of interference filters must be checked occasionally by using a precise spectrophotometer.

$$\text{HBW} = \frac{1-\varrho}{m\pi}\,\lambda_{p(m)}, \qquad\qquad (1.153)$$

where ϱ is the reflectance of metallic layers M_1 and M_2. The peak transmittance of these filters is from 20 to 45%.

More efficient interference filters with HBW < 10 nm and τ_p up to 80% are made of dielectric substances only, and their reflectors M_1 and M_2 are constructed as shown in Fig. 1.69. These are called *dielectric or multidielectric interference*

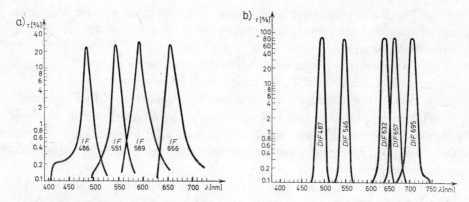

Fig. 1.76. Transmittance curves of typical metal–dielectric *IF* (a) and multidielectric *DIF* (b) interference filters.

filters. For comparison, Fig. 1.76 shows the transmittance curves of some typical metal–dielectric and multidielectric interference filters. It will be seen that the slope of the multidielectric filters is steeper than that of the metal–dielectric filters. The latter are thus less monochromatic than the former.

Generally metal–dielectric interference filters are in use in microscopy because they are much cheaper than multidielectric filters. The latter are, however, more convenient or even indispensable in exact microinterferometric measurements (see Chapter 16).

A very useful tool for much microscopical research is the *linear variable interference filter*. Its spacer has a continuously variable optical thickness, thus the transmitted peak wavelength changes with position along the length of this filter. It provides a continuous selection of the peak wavelength over a specified spectral range, typically from 400 to 700 nm, when the filter is displaced behind an entrance slit.

1.6.5. *Interference produced by birefringent elements*

It was shown in Subsections 1.2.3 and 1.4.5 that if light traverses a birefringent crystal there are, in general, two emergent waves, an ordinary wave and an extraordinary wave, both linearly polarized in orthogonal planes. These waves acquire a certain optical path difference and under some conditions can interfere in the region of space where they superpose.

The interference conditions for two such waves are summarized in *Fresnel–Arago laws* [1.4] :

(a) Two linearly polarized waves whose vibration directions are orthogonal cannot interfere destructively and constructively, hence they cannot produce interference fringes.

(b) Two linearly polarized waves which originate from orthogonally polarized components of unpolarized light cannot interfere destructively and constructively, even if their vibration directions have been rotated until they coincide.

(c) Two linearly polarized waves the vibrations of which are in the same direction can interfere constructively and destructively, provided they originate from the same linearly polarized wave or from the same linearly polarized component of unpolarized light.

An illustration that will help to make these laws more comprehensible is given in Fig. 1.77. An incident ray is split inside a double refracting plate D into two rays, o and e, whose vibration directions are at right angles (Fig. 1.77a). According to the first Fresnel–Arago law, these rays cannot interfere constructively and destructively with each other because they are incoherent. Neither does interference occur if the directions of vibration of the rays o and e are set parallel to each other by means of a polarizer (analyser) A, as shown in Fig. 1.77b related to the second Fresnel–Arago law.

It was mentioned earlier in this chapter that a ray of unpolarized light is equivalent to two mutually incoherent orthogonal components of equal amplitude. Thus in Figs. 1.77a and b the o-ray and the e-ray have no permanent phase difference between their vibrations; hence they cannot produce stable interference fringes. In order to make these two rays coherent, another polarizer P must be placed across the incident light beam so that only a single one of their linearly polarized components falls on the birefringent plate D (Fig. 1.77c). If the vibration directions of the polarizer P and analyser A are at 45° to the principal directions of light vibration in the plate D, the two interfering beams have the same amplitudes (a) and the most contrasty interference occurs. If the vibration directions of the polarizers, P and A, are adjusted to be parallel to each other, the intensity (I) in the interference pattern is given by Eq. (1.112), which in this case may be rewritten as

$$I = 4a^2\cos^2\frac{\varphi_d}{2} = 4a^2\cos^2\frac{\pi\delta_d}{\lambda}, \qquad (1.154a)$$

whereas between crossed polarizers the intensity is expressed by

$$I = 4a^2\sin^2\frac{\varphi_d}{2} = 4a^2\sin^2\frac{\pi\delta_d}{\lambda}, \qquad (1.154b)$$

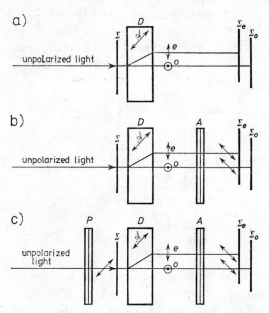

Fig. 1.77. An illustration of the Fresnel–Arago laws.

where φ_d and δ_d are, respectively, the phase difference and the optical path difference between the ordinary and extraordinary rays, o and e. These quantities are defined by Eqs. (1.61a) and (1.61b).

In the first case (Eq. (1.154a)) maximum values of intensity, $I = 4a^2$, occur for $\delta_d = 0, \lambda, 2\lambda, \ldots$, and minimum values, $I = 0$, for $\delta_d = \lambda/2, 3\lambda/2, 5\lambda/2, \ldots$ Whereas in the second case (Eq. (1.154b)) the converse is true: maximum values of I are for $\delta_d = \lambda/2, 3\lambda/2, 5\lambda/2, \ldots$, and minimum values for $\delta_d = 0, \lambda, 2\lambda, \ldots$

The first of the Fresnel–Arago laws should not, however, be taken to mean that there is no interaction between two superposed orthogonally polarized beams which are derived from the same linearly polarized beam. This law only states that interference fringes cannot be produced. Depending on the optical path difference δ_d between rays o and e, their superposition gives a resultant beam, which is, in general, elliptically polarized with right-handedness or left-handedness. In particular instances, when the phase difference $\varphi_d = 0, \pi, 2\pi, \ldots$, elliptical polarization becomes transformed into linear polarization, and if $\varphi_d = \pi/2, 3\pi/2, 5\pi/2, \ldots$ and the amplitudes of the component beams are the same, the resultant beam has a circular polarization.

Interference devices including birefringent elements as beam-splitters and beam-recombiners are said to be *polarization interferometers*. The ¦*Wollaston prism*, in particular, is one such element. It is made of two right-angle wedges of either quartz crystal or calcite cemented together so as to form a plane parallel plate (Fig. 1.78). The optic axes $\mathscr{A}$ of both wedges are at right angles to each other

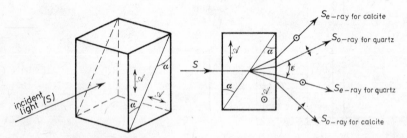

Fig. 1.78. Wollaston prism made of quartz or calcite and its principal cross section illustrating the splitting-up of an incident ray S into two rays: ordinary (S_o) and extraordinary (S_e).

and parallel to the external surfaces of the prism. At the cemented interfaces the incident ray S is angularly split into two components, S_o and S_e. If quartz is used, the S_e-ray is refracted towards the base and the S_o-ray towards the apex of the second wedge. For calcite the opposite holds good. The angular splitting ε of the rays S_o and S_e depends on the apex angle α of the wedges and on the birefringence $n_e - n_o$ of the crystal used. This dependence is given by

$$\varepsilon = 2|n_e - n_o| \tan \alpha. \qquad (1.155)$$

Later on we shall consider a Wollaston prism made of quartz, where the birefringence $n_e - n_o$ will be positive.

The Wollaston prism allows us to obtain both fringe interference and uniform interference. The former occurs when an extended light beam, in particular a collimated beam, strikes the prism placed between two polarizers, P and A (Fig. 1.79). Let Σ denote a plane wavefront of the linearly polarized beam $S - S'$. When passing through the Wollaston prism, the wavefront is sheared into two components, Σ_o and Σ_e, linearly polarized at right angles and deflected in the opposite sense. Hence, a continuously variable optical path difference Δ occurs between these components along the direction perpendicular to the apex edges of the prism. There is, however, a central line (C), along which the two sheared wavefronts intersect and where $\Delta = 0$. This line is determined by a transverse section of the Wollaston prism, where its two wedges have the same thickness.

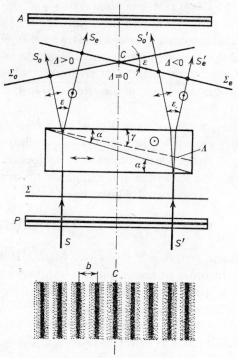

Fig. 1.79. Illustrating the formation of interference fringes by a Wollaston prism made of a positive birefringent crystal (e.g., quartz).

If on the one side of the central line C the optical path difference Δ is positive ($\Delta > 0$), then on the other side it is negative ($\Delta < 0$). When looking through the polarizer (analyser) A, one can observe an interference pattern inside the prism. This pattern consists of straight dark and bright fringes, whose direction is parallel to the apex edges of the prism wedges. If the polarizer P and analyser A are crossed, the dark fringes appear, where $\Delta = 0, \pm \lambda, \pm 2\lambda, ...$, whereas bright fringes occur at $\Delta = \pm \lambda/2, \pm 3\lambda/2, \pm 5\lambda/2, ...$ Conversely, dark fringes occur at $\Delta = \pm \lambda/2, \pm 3\lambda/2, \pm 5\lambda/2, ...$, and bright ones at $\Delta = 0, \pm \lambda, \pm 2\lambda, ...$, when the vibration direction of the analyser A is adjusted to be parallel to that of the polarizer P. A fringe for which $\Delta = 0$ is said to be the *fringe of the zero interference order*. The interfringe spacing (b) is given by

$$b = \frac{\lambda}{\varepsilon} = \frac{\lambda}{2(n_e - n_o)\tan\alpha}.$$

(1.156)

In white light the interference fringes, with the exception of the zero order fringe, are coloured (see Fig. IId). The variation of colours in a function of Δ is the same as for white light interference in isotropic thin films (see Table 1.8). Between crossed polarizers the interference colouration corresponds with that of isotropic thin films observed in reflected light, whereas interference colours occurring between parallel polarizers correspond with those of isotropic thin films observed in transmitted light. In the first case the fringe of the zero order is dark while in the second case it is bright.

A drawback of the Wollaston prism is an optical asymmetry which is responsible for the fact that interference fringes are not localized in a plane parallel to the external faces of the prism. In Fig. 1.79 this plane is marked by the pecked line (Λ). A simple trigonometrical calculation shows that it is inclined to the exit face of the prism at an angle γ which is approximatively given by

$$\gamma = \frac{1}{2}\left(\frac{1}{n_e} + \frac{1}{n_0}\right)\alpha.$$

(1.157)

If the Wollaston prism is made of quartz crystal, $\gamma \approx 0.65\alpha$ (see Section 16.4 in Vol. 3 and Ref. [1.35] for details). In order to cancel out this drawback, a combination of two identical Wollaston prisms may be used (Fig. 1.80a). Here the plane of localization of interference fringes (Λ) lies between the two prisms and is parallel to their external faces. Instead of this combination a single birefringent prism, as shown in Fig. 1.80b, can be used as well. This prism is called the *symmetrical Wollaston prism*.

Another drawback of the Wollaston prism is the localization of fringes inside it, which makes it difficult to use in interference microscopy (see Chapters 7 and 16). Nomarski devised a birefringent prism, whose plane of localization of interference fringes is situated outside the prism [1.36]. This he did by cutting one of the Wollaston prism wedges at an angle β to the optic (crystallographic) axis (Fig. 1.81). The plane of localization of interference fringes (Λ) is not, however, parallel to the external faces of this prism.

When a convergent light beam (S–S') strikes the Wollaston prism, and its convergence point B coincides with the plane of the localization of interference fringes, then a uniform interference arises, which can be observed, for instance,

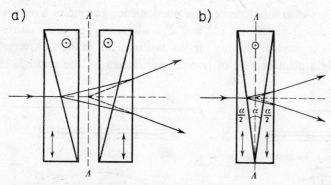

Fig. 1.80. A combination of two Wollaston prisms (a) and a symmetrical Wollaston prism (b) in which the plane Λ of localization of interference fringes is parallel to the external faces of the prisms.

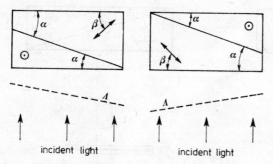

incident light   incident light

Fig. 1.81. Nomarski's birefringent prism.

on a screen E placed behind the analyser A. In this case the same optical path difference Δ occurs between the sheared wavefronts Σ_o and Σ_e, which are derived from a single incident wavefront Σ. This optical path difference may be varied from zero to some extreme values by shifting the prism in the transverse direction as shown in Fig. 1.82a by an arrow (p). When Δ is varied, the intensity of light on the screen changes in accordance with Eq. (1.154a) or (1.154b), where Δ should be used instead of δ_d. If the beam S–S' consists of white rays, then *uniform tint interference* occurs ; varying Δ causes variation of interference colours according to Table 1.8.

If, however, the Wollaston prism is shifted vertically, as shown in Fig. 1.82b, then its plane of the localization of interference fringes (Λ) will be beyond the convergence point B of the light beam, and a variable optical path difference Δ will occur between the wavefronts Σ_o and Σ_e. In this case fringe interference will appear instead of uniform interference. The phenomenon resembles Young's interference of two spherical wavefronts (Σ_o and Σ_e), whose centres of curvature (B_1 and B_2) are slightly separated. Shifting the Wollaston prism in the transverse direction p causes only a displacement of interference fringes on the screen.

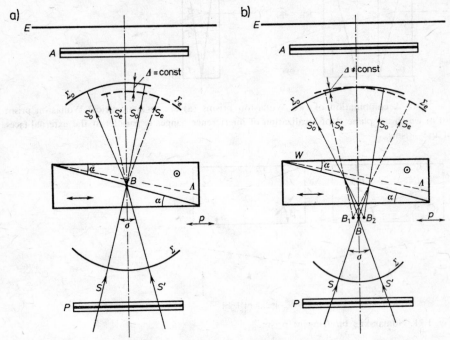

Fig. 1.82. Wollaston prism in a convergent light beam: a) formation of homogeneous interference, b) the prism produces fringe interference.

Extended uniform interference occurs only when the convergence angle σ (Fig. 1.82a) is small. Otherwise the interference pattern has the form of a double system of equilateral hyperbolic fringes (Fig. 1.83). The interference can be regarded as uniform only in the centre of the pattern. The hyperbolic appearence becomes more evident as the apex angle α of the Wollaston prism increases. This prism should therefore be used with light beams of a small angular aperture.

There are many other birefringent elements for double refracting interference systems [1.37], but the Wollaston prism and its modifications are the most useful in interference microscopy and microinterferometry.

Further details relating to the material presented in this subsection may be found in Refs. [1.32], [1.33] and [1.37].

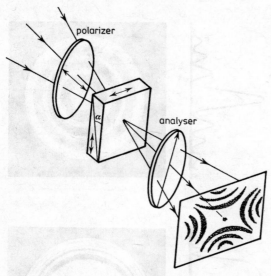

Fig. 1.83. Wollaston prism in a convergent light beam. Formation of hyperbolic fringes in the far field.

1.7. Diffraction of light

Light diffraction, like interference, is one of the fundamental phenomena inherent in microscopy. It always occurs when light meets an obstacle or aperture, in particular small particles, pinholes, slits, gratings, and like objects. In this section only grating-like objects are taken into consideration, and some basic conceptions relating to diffraction phenomena are explained. The diffraction approach to image formation in the microscope will be discussed in Chapter 3.

1.7.1. Fresnel and Fraunhofer diffraction

When light waves pass through apertures or meet obstacles, they are usually able to bend beyond the limit of the geometric shadow of the apertures and obstacles. This bending of waves is known as *diffraction*. Diffraction phenomena are easily explained by the *Huygens principle*, which states that each point on a wavefront can be considered as a new source of a wave. Figure 1.84a shows a portion of a collimated light beam being blocked by a baffle *B*, in which there is a pinhole. According to the Huygens principle, this same pinhole can be con-

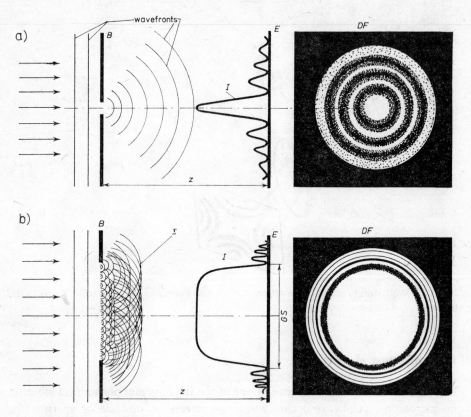

Fig. 1.84. Passage of a light wave through a pinhole (a) and a large aperture (b).

sidered as a new source of a spherical wave (in reality, semispherical wave). Similarly, an aperture (Fig. 1.84b) can be regarded as a source of many spherical wavelets whose superposition gives a new wavefront Σ beyond the baffle B. If a screen E is placed at a distance over the aperture, the light is observed on the screen beyond the geometric shadow GS of the aperture. The intensity distribution (I) at the geometric shadow limit and outside it is, however, not uniform. When the incident light is coherent, the diffracted rays are able to interfere with each other and produce interference fringes (DF), which in this case are called *diffraction fringes*. The appearance of the diffraction pattern observed on the screen E depends upon its distance z.

One distinguishes two basic kinds of *diffraction patterns*: *Fresnel* or *near-field diffraction patterns*, and *Fraunhofer* or *far-field diffraction patterns*. There is no sharp limit between either kind of diffraction, and they are specified by different criteria. In lensless systems a far-field diffraction pattern arise, theoretically at infinity but practically at a distance[36]

$$z > \frac{d^2}{\lambda},$$ (1.158a)

where λ is the wavelength of diffracted light and d is the diameter of the diffracting object. In lens systems the images formed in a plane optically conjugate to the light source are Fraunhofer diffraction patterns (they were observed by Fraunhofer in 1821). Anything else is Fresnel diffraction. In general, the Fraunhofer pattern differs from the Fresnel pattern, but sometimes both patterns are similar. The difference becomes more evident with increasing size of the diffracting obstacles or apertures. As a rule, characteristic features of the Fresnel and Fraunhofer

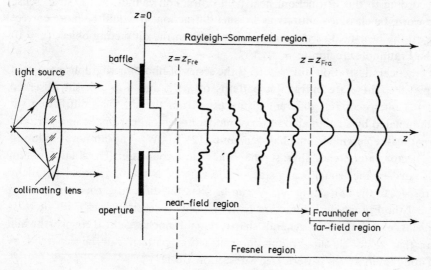

Fig. 1.85. Designation of various regions of light diffraction.

[36] Some authors even accept a smaller distance

$$z > \frac{d^2}{2\lambda}.$$ (1.158b)

diffraction are diffraction fringes at the edges of the geometrical shadows of objects for the former and fringed images of the light source for the latter [1.4].

This rather traditional classification of diffraction patterns presented here can be found in the majority of books on optics, although it is not entirely correct. A more precise classification has been proposed by Gaskill [1.8], who distinguishes three *regions of diffraction* (Fig. 1.85):

(a) the *Rayleigh–Sommerfeld region*, which comprises the entire space of diffraction from $z = 0$ to $z = +\infty$;

(b) the *Fresnel region*, which is a portion of the Rayleigh–Sommerfeld region; it extends from a distance $z = z_{Fre} > 0$ to $z = +\infty$;

(c) the *Fraunhofer region*, which is a portion of the Fresnel region; it extends from $z = z_{Fra} > z_{Fre}$ to $z = +\infty$.

In the last region the size of the diffraction pattern can increase with an increasing distance z, but the pattern shape remains the same, whereas in the intermediate region, between $z = z_{Fre}$ and $z = z_{Fra}$, both the size and the shape of the diffraction patterns change as the distance z increases.

According to this distinction, near-field diffraction cannot, in the strict sense of the word, be classified entirely as Fresnel diffraction. It would be more correct to regard the near-field as the region lying between the diffracting object ($z = 0$) and the Fraunhofer region ($z = z_{Fra}$).

A practical illustration to Fig. 1.85 is the series of photomicrographs presented in Fig. 1.86. These are diffraction patterns of a slit and a double-slit observed through a microscope at different distances z from the slit. The diffracting slit was illuminated by a collimated beam of moderately monochromatic and partially coherent light emerging from a halogen lamp (12V/100W). The lamp was followed by a collector, interference filter, slit diaphragm, and condenser. The slit diaphragm was located in the front focal plane of the condenser and orientated parallelly with respect to the diffracting slit. As can be seen, the diffractograms at $z \geqslant 8\ \mu m$ represent the Fraunhofer diffraction patterns; only their size varies with increasing distance z but their general shape remain unchanged. It is worthwhile noting that a double-slit Fraunhofer diffraction pattern can be regarded as the Young interference pattern (see Subsection 1.6.1).

Another example of the diffraction patterns is shown in Fig. 1.87. These are the Fresnel diffractograms of a circular aperture illuminated by a divergent beam emerging from a point source of highly monochromatic light. The distance (d) between the light source and the diffracting aperture was approximately equal to 500 mm. The diffractograms were recorded at a constant distance $z = 750$ mm, but the distance d was slightly changed. The diameter of the aperture was equal

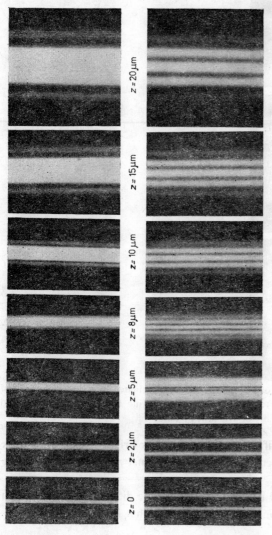

Fig. 1.86. Diffraction patterns of a slit (top) and a double-slit (bottom) observed through a microscope. The slit width is equal to 2 μm (top) and 0.8 μm (bottom). The distance between two slits (bottom) is equal to 3.5 μm.

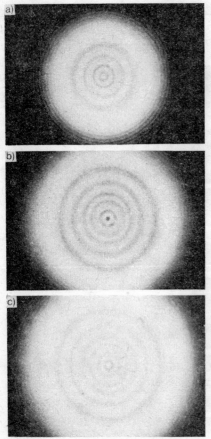

Fig. 1.87. Fresnel diffraction patterns of a circular aperture illuminated by a divergent beam of highly coherent (laser) light (by courtesy of K. Patorski, Institute of Design of Precise and Optical Instruments, Technical University of Warsaw, Poland).

to 1 mm. As can be seen, the diffractograms show their bright (a and c) or dark (b) centres due to an odd or even number of the Fresnel zones which enter the diffracting aperture [1.4].

The following discussion will deal primarily with the diffraction phenomenon in the Fraunhofer region, and only occasionally with that in the Fresnel region.

1.7.2. Diffraction gratings—basic objects for microscopy

An exceptionelly useful tool for the study of theoretical and practical problems of light microscopy is the *diffraction grating*. Its simplest form is shown in Fig. 1.88.

This is an *amplitude grating* with a rectangular profile of transmittance obtained, e.g., by vacuum deposition of many opaque (metallic) strips on a glass plate. The equidistant strips of equal width form narrow slits. The spacing (p) between the centres (or two homologous points) of two neighbouring slits is called the *period*, and the reciprocal of p is known as the *spatial frequency*. This quantity is denoted by u, thus $u = 1/p$.

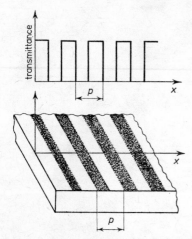

Fig. 1.88. Amplitude grating with rectangular profile of transmittance.

If a parallel light beam of wavelength λ strikes normally such a grating (Fig. 1.89a), then each slit diffracts light over a range of angles ϑ. Theoretically this range extends from $\vartheta = -90°$ to $\vartheta = +90°$. The diffracted beams from successive slits interfere constructively and destructively because they are derived from the same wavefront Σ. Constructive interference occurs for the diffraction angles ϑ_m which satisfy the following equation:

$$p\sin\vartheta_m = m\lambda,$$

(1.159)

where $m\lambda = \delta$ is the optical path difference between two parallel beams issuing from neighbouring slits (Fig. 1.89b) and m is an integer ($m = 0, \pm 1, \pm 2, ...$), called the *diffraction order*.

Diffraction and interference produce the *diffraction spectrum*, which consists of a number of parallel beams ..., $S_{-2}, S_{-1}, S_0, S_{+1}, S_{+2}, ...$, whose diffraction angles are ..., $\vartheta_{-2}, \vartheta_{-1}, \vartheta_0, \vartheta_{+1}, \vartheta_{+2}, ...$ Firstly, there is the direct or zero order beam S_0, for which $\vartheta_0 = 0$; it propagates as if the grating did not exist. Then, there are two beams, S_{-1} and S_{+1}, of the first order, for which $\vartheta_{\pm 1}$

$= \text{arc sin}(\pm \lambda/p)$, two beams, S_{-2} and S_{+2}, for which $\vartheta_{\pm 2} = \text{arc sin}(\pm 2\lambda/p)$, and so on. The beams S_{-1} and S_{+1}, S_{-2} and S_{+2}, and so on, occur at symmetrical intervals on both sides of the direct beam. This effect occurs when the slits are very narrow. Otherwise, some symmetrical beams of higher orders depending on the relation between the width of slits and the period do not exist in the diffraction spectrum.

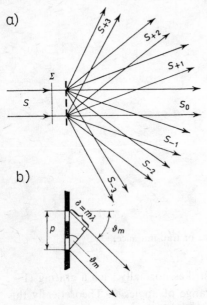

Fig. 1.89. Diffraction of light by an amplitude grating with rectangular profile of transmittance.

The diffracted beams can be focused in the focal plane of a convergent lens L (Fig. 1.90) as bright spots whose intensity decreases as diffraction orders increase. As shown in Fig. 1.90, only beams up to the second order are able to enter the lens because its aperture is limited and cannot comprise all the diffracted beams. The *diffraction spots*, usually called the *diffraction maxima*, are located in a direction defined by the angle ϑ_m, which results from Eq. (1.159).

From the triangle CQ_0Q_{+1} (C is the centre or the image-side nodal point of the lens) it results that the distance between the first orders and the zero order maxima, Q_{-1} and Q_0 or Q_{+1} and Q_0, can be expressed by

$$\zeta_1 = f' \tan \vartheta_1 = f' \frac{\sin \vartheta_1}{\cos \vartheta_1}, \tag{1.160a}$$

where f' is the focal length of the lens L (Fig. 1.90). Combining Eqs. (1.159) and (1.160a) shows that

$$\zeta_1 = \frac{f'\lambda}{p\cos\vartheta_1} = \frac{f'u\lambda}{\cos\vartheta_1}. \tag{1.160b}$$

If the diffraction angles are small, then $\cos\vartheta_1 \approx 1$, and Eq. (1.160b) reduces simply to

$$\zeta_1 = f'u\lambda. \tag{1.160c}$$

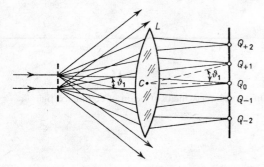

Fig. 1.90. Formation of a diffraction spectrum in the focal plane of a convergent lens.

Similarly, the distance ζ_m of the diffraction maxima of higher orders may be expressed as

$$\zeta_m = mf'u\lambda. \tag{1.161}$$

The above equations are basic to Fourier transform microscopy (see Chapters 3 and 17).

If a diffraction grating is struck obliquely by a collimated beam of light (Fig. 1.91), then Eq. (1.159) takes the form

$$p(n_2\sin\vartheta_m - n_1\sin\theta) = m\lambda, \tag{1.162}$$

where θ is the angle of incidence, n_1 is the refractive index of the medium from which the light falls on the grating, while n_2 is that of the medium into which the diffracted light enters. It is worth noting that for $m = 0$, Eq. (1.162) may be written as $n_2\sin\vartheta_0 = n_1\sin\theta$. This is simply Snell's law of refraction (see Eq. (1.65)). If the grating is surrounded by air, $n_2 = n_1 = 1$, and if $\theta = 0$, then Eq. (1.162) reduces simply to Eq. (1.159).

In optics, and especially in microscopy, there are objects similar to gratings whose transmittance distribution obeys a $\sin^2 x$ (or $\cos^2 x$) law (Fig. 1.92). These are called *sinusoidal gratings*. When illuminated by a parallel beam, such a grating

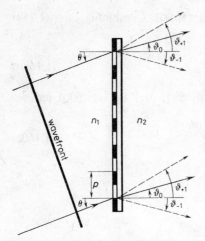

Fig. 1.91. Amplitude diffraction grating in an obliquely incident beam of collimated light.

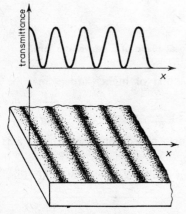

Fig. 1.92. Amplitude grating with sinusoidal profile of transmittance.

produces only three diffracted beams (Fig. 1.93) : the direct beam (S_0) and two beams of the first order (S_{-1} and S_{+1}). This kind of diffraction is very useful in applied optics, especially in holography. A sinusoidal amplitude grating may be easily obtained by a suitable photographic recording of fringes occurring in the two beam interference of highly monochromatic light [1.38].

In its plane the amplitude grating modifies the amplitude (intensity) of incident light. In microscopy there are also objects which are similar to *phase gratings*. They modify the phase of incident light while the amplitude remains unchanged.

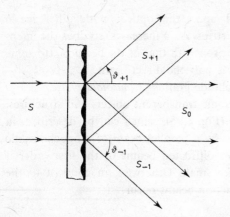

Fig. 1.93. Diffraction of light by an amplitude grating with sinusoidal profile of transmittance.

The simplest form of such a grating is shown in Fig. 1.94. This is a glass plate on which a series of narrow strips has been deposited, e.g., by vacuum evaporation of a transparent (dielectric) material. The strips produce a phase variation φ $= 2\pi\delta/\lambda = 2\pi(n_M - n)t/\lambda$, where λ is the wavelength of incident light, n and t

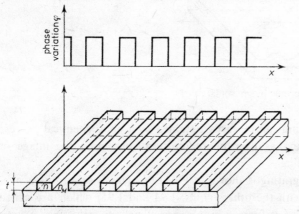

Fig. 1.94. Phase grating with a profile in rectangular relief.

are, respectively, the refractive index and the thickness of the strips, and n_M is the refractive index of the gaps between the strips (for air gaps, $n_M = 1$). The phase variation φ has the form of a rectangular function of x. If such a grating is used instead of the amplitude one (Figs. 1.88 to 1.91), the same diffraction patterns as before are observed. The intensities of the direct beam S_0 and the diffracted beams $S_{\pm 1}$, $S_{\pm 2}$, ... depend upon the optical path difference $\delta = (n_M - n)t$ produced by the transparent strips. As δ increases, so does the intensity of the beams $S_{\pm 1}$, $S_{\pm 2}$, ... by comparison with that of the beam S_0 (to some limits). It is evident that for $\delta = 0$ there is only the direct beam.

On the analogy of the sinusoidal amplitude gratings, *sinusoidal phase gratings* are also used in microscopy to describe some transparent objects and structures. Such a grating has a sine surface profile (Fig. 1.95), which can be described, in particular, by $\sin^2 x$ function. Illuminated by a parallel beam of monochromatic light, it produces a direct beam and two diffracted beams of the first order if the phase variations of incident light are small. Otherwise, in contrast to the sinusoidal amplitude grating, many diffraction beams occur.

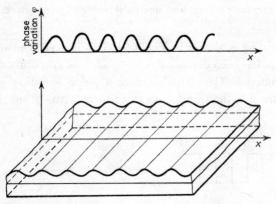

Fig. 1.95. Phase grating with a profile in sinusoidal relief.

A sinusoidal phase grating can be easily obtained by a photochemical process which depends on bleaching a silver photoemulsion containing the image of a sinusoidal amplitude grating.

Another kind of phase grating is shown in Fig. 1.96. This is said to be a *non-relief grating* by contrast with the others (Figs. 1.94 and 1.95) which are called *relief phase gratings*. This last is formed by a series of narrow strips which have the same thickness but two refractive indices of alternate values n_1 and n_2.

The diffraction gratings as described above work in transmitted light. In microscopy, especially in metallography, there are objects which simulate gratings working in reflection (Fig. 1.97). In such a case the diffraction formula (1.162) takes the form

$$p(\sin \theta + \sin \vartheta_m) = m\lambda. \tag{1.163}$$

This equation gives the angular direction ϑ_m in which the m-th order diffracted beam of wavelength λ propagates for an angle of incidence θ. For the zero order beam, Eq. (1.163) reduces simply to the ordinary law of reflection ($\vartheta_0 = -\theta$).

When using Fourier analysis, any microscopical object can be regarded as a particular diffraction grating or as a collection of amplitude and/or phase

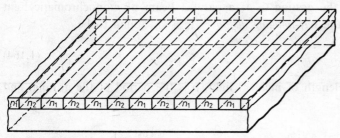

Fig. 1.96. Phase grating with a flat surface.

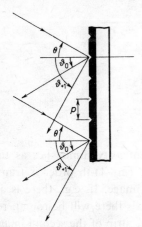

Fig. 1.97. Reflection grating.

gratings of different periods and of various orientations of periodic or quasi-periodic structures. The Fourier analysis of microstructures is covered by Chapter 17.

At the end of this subsection it is worth noting that the Fraunhofer diffraction spectrum, ... Q_{-2}, Q_{-1}, Q_0, Q_{+1}, Q_{+2}, ... , as shown in Fig. 1.90, is said to be the *Fourier spectrum*. It can be observed in the exit pupil of the microscope objective.

1.7.3. Talbot's phenomenon

Diffraction gratings, as well as some other periodic structures, can produce their images by themselves. This phenomenon of *self-imaging*, discoverd by Tabolt in 1836, results from the Fresnel diffraction, and manifests itself as a series of grating images G_1, G_2, G_3, ... (Fig. 1.98), which re-occur at some distances z_1, z_2, z_3, ... from the grating G. If a parallel beam of monochromatic light strikes the grating, the *self-imaging distances* (z_m) are found to be

$$z_m = m \frac{2p^2}{\lambda},$$

(1.164)

where λ is the wavelength of incident light, p is the period of the grating, and $m = 1, 2, 3, ...$

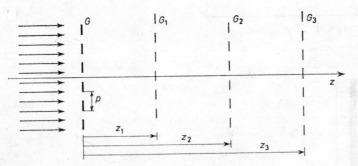

Fig. 1.98. Grating self-imaging (Talbot's phenomenon).

The images G_1, G_2, G_3, ... are alike, but their intensity diminishes as the imaging order m increases. Besides, the image of the $(m+1)$-th order is transversely shifted by $p/2$ with respect to the m-th order image. If, e.g., there is an opaque strip of the grating G at axis z, then at this axis there will be found, respectively, a clear strip of the first image G_1, an opaque strip of the second image G_2, a clear strip of the third image G_3, and so on.

Talbot's phenomenon can be easily observed by means of a microscope if a linear grating of a suitable spatial frequency is placed obliquely to the optical axis of the objective on the object table. It is then possible to distinguish several sharp grating images across the field of view of the microscope (see Fig. III). Sometimes this phenomenon is so very evident that it is difficult to focus correctly the proper image of an object with a fine periodic structure. However, in partially coherent or incoherent light secondary Talbot images are generally weakly contrasted and do not disturb the proper microscopic image.

1.7.4. Fresnel zone plates—are they of interest for X-ray microscopy?

There are diffraction gratings which can focus plane or spherical waves and produce images. The principle of such a grating, called the *Fresnel zone plate*, is shown in Fig. 1.99. The plate *FP*, orientated normally to the axis (*z*) of a light beam, contains a series of transparent and opaque circular zones whose diameters

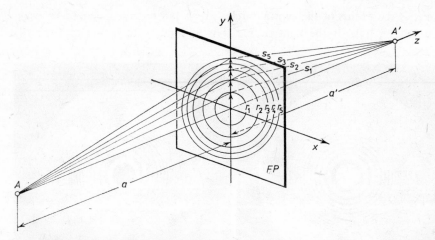

Fig. 1.99. Principle of the Fresnel zone plate.

are fixed so as to achieve a constructive interference of diffracted beams bent towards an axial point A', which constitutes the image of an axial object point A. This condition is fulfilled by the following relations:

$$s_1 = a' + \tfrac{1}{2}\lambda, \quad s_2 = a' + \tfrac{1}{2}2\lambda, \quad s = a' + \tfrac{1}{2}3\lambda, \ldots, s_m = a' + \tfrac{1}{2}m\lambda, \quad (1.165)$$

where a' is the distance between *FP* and A', and $s_1, s_2, s_3, \ldots$ are the path lengths of rays emerging from the borders of the successive zones of radius $r_1, r_2, r_3, \ldots$ and arriving at the image point A'. As can be seen, between the path lengths $s_1, s_3, s_5, \ldots$ as well as between $s_2, s_4, s_6, \ldots$ there are optical path differences equal to $\lambda, 2\lambda, 3\lambda, \ldots$ Hence the rays diffracted in this way will interfere constructively at point A'.

Applying Pythagoras theorem to the geometry of Fig. 1.99, the radius of the *m*-th zone is

$$r_m = \sqrt{\left(a' + \frac{m\lambda}{2}\right)^2 - a'^2} = \sqrt{a'm\lambda}\,\sqrt{1 + \frac{m\lambda}{4a'}}. \qquad (1.166a)$$

The quantity $m\lambda$ is usually small in comparison with a', and Eq. (1.166a) reduces simply to

$$r_m = \sqrt{m\lambda a'}, \tag{1.166b}$$

or

$$r_m = r_1 \sqrt{m}, \tag{1.166c}$$

where r_1 is the radius of the central circle, which can be either opaque or transparent (Fig. 1.100). If the light source A is at infinity $(a = -\infty)$, a parallel

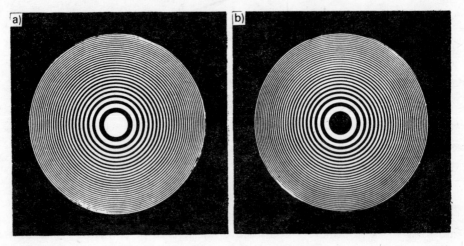

Fig. 1.100. Fresnel zone plates: a) with clear centre, b) with black centre.

light beam falls upon the Fresnel plate FP (Fig. 1.99), and in this case the image distance a' simply represents the focal length f'. In particular, Eqs. (1.166 b,c) yield

$$f' = \frac{r_1^2}{\lambda}. \tag{1.167}$$

A typical Fresnel zone plate has a rectangular profile of transmittance (Fig. 1.101a). If such a circular grating is illuminated by a collimated beam of monochromatic light, its complete diffraction spectrum is more complicated than that as shown in Fig. 1.99. Firstly, we have a direct beam B_0, a series of real images $B_{+1}, B_{+2}, B_{+3}, \dots$ of the light source B, and another series of virtual images $B_{-1}, B_{-2}, B_{-3}, \dots$ (Fig. 1.101b). All these images lie on the grating axis z, as does the light source itself. The images B_{+1} and B_{-1}, B_{+2} and $B_{-2}, \dots$ are strictly related to the diffraction maxima Q_{+1} and Q_{-1}, Q_{+2} and Q_{-2}, ... of a typical line grating such as shown in Fig. 1.90.

The situation differs, however, if the circular grating has a "sinusoidal" profile of transmittance such as shown in Fig. 1.102a. In this case all we get are the

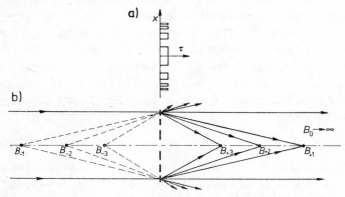

Fig. 1.101. Rectangular profile of transmittance of a circular zone plate (a) and its diffraction spectrum (b).

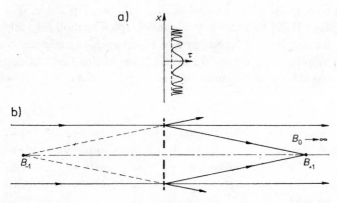

Fig. 1.102. "Sinusoidal" profile of transmittance of a circular zone plate (a) and its diffraction spectrum (b).

direct beam B_0, the diffracted real image B_{+1}, and the virtual image B_{-1} of the light source B (Fig. 1.102b). This is analogous with the line sinusoidal grating (Fig. 1.93). Sinusoidal zone plates may be obtained by a suitable photographic recording of interference rings occurring between two superposed coherent wavefronts one of which is plane while the other is spherical.

On the analogy of linear diffraction gratings, which were described in Sub-section 1.7.2, both amplitude and phase zone plates[37] exist in practice. So far they have had no practical application in light microscopy, but at the present time they are of great interest for X-ray microscopy [1.39–1.43]. Moreover, it is worth noting that there is a close similarity between holographic images, real and virtual, and those formed by a sinusoidal zone plate.

In considering the focusing properties of a Fresnel zone plate for X-ray microscopy, we see from Eq. (1.167) that the product $f'\lambda$ appears to be very small as is generally required in microscopy. For an X-ray microscope, one might consider $f' = 10$ mm, If, for instance, the wavelength λ of X-rays is equal to 0.1 nm, then $f'\lambda = 10^6$ nm^2. This requires $r_1 = 1$ µm. From Eq. (1.166c) it appears that the width of the m-th zone is

$$w_m = r_1 \left(\sqrt{m+1} - \sqrt{m} \right).$$

(1.168)

If, for instance, $m = 36$ and $r_1 = 1$ µm, then $w_{36} = 0.08$ µm. A zone plate made to these specifications appears to be out of the question, so that the product $f'\lambda$ needs to be considerably larger. Fortunately, there is now a trend towards soft X-ray microscopy, which uses synchrotron radiation of a wavelength approaching 10 nm. Under these conditions, $r_1 \approx 10$ µm and $w_{36} \approx 0.8$ µm for $f' = 10$ mm. Zone plates made to these specifications do seem possible. The authors of the cited paper [1.39] have developed a holographic method for making microzone plates for soft X-ray microscopy and set up an X-ray microscope by means of which a resolution of up to 70 nm has been obtained on wet and unstained cells with a synchrotron radiation of wavelength $\lambda = 4.6$ nm.

*

*   *

It is hoped that the contents of this chapter is sufficient to give the reader a suitable knowledge of physical optics without the necessity to dip into various other books on optics. The interested microscopist who needs more details, however, regarding this and related matters is referred to the works of the other authors cited in the references.

---

[37] It is worth noting here that phase zone plates should not be confused with the Fresnel lens largely used in lighting instruments or with Fresnel screens which constitute standard equipment for rear projection microscopy or for some types of overhead projectors used by lecturers,

2. Light Microscopes within the Scope of Geometrical Optics, Their Construction, Standard Parameters, and Basic Properties

The principles of microscopical imaging within the scope of geometrical optics are based on the refraction and specular reflection of rays as well as on the ability of lenses and mirrors to focus and change the convergence or divergence of light beams. Therefore, a knowledge of basic optical elements and their fundamental image-forming properties is necessary to understand clearly the geometrical theory of image formation in the microscope. But before that, it will be useful to consider some definitions and concepts relating to microscopic objects.

2.1. Microscopic objects: classification and notions

Microscopical studies cover a great variety of objects and structures. These are classified in different ways, but a classification that takes into account the interaction of light with matter would appear to be the most useful. The light–matter interaction manifests itself in different forms of which the most important for light microscopy were described in the preceding chapter. These phenomena belong to pre-laser physical optics. Today we know that the light emitted by high-power or high-energy lasers is able to create some new forms of light-matter interactions known as non-linear phenomena. These are the subject of non-linear optics which is beyond the scope of this book.

2.1.1. Amplitude and phase objects

Objects which change the intensity of trasmitted or reflected light are called *amplitude objects* (Fig. 2.1a). They are observed in a conventional bright-field microscope, as a consequence of modulation in the amplitude of the light waves which produce the image. As is well known, the intensity of light is proportional to the square of the amplitude of the light wave.

Objects which do not alter light intensity, but only shift the phase of the light wave are called *phase objects* (Fig. 2.1b). They are normally invisible because the human eye, as well as other light receptors, is insensitive to changes in the

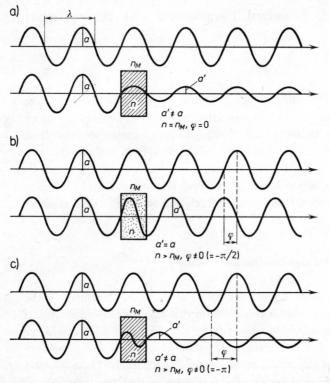

Fig. 2.1. Amplitude (a), phase (b) and phase–amplitude (c) objects; a—amplitude of incident light wave, a'—amplitude of exiting light wave, λ—wavelength, φ—phase shift, n—refractive index of the object, n_M—refractive index of a surrounding medium.

relative phase shifts of light waves. For transparent specimens the phase shifts are mainly due to different refractive indices and/or the varying thickness of the objects and their surrounding medium. In particular, living cells and their organelles usually belong to this class of objects.

Consider a parallel light beam with the plane wavefront Σ (Fig. 2.2) passing through a transparent particle O of refractive index n and thickness t. This object is surrounded by a medium M having equal transparency but a different refractive index n_M. If $n_M > n$ (Fig. 2.2a), the wavefront Σ' immediately above the object O is retarded in phase by an amount

$$\varphi = \frac{2\pi\delta}{\lambda}, \text{ expressed in terms of radians,} \qquad (2.1a)$$

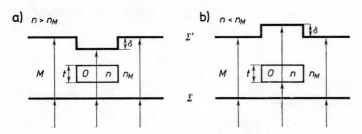

Fig. 2.2. Phase-retarding (a) and phase-advancing (b) transparent objects.

or

$$\varphi = \frac{360°\delta}{\lambda}, \text{ expressed in terms of degrees of arc,} \qquad (2.1b)$$

where λ is the wavelength and δ is the optical path difference defined as

$$\delta = (n_M - n)t. \qquad (2.2)$$

Conversely, when $n_M < n$, the wavefront is advanced in phase (Fig. 2.2b). Consequently, objects with $n > n_M$ are called *phase-retarding objects* ; physically they give a negative phase shift ($\varphi < 0$), whereas those with $n < n_M$ are called the *phase-advancing objects*, and produce a positive phase shift ($\varphi > 0$).

In the case of light-reflecting specimens the phase shifts of the wavefront are, in particular, due to microsteps and other variations in specimen flatness, as well as phase jumps at different grain areas. Depressions behave as phase-retarding objects, elevations as phase-advancing objects (Fig. 2.3). If there are no phase jumps, they produce an optical path difference

$$\delta = 2t, \qquad (2.3)$$

where t is the height of an elevation (Fig. 2.3b) or the depth of a depression (Fig. 2.3a).

In reality, there are no ideal amplitude or phase objects. Real specimens show, to some degree, both amplitude and phase properties (Fig. 2.1c). However, it is convenient to categorize microscopic specimens into one or the other group, according to whether amplitude or phase changes of light are predominant.

Both amplitude and phase objects are *isotropic* or *anisotropic*. The latter not only produce amplitude or phase modulation of a light wave but also split its wavefront into two components by double refraction and change the state of light polarization.

Phase objects are also frequently referred to as *transparent specimens* and amplitude objects as *light-absorptive specimens*.

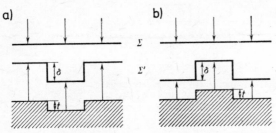

Fig. 2.3. Light-reflecting object with a depression (a) and that with an elevation (b).

2.1.2. Self-luminous and non-self-luminous objects

An object is *self-luminous* when it emits its own light as does a glowing filament, for example. In microscopy *fluorescing specimens* constitute a large class of such self-luminous objects which are studied by means of fluorescence microscopes. However, the most common class of microscopic specimens falls within the scope of ordinary microscopy, and consists of thin partially transparent objects transilluminated by light from an external source. Such objects are said to be *non-self-luminous*. There is a fundamental difference between these two classes of objects. Namely, each point of a self-luminous object emits light waves which are quite independent of those emitted by every other point, while different points of a non-self-luminous object may be illuminated by light from a single point of the external source. The waves propagating from different points of the latter object will not now be completely independent of each other, as they originate from the same source and can produce some interference effects. These effects become more predominant with an increase in the coherence degree of incident light. Therefore, non-self-luminous objects can be regarded as illuminated coherently, incoherently, and partially coherently. In brief, they are called *coherent*, *incoherent*, and *partially coherent objects*.

2.1.3. Point and extended objects

In microscopy the terms *point object* and *extended object* are also in use. The first term means that a single object under consideration is theoretically so small that it can be regarded as a geometrical point, but in practice all objects whose size is comparable with the resolving power of microscope objectives are numbered among the class of point objects. A single self-luminous point object is coherent, but any two such objects are mutually incoherent.

From what has been said above it follows that extended objects are all those whose size is greater than the resolving power of the microscope. They are classified as *periodic, quasi-periodic*, and *non-periodic objects*.

A periodic object consists of identical elements which occur, stricktly speaking, at regular intervals from $-\infty$ to $+\infty$. Otherwise, if the periodicity repeats itself along a finite length, we have a quasi-periodic object. If the periodicity length occupies a large portion of the linear field of view of the microscope, quasi-periodic objects can be treated as periodic objects. When the periodicity occurs along a single direction, we speak of *one-dimensional periodic objects*. Linear diffraction gratings, which were discussed in Subsection 1.7.2, are among such objects. If, however, the periodicity occurs along two directions in the same plane, the periodic object is said to be *two-dimensional*. Two crossed linear plane gratings, for instance, constitute an object of this kind. Similarly, objects with periodicity occuring along three Cartesian coordinates are said to be *three-dimensional periodic objects*. Some diatoms, for instance, have a structure which places them among the last group of objects.

Within the scope of Fourier optics any extended object, quasi-periodic or non-periodic, can be regarded and analysed as being composed of a large number of periodic components.

2.1.4. Object contrast

One of the most important factors in microscopical imaging is *optical contrast*. This quantity is specified as brightness difference or colour variations between the object and its surround and within different details of a specimen. The properties of the specimen producing the brightness or colour differences with which we see the microscopical image are light absorption or reflection, spatial variations in refractive indices, light scattering, and other optical phenomena. The absorption (or reflection) may be neutral or selective within the visible spectrum. In the first case it gives gray or black images of objects or their details and we then speak of *brightness contrast*. The same kind of contrast is created by the interference of monochromatic and, in some circumstances, white light.

Brightness contrast (in brief, *contrast*) specifies the differences in light intensities between the object and its surround. Actually, this quantity is defined in various ways, but for the purpose of light microscopy the following definitions are the most convenient:

$$C = \frac{I_b - I_o}{I_b + I_o}, \qquad (2.4a)$$

$$C = \frac{I_b - I_o}{I_b}, \qquad (2.4b)$$

where I_b is the light intensity (or brightness) of the background and I_o is that of the object. If the latter is darker than the background ($I_o < I_b$), then the contrast is *positive* ($C > 0$). Conversely, when the background is darker than the object ($I_o > I_b$), the contrast is *negative* ($C < 0$).

If the light intensity (brightness, luminance) is changed by selective spectral absorption or reflection, which modifies the colour of light, then *colour contrast* occurs. This kind of contrast is also produced by the interference of white light. The definition and measurement of colour contrast are more complicated than those of brightness contrast (see, e.g., Ref. [5.27]).

In transmitted light the contrast of a plate-like non-self-luminous object can be simply determined by measuring its transmittance (τ_o) and that (τ_b) of the surrounding medium. Substituting τ_o and τ_b, instead of I_o and I_b, into Eqs. (2.4) gives the value of the object contrast. For phase objects that value is, of course, equal to 0 because $\tau_o = \tau_b$.

Consequently, in reflected light the contrast of an object detail can be determined by measuring its reflectance (ϱ_o) and that (ϱ_b) of the surrounding area. The transmittance or reflectance of small objects is measured by means of microphotometers (see Chapter 14).

The contrast of grating-like objects (see Subsection 1.7.2) results from their profile of transmittance (or reflectance). Here I_o and I_b in Eqs. (2.4) denote the minimum and maximum values of transmittance (or reflectance), respectively.

Sometimes, and especially in coherent or Fourier optics, contrast is expressed by using the concepts of amplitude transmittance or amplitude reflectance instead of the intensity or brightness quantities.

When looking into the microscope, we do not observe the contrast of objects but the contrast of their images. Object contrast cannot be, in general, identified with *image contrast*. The latter and its relation to the former will be discussed later in the next chapter (see Section 3.8).

2.2. Lenses and their basic properties

When the fundamentals of lens optics are understood, it is easier to tackle both the geometrical theory of image formation in the microscope and the diffraction approach to microscopical imaging. Although mirrors, prisms, and other optical elements play important roles in microscopy, particular emphasis is given here to optical systems containing only lenses; the imaging properties of mirrors, reflecting prisms, and plane parallel plates are readily deduced from the study of lenses. There are several lens shapes—spherical, aspherical, and prismatic—but the most frequently encountered are spherical; these will be discussed in greater detail, while the others will only be mentioned in passing.

2.2.1. Cardinal points of a lens

There are two main groups of single lenses: *convex* (Fig. 2.4) and *concave* (Fig. 2.5). Their configurations are different, and the most common are: *biconvex*,

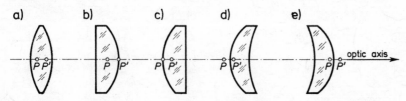

Fig. 2.4. Single positive lenses: biconvex (a), planoconvex (b, c) and meniscus-convex (d, e), and dependence of the position of their principal points (P, P') on lens shape.

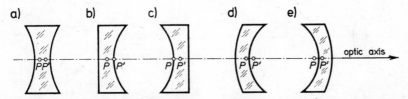

Fig. 2.5. Single negative lenses: biconcave (a), planoconcave (b, c) and meniscus-concave (d, e), and dependence of the location of their principal points (P, P') on lens shape.

planoconvex, meniscus-convex, and *biconcave, planoconcave, meniscus-concave.* Convex and concave lenses are also called *positive* and *negative lenses,* respectively. But this nomenclature is true only when a lens is surrounded by a less dense medium. Otherwise, if the medium is optically denser than the lens material, the convex lens is negative and the concave lens is positive. The former situation (lens optically denser than its surround) is typical: as a rule lenses are surrounded by air. In microscopy there are, however, optical systems (so-called *immersion objectives* and *condensers*) which contain front lenses that are one-sidedly in contact with an immersion liquid.

A positive lens is one that causes a convergent light beam to converge more rapidly (Fig. 2.6a), or causes a divergent beam either to converge (Fig. 2.6c and d)

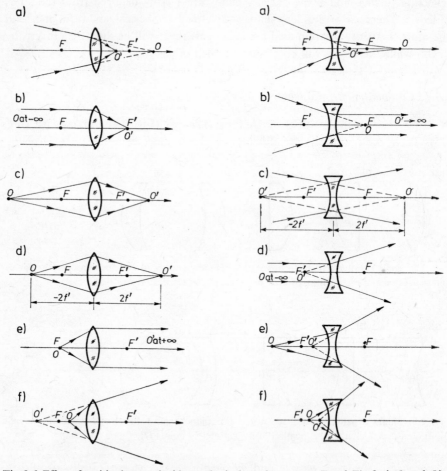

Fig. 2.6. Effect of positive lens on incident spherical or plane wave; *F* and *F'*—foci, *O* and *O'*—point light source and its image (sometimes the image of an object point is called the antipoint).
Fig. 2.7. Effect of negative lens on incident spherical or plane wave; *F* and *F'*—foci, *O* and *O'*—point light source and its image.

or diverge less rapidly (Fig. 2.6*f*). A particular situation occurs when a parallel light beam strikes the positive lens axially: the beam is focused at an axial point F' (Fig. 2.6b). This point is called the *second focus* or *image-side focus*.[1] It is on the right side of the lens if light strikes its left side. Whereas the *first* or *object-side focus* F is determined by the particular axial point from which the light emerging leaves the lens as a parallel beam (Fig. 2.6e).

A negative lens is one that causes a convergent light beam either to converge less rapidly (Fig. 2.7a) or to diverge (Fig. 2.7c), or causes a divergent beam to diverge more rapidly (Figs. 2.7e and f). However, in a particular situation a convergent beam can emerge from the negative lens as a parallel beam (Fig. 2.7b). In this case extensions (pecked lines in Fig. 2.7) of incident rays intersect at the first focus F of the lens. The opposite happens when a parallel beam strikes the lens (Fig. 2.7d): the emergent beam is divergent, but its extensions intersect at the second focus F'. As can be seen, the first focus of the negative lens is, in contrast to the positive lens, behind the negative lens and the second focus is before it.[2]

An incident ray S_1 (Fig. 2.8a) passing through the first focus F of a positive lens leaves the lens as a ray S_1' parallel to the optic axis of the lens. The light

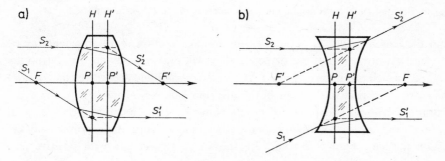

Fig. 2.8. Construction of principal planes, H and H', for positive (a) and negative (b) lenses.

is refracted at both surfaces of the lens. If, however, extensions of the incident ray S_1 and the emergent ray S_1' are depicted, they intersect at a surface H, which is called the *first* (or *object-side*) *principal surface*. Next, let us consider an incident ray S_2 travelling parallel to the optic axis of the lens. After refraction, it passes

---

[1] The term "focal point" is also used instead of "focus".

[2] Sometimes the first focus is called the *front focus* and the second focus is called the *back focus*, but this nomenclature is largely true for positive lenses only.

through the second focus F' as a skew ray S'_2. Extensions of the latter and of the former intersect at a surface H', which is said to be the *second* (or *image-side*) *principal surface*. In the vicinity of the lens axis the surfaces H and H' are nearly plane surfaces, and are thus generally referred to as *principal planes*. The inter-sections, P and P', of these planes with the optic axis of the lens are known as the *first* (or *object-side*) and the *second* (or *image-side*) *principal points*, respect-ively. A similar construction of the principal planes for a negative lens is shown in Fig. 2.8b.

Apart from the foci and principal points, the third pair of useful points is determined by a particular skew ray S (Fig. 2.9), which after refraction emerges

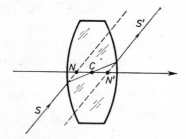

Fig. 2.9. Determination of the location of nodal points N and N' (C—centre of lens).

from the lens as a ray S' parallel to the initial direction of the incident ray S. If extensions of both rays are depicted, they will intersect the lens axis at points N and N' which are known as the *first* (or *object-side*) and *second* (or *image-side*) *nodal points*, respectively. If on both sides the lens is surrounded by the same medium, the nodal and principal points coincide. Their location depends upon the lens configuration, and sometimes either one or both these points may lie outside the lens. Figures 2.4 and 2.5 show how this occurs with positive and negative lenses of different configuration.

The three pairs of points: F and F', P and P', N and N', are called the *cardinal points*. They lie at the optic axis of the lens, i. e., at a line which passes through the centres of curvature of lens surfaces (Fig. 2.10). On paper its positive direction and the direction of light propagation are conventionally shown from left to right. If one of the lens surfaces is plane (Figs. 2.4b, c and 2.5b, c), the optic axis of the lens is determined by a line passing through the centre of curvature of the spherical surface and normally to the plane surface.

When either the location of the points F, F', P, and P' or the location of F, F', N, and N' are known, the geometrical construction of lens imaging can be determined without taking into consideration the refraction of rays at each surface of the lens. Thus any lens can be simulated by their foci and their principal planes or nodal points. It is worth noting that in such a situation all rays travelling the lens behave as if they strike the first principal plane, then pass parallel

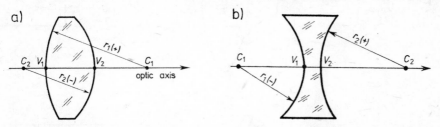

Fig. 2.10. Definition of optic axis and sign convention for radii of curvature r_1 and r_2 of positive (a) and negative (b) lenses (C_1 and C_2—centres of curvature of lens surfaces, V_1 and V_2—vertices of lens).

to the lens axis, and emerge from the second principal plane without refraction at the lens surfaces. This ray tracing is indicated in Fig. 2.11, which illustrates a geometrical construction of the image produced by a positive and a negative lens.

Finally, it should be pointed out that the points of primary or secondary rays are called *objects*, whereas the points (regions) of concentration of the rays or their extensions, after refraction or reflection of light, are *images*. When the rays actually intersect, the image is *real*; if only their extensions intersect, the image is *virtual*. By analogy to images, objects too are real and virtual. In Fig. 2.11a both the object AB and its image $A'B'$ are real, while in Fig. 2.11b the object AB is real and its image $A'B'$ is virtual. The totality of the image points defines the *image space* of a lens. Consequently, the *object space* is defined by all object points. The corresponding points in these two spaces are called the *conjugate points*.

2.2.2. Diaphragms, pupils, and windows

In practice, there are no lenses and optical systems without *diaphragms*. A diaphragm (or *stop*) is an opaque screen with an opening, usually circular. In particular, the edges of lens mounts can act as diaphragms.

One distinguishes two basic kinds of diaphragms. The one which affects the aperture of an optical system is called the *aperture diaphragm*, while that which determines what field of an extended object is imaged by a lens or optical

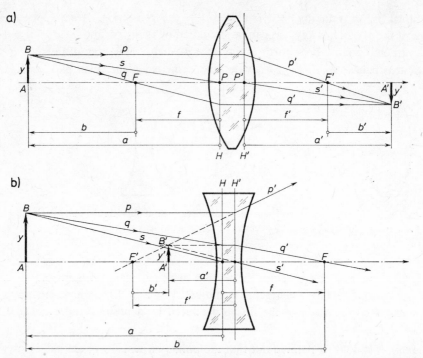

Fig. 2.11. Geometrical construction of the image produced by positive (a) and negative (b) lenses (p, q, s and p', q', s'—conjugate rays).

instrument is called the *field diaphragm*. Their role is to prevent some severely aberrated rays and stray light from reaching the image plane, to ensure suitable distribution of light in both the object space and the image space, and to produce images of adequate brightness.

Each diaphragm can be considered as a special object whose image is formed, according to the principles of the Gaussian optics, either by the whole optical system or by those parts of it that precede or follow the diaphragm. The image of the aperture diaphragm produced by the part of the optical system that follows this particular diaphragm is called the *exit pupil*, whereas the image formed by the part of the system which precedes the aperture diaphragm is known as the *entrance pupil*. By analogy to the pupils, the respective images of the field diaphragm are called the *exit window* and *entrance window*.

A general relation between the pupils and windows is schematically shown in Fig. 2.12. A multiple lens system contains an aperture diaphragm and a field diaphragm. A spherical wavefront is emitted by an axial light source A. Its image is at point A'. From the incident light the optical system accepts, due to the aperture diaphragm, a cone the solid angle of which (2σ) is subtended at A by the diameter of the entrance pupil E. Similarly, the image-forming cone of light is determined by the angle $2\sigma'$ subtended at A' by the diameter of the exit pupil E'. The angle 2σ is called the *object-side angular aperture* or, in short, *angular aperture*. Consequently, the angle $2\sigma'$ is called the *image-side angular aperture*.

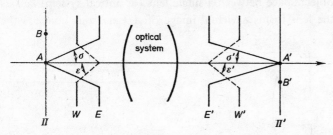

Fig. 2.12. Pupils (E, E') and windows (W, W') of an optical system.

If other parameters remain fixed, the angular aperture is the quantity which mainly determines image brightness. In microscopy, however, there is another parameter, viz., the *numercial aperture*, which is generally used as a measure of the light-gathering power of the microscope. This quantity relates to microscope objectives, and is usually denoted by A and specified as a product of the sine of the angular semi-aperture (σ) and the refractive index (n) of the object space, i.e., $A = n\sin\sigma$. This *object-side numerical aperture* has its conjugate counterpart in the image space, viz., the *image-side numerical aperture* $A' = n'\sin\sigma'$. In microscopy the refractive index $n' = 1$ since the image space is always the air; hence $A' = \sin\sigma'$.

The quite different characteristic angles 2ε and $2\varepsilon'$ are determined by the windows and centres of the pupils. The angle 2ε subtended at the centre of the entrance pupil E by the diameter of the entrance window W is called the *angular field of view* or the *field angle*. Consequently, the angle $2\varepsilon'$ subtended at the centre of the exit pupil E' by the diameter of the exit window W' is called the *image field angle* or the *angle of view*. The field angle determines that portion of the object plane Π which can be imaged by the optical system in its image plane Π'. If other parameters remain fixed, the imaged portion of a specimen under examination is determined by the angle of view of the eyepieces.

Let us now consider an arbitrary off-axis object point B from the angular field of view (Fig. 2.12). In the cone of rays that proceeds from this point, there will be a special ray which passes through the centre of the entrance pupil E.

In the theory of optical imaging this ray is known as the *chief* or *principal ray*. When the optical system is free from aberrations, the chief ray will also traverse the centre of the aperture diaphragm and the centre of the exit pupil.

Figure 2.12 represents a general configuration of pupils and windows. Now, let us look at some specific situations shown in Fig. 2.13, where the same notation is applied as in Fig. 2.12. Diagram a illustrates a system where an aperture diaphragm *AD* is in the object space between a single lens (or optical system) and its front focus *F*. Here the lens forms a virtual image of *AD* and this image is the

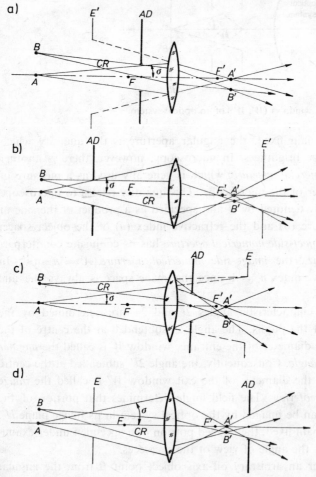

Fig. 2.13. Pupils of an optical system whose aperture diaphragm (*AD*) is ahead of (a, b) and behind (c, d) the system.

exit pupil E'. It is localized before the lens, while a separate entrance pupil does not exist and its function is simply fulfilled by the aperture diaphragm itself. If this diaphragm is moved to the left of F, the exit pupil E' becomes real and its localization is behind the lens (diagram b). In the specific case that the aperture diaphragm is situated in the front focal plane, the exit pupil will be at infinity, and the chief ray (CR) in the image space will be parallel to the optic axis of the lens. Such a configuration is said to be *telecentric on the image side*. In microscopy this may be exemplified by a microscope condenser, whose aperture diaphragm is situated in its front focal plane.

In contrast to the previous diagrams, Figs. 2.13c and d illustrate systems where the image of AD is the entrance pupil E, while the exit pupil is the aperture diaphragm itself. If this diaphragm is placed between the lens and its back focus F' (diagram c), the entrance pupil E is the virtual image of AD and is localized behind the lens. If, however, the diaphragm AD is moved to the right of F', the pupil E is the real image of AD and is localized before the lens. In an intermediate case, when the aperture diaphragm is situated in the back focal plane of the lens, the entrence pupil will be at infinity and all the chief rays in the object space will be parallel to the optic axis of the lens. Such a configuration is said to be *telecentric on the object side*.

A situation rather more similar to that occurring in the common microscope is shown in Fig. 2.14. The optical system consists of two positive lenses, L_1 and L_2, representing a two-stage imaging system. The object O to be imaged is placed in the object plane Π, the lens L_1 forms an intermediate image O' at Π', and the lens L_2 produces a final image O'' at Π''. Two natural diaphragms, D_1 and D_2, are formed by the edges of the lens mounts, and the third one, D_3, is localized in the image plane Π'. The images (entrance windows) of D_2 and D_3 in the object space are marked by W_2 and W_3, respectively.

From Fig. 2.14a the following facts ensue: the diaphragm D_1 is simultaneously the entrance pupil for the system; rays *1* and *2* emerging from the axial point A of the object O are the limits for a light beam which is transmitted by the system from point A; angle 2σ subtended at A by the diameter of D_1 is the object-side angular aperture; rays *3* and *4* are the limits of the beam which emerges from the off-axis object point B and enters the system; angle 2ε subtended at the centre of D_1 by the diameter of the entrance window W_2 is the angular field of view; no ray emerging from an object point beyond C can pass the system. Figure 2.14a also shows how various diaphragms cooperate to limit the size of the transmitted light beams emerging from different object points. For the point B the meridional section of the beam limited by rays *3* and *4* is nearly as little as half

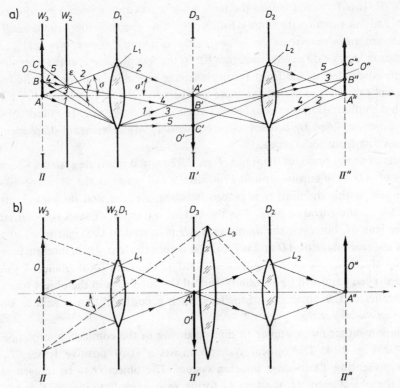

Fig. 2.14. Vignetting in two-stage imaging system: a) without field lens, b) with field lens.

the beam emerging from the axial point. A further, and even more rapid, reduction of beam sizes occurs for points between B and C, while for C only a single ray (5) is transmitted by the system. In the image planes, II' and II'', this reduction of light implies a gradual fading of the object image. This injurious phenomenon is known as *vignetting*.

The most obvious way of eliminating vignetting would be to enlarge the diameter of the lens L_2 until the entrance window W_2 no longer interferes with any object ray which can be accepted by the entrance pupil D_1. This remedy is, however, impractical and another solution illustrated in Fig. 2.14b is in general use. An additional lens, L_3, is inserted between L_1 and L_2 near (before or behind) the image plane II'. The lens L_3, called a *field lens*, prevents lens L_2 from acting as the field stop. Now the entrance window is nearly coincident with the entrance pupil D_1. The role of the field lens is to greatly enlarge the angular field of view

2ε (by a factor of 4) compared with that in the system of Fig. 2.14a. In the following pages it will be shown that lenses L_2 and L_3 constitute a microscope ocular, while L_1 is a microscope objective.

The selection of the position of a diaphragm depends on the function of an optical system and in part on mechanical considerations. Frequently the aperture and field stops have the form of an *iris diaphragm* with an opening that can be continuously varied. However, it should be pointed out that the diameter and axial position of the aperture diaphragm affect the aberrations of the system and alter their relative proportions [2.1].

In general, field diaphragms are of lesser optical importance than aperture diaphragms since they do not affect aberrations and can only cutoff a strongly aberrated marginal zone of the image field. In microscopy, however, both aperture and field diaphragms are equally significant (see Subsection 2.3.4).

2.2.3. Basic equations of lens optics

The basic principles of lens optics are derived from Snell's law of refraction (see Subsection 1.5.1) and *Fermat's principle of least time*. The latter states that for any physically possible path of rays travelling through an optical system the time of travel of the light is a minimum [1.1, 2.1, 2.2]. Frequently this principle is expressed more precisely in the form

$$\int_{O}^{O'} n\,dl \quad \text{is stationary,} \tag{2.5}$$

where dl is a differential segment of length along any one of the possible paths from O to O' (Fig. 2.15). The integral of ndl expresses the *optical path length*, and is proportional to the time of travel of the light.

From Fermat's principle it results that in isotropic media, with which we are dealing in this chapter, rays are normal to wavefronts. This relation enables

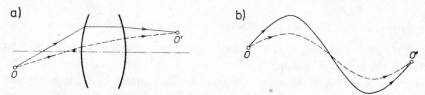

Fig. 2.15. Fermat's principle: a) for lenses, b) for a medium of continuously varying refractive index. The solid line shows a physically possible ray path from points O to O', whereas broken lines represent any other possible path.

an analogue for ideal image formation to be constructed by using the concept of rays. Physically the point image O' (Figs. 2.6 and 2.7) is perfect when all the image-forming wavefronts are spherical. Consequently, if after refraction through a lens the rays of a light beam intersect at a single point, then the wavefronts are spherical, and vice versa. Thus a point image is perfect if all the image-forming rays converge at a single point. This analogue is valid only within the scope of *Gaussian optics* or *paraxial optics*, i.e., it is true for a region close to the optic axis of the lens. Some useful formulae related to this region are given in this section, whereas departures from the Gaussian optics or *optical aberrations* will be discussed in the next part of the chapter.

Single lens. The most important parameter of every lens is its *focal lenght*. We distinguish between the *image-side focal length*, denoted by f', and the *object-side focal length*, denoted by f. These are defined by the distances from either principal point to the corresponding focus, i.e., $f' = P'F'$ and $f = PF$ (Figs. 2.8 and 2.11). These magnitudes take signs that are either positive or negative defined by the order of letters P', F' and P, F in relation to the positive direction of the optic axis of the lens, i.e., to the direction of light propagation. Consequently, in Fig. 2.16a the length f' is positive and f is negative, while in Fig. 2.16b the length f' is negative and f is positive. As can be seen, the focal lengths f and f' of a given lens are always opposite in sign, and

$$\frac{f}{f'} = -\frac{n}{n'}, \tag{2.6a}$$

where n and n' are the refractive indices of media before and behind the lens, respectively. If these media are the same, e.g., the lens is surrounded by air, then $n = n'$ and

$$f = -f'. \tag{2.6b}$$

The quantity

$$K = \frac{n'}{f'} = -\frac{n}{f} \tag{2.7}$$

is called the *power of lens*. It is expressed in m^{-1} or in dioptres. A lens of one dioptre has the power $K = 1\ m^{-1}$.

The reciprocal of K is the *equivalent focal length*, denoted by f_e. It is obvious that $f_e = f'/n' = -f/n$ or, if a lens is surrounded by air, $f_e = f' = -f$.

Let the first and second surfaces of a lens have radii of curvature r_1 and r_2, respectively. These radii may be positive or negative. There are different con-

ventions for defining their sign, but the most common is as indicated in Fig. 2.10:
the radius of curvature (r) of a lens surface is positive if the centre of curvature
(C) of the surface lies to the right of its vertex (V), and negative otherwise (on
the assumption that light propagates from left to right). If the lens has a refractive
index n_L and an axial thickness t, then its power K is determined by the expression

$$K = \frac{n_M}{f'} = (n_L - n_M)\left(\frac{1}{r_1} - \frac{1}{r_2}\right) + \frac{(n_L - n_M)^2 t}{n_L r_1 r_2}, \tag{2.8a}$$

where n_M is the refractive index of a surrounding medium. If this medium is
air, $n_M \approx 1$ and Eq. (2.8a) takes the form

$$K = \frac{1}{f'} = (n_L - 1)\left(\frac{1}{r_1} - \frac{1}{r_2}\right) + \frac{(n_L - 1)^2 t}{n_L r_1 r_2}. \tag{2.8b}$$

For a thin lens t is small and the above expression reduces simply to

$$K = \frac{1}{f'} = (n_L - 1)\left(\frac{1}{r_1} - \frac{1}{r_2}\right). \tag{2.8c}$$

The distance from the front vertex V_1 of a lens to the first focus F (Fig. 2.16)
and that from the back vertex V_2 to the second focus F' are called, respectively,
the *front focal length*, denoted by f_v, and the *back focal length*, denoted by f_v'. If
the lens is surrounded by air, these quantities are determined by the expressions

$$f_v = -(f' - v_1) = -f'\left(1 + \frac{n_L - 1}{n_L}\frac{t}{r_2}\right), \tag{2.9a}$$

$$f_v' = f' - (-v_2) = f'\left(1 - \frac{n_L - 1}{n_L}\frac{t}{r_1}\right), \tag{2.9b}$$

where v_1 and v_2 are the distances of the principal planes from the vertices, measured as indicated in Fig. 2.16. These distances are expressed by

$$v_1 = -f'\frac{n_L - 1}{n_L}\frac{t}{r_2}, \tag{2.10a}$$

$$v_2 = -f'\frac{n_L - 1}{n_L}\frac{t}{r_1}. \tag{2.10b}$$

For a thin lens t is small and $f_v \approx f_v' \approx f'$, and $v_1 \approx v_2 \approx 0$.

The reciprocal of f_v and that of f_v' are called the *front vertex power* and the
back vertex power, respectively. These quantities are used mainly for descriptions
of spectacle glasses.

From Eqs. (2.10) it results that the distance i (Fig. 2.16) between the principal points P and P' is found to be

$$ i = t - (-v_1) - v_2 = t\left[1 - f'\,\frac{n_L - 1}{n_L}\left(\frac{1}{r_1} - \frac{1}{r_2}\right)\right]. \qquad (2.11a) $$

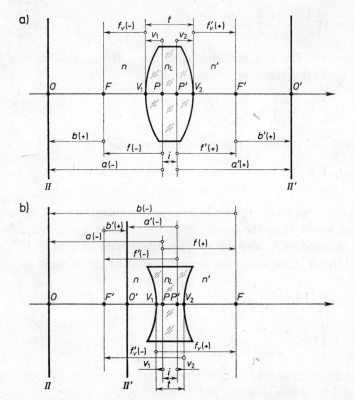

Fig. 2.16. Basic imaging parameters for positive (a) and negative (b) lenses.

By application of Eq. (2.8b) this distance, sometimes called the *interstitum*, may by written in a more convenient form

$$ i = t\left[\frac{n_L - 1}{n_L} + \left(\frac{n_L - 1}{n_L}\right)^2 f'\,\frac{t}{r_1 r_2}\right], \qquad (2.11b) $$

which shows that if the thickness t of a lens is small compared to its radii of curvature r_1 and r_2, then the second term in square brackets is small and the distance i is nearly constant with lens bending at the value $t(n_L - 1)/n_L \approx 0.3\, t$.

When the lens is very thin, its thickness is equal nearly to zero and $i \approx 0$. In this case the principal points (and nodal points) coincide with the *optical centre* of the lens. A ray passig through the latter (C, Fig. 2.9) emerges parallel to the incident ray.

Two axial points, O and O' (Figs. 2.6, 2.7 and 2.16), are said to be *conjugate* when one is the image of the other. The same definition holds good for two planes, II and II', one of which is the image of the other. The relation between the axial conjugates O and O' is given by the *Cartesian conjugate distance equation* (or *Cartesian lens equation*):

$$\frac{f'}{a'} + \frac{f}{a} = 1,$$ (2.12a)

or, if $f = -f'$,

$$\frac{1}{a'} + \frac{1}{a} = \frac{1}{f'},$$ (2.12b)

where a and a' are the *conjugate distances* from the principal points: a is the *object distance* and a' is the *image distance*. They are measured from the corresponding principal points, and take either positive or negative sign defined by the order of letters P, O and P', O' in relation to the positive direction of the optic axis of the lens.

Locations of the object O and its image O' (Fig. 2.16) can also be related to the corresponding foci F and F'. Here the lens equation takes the form of the *Newton conjugate distance equation*

$$bb' = ff',$$ (2.13a)

or, if $f = -f'$,

$$bb' = -f'^2,$$ (2.13b)

where b and b' are the conjugate distances from the foci; b is the object distance and b' is the image distance. They are measured from the corresponding focal points, and take either positive or negative sign defined by the order of letters F, O and F', O' in relation to the positive direction of the optic axis of the lens. Equations (2.12) and (2.13) can be easily derived from the geometry of Figs. 2.11a and 2.11b.

In microscopy lenses serve mainly to produce magnified images. We distinguish three basic kinds of *magnification*: *transverse* or *lateral*, *longitudinal* or *axial*, and *angular*, denoted by M, M_l, and M_a, respectively. Transverse magnification (M) is the most important and is defined by the ratio of the lateral size of the image to that of the object. Referring back to Figs. 2.11 and 2.12 we have

$$M = \frac{y'}{y},$$
(2.14a)

where y and y' are the heights of the object and its image, respectively. These quantities are measured from the lens axis, and are positive upward and negative downward of this axis. From the geometry of Figs. 2.11a and b and by application of Eqs. (2.13) transverse magnification may be expressed as follows:

$$M = -\frac{f}{b} = -\frac{b'}{f'} = -\frac{fa'}{f'a},$$
(2.14b)

or, if $f = -f'$,

$$M = \frac{f'}{b} = -\frac{b'}{f'} = -\sqrt{-\frac{b'}{b}} = \frac{a'}{a}.$$
(2.14c)

In Fig. 2.11a the height y is positive and y' is negative, so that M is negative, and the image $A'B'$ of the object AB is inverted; in Fig. 2.11b, on the other hand, both y and y' are positive, so that M is positive, but the erect image $A'B'$ of the object AB is virtual. By definition, a virtual image is determined by intersections of the extensions of rays, while a real image is formed by the intersections of the rays themselves. In general, if the image is erect, the transverse magnification M is positive, and conversely, if the image is inverted, M is negative. When $|M| > 1$, the image size is greater than the object size, and if $|M| < 1$, the image size is smaller than the object size.

The angular magnification is defined as[3]

$$M_a = \frac{\tan \sigma'}{\tan \sigma},$$
(2.15a)

where σ and σ' are the conjugate angles as indicated in Fig. 2.17: σ is the angle between an object ray S and the optic axis of lens, and σ' is the angle between the image ray S', conjugate with S, and the lens axis. Their signs may be positive or negative according to the usual convention that an anti-clockwise rotation to the optic axis (or to any other reference axis) is positive, while clockwise rotation is negative. Consequently, in Fig. 2.17 the angle σ is negative and σ' is positive. The geometry of Figs. 2.11 and 2.17, and Eq. (2.14b) show that

$$M_a = \frac{a}{a'} = \frac{f}{b'} = \frac{b}{f'}.$$
(2.15b)

---

[3] Frequently this quantity is also defined as $M_a = \sigma'/\sigma$.

Then, just as $a/a' = -f/Mf'$ so also

$$M_a = -\frac{f}{Mf'}, \tag{2.15c}$$

and

$$MM_a = -\frac{f}{f'}, \tag{2.16a}$$

or, if $f = -f'$,

$$MM_a = 1. \tag{2.16b}$$

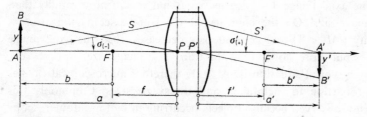

Fig. 2.17. Illustrating the transverse and angular magnifications.

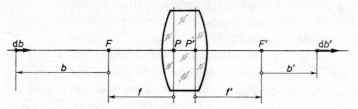

Fig. 2.18. Illustrating the longitudinal magnification.

Longitudinal magnification is defined as a ratio

$$M_l = \frac{db'}{db}, \tag{2.17a}$$

where db and db' are very small (differential) conjugate axial elements (measured along lens axis) of an object and its image, respectively (Fig. 2.18). By differentiation of Eq. (2.13a) it is easy to see that $bdb' + b'db = 0$, then Eq. (2.17a) may be written as

$$M_l = -\frac{b'}{b}. \tag{2.17b}$$

Equation (2.14b) shows that $b' = -f'M$ and $b = -f/M$, so also

$$M_l = -\frac{f'}{f} M^2, \tag{2.17c}$$

or, if $f = -f'$,

$$M_l = M^2. \tag{2.17d}$$

This is a very important conclusion from which it results that the image size of a spatial object is, in general, greatly deformed with respect to the object size. An ideal size fidelity between the object and its image occurs only when $|M| = 1$. If the transverse magnification is greater than unity ($|M| > 1$) and continuously increases, then the axial image size also increases but much more rapidly than the transverse image size. On the contrary, if the transverse magnification is smaller than unity ($|M| < 1$) and continuously decreases, then the axial image size also decreases but more slowly than the transverse image size.

In microscopy a typical situation is $|M| \gg 1$; therefore microscopical images are out of all proportion to their axial and transverse sizes. Commonly this disproportion is not observed because objects occurring in biological microscopy are, in general, very thin and those being examined by means of reflected light microscopes belong to the class of surface objects. Entirely different circumstances occur in stereoscopic microscopy. One of the main properties of any stereoscopic microscope is to show objects in their depth. When looking into this microscope, however, the observer sees the object disproportionally elongated along the optic axis of the microscope. This is, of course, a defect which cannot be overcome within the scope of classical optics. However, some possibilities of producing optical images conforming to object size, i.e., with the same transverse and longitudinal magnifications, are available in holographic microscopy (see Chapter 11).

In microscopical literature the terms: *simple microscope* and *compound microscope* are frequently in use. The former denotes an optical system consisting of a single positive lens. Equations (2.14) show that transverse magnification increases as object distance b decreases. If the latter is sufficiently small compared with the focal length of the lens, it is possible to obtain high transverse magnifications of real images. For instance, if $f' = 10$ mm and $b = -0.1$ mm, $M = -100$. The earliest microscopes, in particular those of Van Leeuwenhoek, were based on this principle. Today the simple microscopes are hardly in use any longer although single lenses or cemented doublets and triplets are widely used as loupes or magnifying glasses. They are, however, beyond the scope of this book.

System of two lenses. Let us now consider a system of two lenses, *1* and *2* (Fig. 2.19). The lenses are set coaxially in series. The distance between their adjacent principal points P_2' and P_2 is denoted by e, whereas the distance between the adjacent foci F_1' and F_2 is denoted by g. These quantities are taken to be positive if the order of the letters $P_1'P_2$ and $F_1'F_2$ conforms to the positive direction of the optic axis

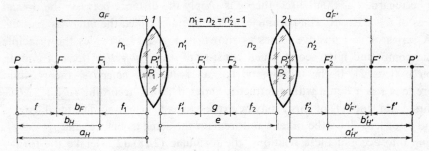

Fig. 2.19. System of two lenses arranged coaxially in series (n_1, n_1', n_2 and n_2'—refractive indices of media before and behind the lenses).

of the lenses, i.e., left-to-right as indicated in Fig. 2.19. If the lenses are surrounded by air, we may arrive at the following relations:

$$f' = -\frac{f_1'f_2'}{g}, \quad f = \frac{f_1 f_2}{g}; \tag{2.18}$$

$$\frac{1}{f'} = \frac{1}{f_1'} + \frac{1}{f_2'} - \frac{e}{f_1'f_2'}, \quad \frac{1}{f} = \frac{1}{f_1} + \frac{1}{f_2} + \frac{e}{f_1 f_2}; \tag{2.19}$$

$$a_H = \frac{ef'}{f_2'}, \quad a_{H'}' = -\frac{ef'}{f_1'}; \tag{2.20}$$

$$b_F = -\frac{f_1'^2}{g}, \quad b_{F'}' = \frac{f_2'^2}{g}, \tag{2.21}$$

where f_1 and f_1' are the first and second focal lengths of lens *1*, f_2 and f_2' are those of lens *2*, f is the first resultant focal length of the system of two lenses $(1+2)$, f' is the second resultant focal length of this system, a_H is the distance between the first principal point P_1 of lens *1* and the first principal point P of the system of two lenses, $a_{H'}'$ is the distance between the second principal point P_2' of lens *2* and the second principal point P' of this system, b_F is the distance between the first focus F_1 of lens *1* and the first focus F of the system of two lenses, and $b_{F'}'$ is the distance between the second focus F_2' of lens *2* and the second focus F of the system. The quantities a_H, $a_{H'}'$, b_F, and $b_{F'}'$ can be either positive or negative. They are positive if the order of letters P_1P, $P_2'P'$, F_1F, and $F_2'F'$ is conformable to the positive direction of the optic axis of a lens system; otherwise their sign is negative.

It is worth noting that Eqs. (2.19) result from Eqs. (2.18) if the substitution $g = e - f_1' + f_2$ is applied into (2.18). Moreover, when lenses *1* and *2* (Fig. 2.19) may be regarded as thin lenses, then e is simply the distance between the optical centres of these lenses. The parameter e is sometimes called the *interlens spacing*.

A series of two positive lenses as shown in Fig. 2.19 provides the principle of the compound microscope which consists of a *dry objective* (lens *1*) and an eyepiece (lens *2*). If the microscope has an *immersion objective*, the medium before lens *1* is a liquid with a refractive index n_1, and, according to Eq. (2.6a), the first focal length f_1 of lens *1* may be expressed as $f_1 = -n_1f_1'$. The focal lengths f_2 and f_2' of lens *2*, on the other hand, remain the same as before, and $f_2 = -f_2'$. Taking into account these relations, the formulae (2.18)–(2.21) take the form

$$f' = -\frac{f_1'f_2'}{g}, \quad f = -n_1f'; \tag{2.22}$$

$$\frac{1}{f'} = \frac{1}{f_1'} + \frac{1}{f_2'} - \frac{e}{f_1'f_2'}, \quad \frac{1}{f} = -\frac{1}{n_1f'}; \tag{2.23}$$

$$a_H = n_1\frac{ef'}{f_2'}, \quad a_{H'}' = -\frac{ef'}{f_1'}; \tag{2.24}$$

$$b_F = -\frac{n_1f_1'^2}{g}, \quad b_{F'}' = \frac{f_2'^2}{g}. \tag{2.25}$$

Equations (2.18)–(2.21) and (2.22)–(2.25) enable the basic optical parameters of a compound microscope to be determined. The first set of these equations relates to the microscope with dry objectives and the second to those with immersion objective.

Equations (2.18)–(2.25) can be extended to a series of three or more lenses because the first two lenses and the next three, four, and so on, can be treated as single systems to which the next lens is added. If M_{l1}, M_{l2}, ... ,M_1,M_2, ..., and M_{a1}, M_{a2}, ... denote, respectively, the longitudinal, transverse, and angular magnifications of the component lenses, then the longitudinal (M_l), transverse (M), and angular (M_a) magnifications of the multilens system are given by

$$M_l = M_{l1}M_{l2} \dots M_{lm} = \prod_{k=1}^{m} M_{lk}, \tag{2.26}$$

$$M = M_1 M_2 \dots M_m = \prod_{k=1}^{m} M_k, \tag{2.27}$$

$$M_a = M_{a1} M_{a2} \dots M_{am} = \prod_{k=1}^{m} M_{ak}. \tag{2.28}$$

The theory presented above relates to paraxial geometrical optics. It is based on the assumption that rays traverse an optical system near its optical axis and that the angles at which the rays strike lens surfaces are small. An angle φ expressed in radians can be regarded as small if the following approximations are valid: $\sin\varphi \approx \varphi$, $\cos\varphi \approx 1$, and $\tan\varphi \approx \varphi$. In this case Snell's law of refraction (Eq. 1.65) may be written as

$$n\varepsilon = n'\varepsilon', \tag{2.29}$$

where ε, ε', n, and n' are the same quantities as in Eq. (1.65) and Fig. 1.33a.

Remembering Eqs. (2.6a), (2.14a) and (2.15a), the relation (2.16a) may be written as

$$ny\tan\sigma = n'y'\tan\sigma', \tag{2.30a}$$

except that for the paraxial region $\tan\sigma \approx \sigma$ and $\tan\sigma' \approx \sigma'$, thus the above equation reduces to

$$ny\sigma = n'y'\sigma', \tag{2.30b}$$

where n and n' are the refractive indices of object and image space, respectively. Equation (2.30b) is of great importance in paraxial optics and is known as the *Lagrange–Helmholtz invariant*.

2.2.4. Aberrations

We say that a lens or optical system exhibits ideal imaging behaviour if it causes point, line, and plane objects to be imaged as point, line, and plane images, respectively. In practice, however, such imagery is impossible because lenses exhibit some optical defects or aberrations that cause nonideal images to be formed. The subject of optical aberrations is large and complex. Its detailed discussion is beyond the scope of this book, and only the causes and consequences of the most frequently encountered aberrations will be mentioned here. For the interested reader more details regarding this subject may be found in Ref. [2.1] or in the more advanced books on physical and geometrical optics cited in Chapter 1.

Optical aberrations are commonly divided into two categories: *monochromatic* and *chromatic aberrations* ; the first category is referred to monochromatic while the other to polichromatic light. They can be regarded as undesirable deformations of wavefronts or as deviations of rays from desirable paths. Consequently, we speak of *wave aberrations* in the first case and of *geometrical* (or *ray*) *aberrations* in the second. The latter can be derived from the former, and vice versa. Furthermore, it is useful to distinguish between aberrations of the image of a point object (*point-imaging aberrations*) and aberrations of the image of extended objects where each object point is imaged as a true point (with diffraction effects accounted for) but object and image shapes do not entirely correspond. The point-imaging aberrations in monochromatic light are *spherical aberration*, *coma*, and *astigmatism*, while *field curvature* and *distortion* are *image shape aberrations*.

Spherical aberration. The nature of this aberration is illustrated in Fig. 2.20. Rays emerging from an axial point object A and traversing a lens near its optic axis are brought to an image point A' at a different position than rays passing through the lens near its edge and producing another image $\tilde{A}'$. The point A' is said to be the *paraxial* or *Gaussian image*, while $\tilde{A}'$ is called the *marginal image*. Between these two points there are many point images formed by intermediate rays. Thus, in reality, in any plane perpendicular to the lens axis no single point image occurs between A' and $\tilde{A}'$ but a patch, called the *circle of confusion*. The rays near the Gaussian and marginal images touch a surface of revolution, called the *caustic*, the intersection of which with the divergent (postfocus) cone of marginal rays demarcates the *circle of least confusion* (Fig. 2.20b). The latter is frequently regarded as the best image of a point object when spherical aberration occurs in the optical system; thus the plane of the best image no longer coincides with the paraxial image plane.

If a parallel beam strikes the lens (Fig. 2.20c), the paraxial rays form the *paraxial* (or *Gaussian*) *focus F'* and marginal rays are brought to a *marginal focus $\tilde{F}'$*.

We distinguish between *longitudinal* (or *axial*) *sphericial aberration* and *transverse* (or *lateral*) *spherical aberration*. The former is expressed by a difference $\Delta s'$ in vertex image distances $\tilde{s}'$ and s', i.e.,

$$\Delta s' = \tilde{s}' - s', \tag{2.31}$$

whereas the latter is defined by the radius ϱ of the circle of confusion in the paraxial image plane. Between ϱ and $\Delta s'$ the following relation holds:

$$\varrho = -\Delta s' \tan \sigma', \tag{2.32}$$

where σ' is the angle at which the most marginal ray intersects the optic axis of the lens in the image space. This ray and its conjugate counterpart in the object space are called the *image-side* and *object-side aperture rays*. Consequently, the conjugate angles σ and σ' are said to be the *object-side aperture angle* and the *image-side aperture angle*.

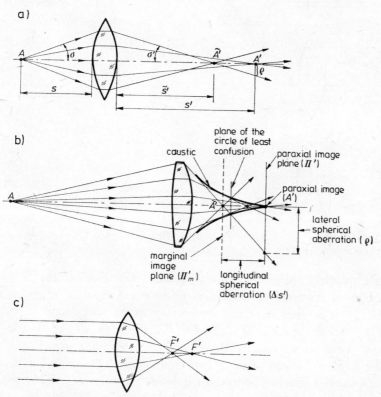

Fig. 2.20. Spherical aberration.

If $\tilde{s}' < s'$ (Fig. 2.20a), then $\Delta s' < 0$ and the spherical aberration is *negative*; on the contrary, this aberration is *positive* if $\Delta s' > 0$ or $\tilde{s}' > s'$. An optical system with negative $\Delta s'$ is said to be *undercorrected*, whereas one with positive $\Delta s'$ is said to be *overcorrected*.[4] Putting aside some exceptions (e.g., aplanatic menisci), positive lenses exhibit negative spherical aberration, whereas negative lenses

---

[4] This convention is not generally accepted. Some authors take $\Delta s' > 0$ for undercorrected systems and $\Delta s' < 0$ for overcorrected systems.

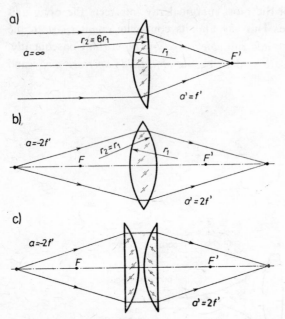

Fig. 2.21. Examples of lens shapes to minimize spherical aberration when the object distance $a = -\infty$ (a), and $a = -2f'$ (b, c).

exhibit positive spherical aberration. Thus it is clear that spherical aberration of optical systems is corrected by combining positive and negative lenses.

In general, $\Delta s'$ is a function of the third power of the lens diameter, and spherical aberration can therefore be reduced by decreasing the size of the aperture diaphragm of the optical system. Moreover, if the object and image conjugate planes are fixed, the effects of the spherical aberration of a single lens can be reduced by a suitable selection of lens shape. Some examples of such a procedure are shown in Fig. 2.21. A general and very simple way of reducing the effects of the spherical aberration of a single lens with surfaces of different curvature is as follows: if the object-side and image-side conjugate planes, Π and Π' (Fig. 2.22), are fixed and one of them is farther away than the other, then that lens surface which is more bulgy should be turned towards the more distant conjugate plane. For simple multiple lens systems spherical aberration is minimized when deviations of rays are spread over several lens surfaces and attains a minimum value if the deviations of conjugate rays are the same at each surface (see Fig. 2.21c). Some microscope eyepieces, e.g., Huygens oculars, are devised in accordance with this principle.

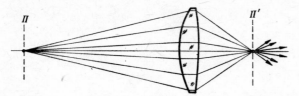

Fig. 2.22. Proper orientation of lens to minimize spherical aberration.

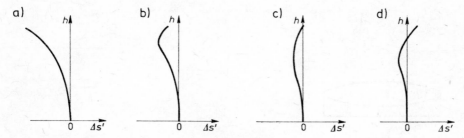

Fig. 2.23. Typical graphs of uncorrected (a) and corrected spherical aberration (b–d).

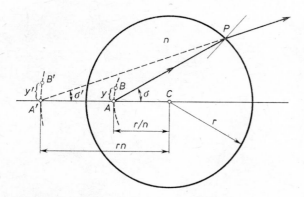

Fig. 2.24. Ball lens and its aplanatic points A and A'.

Spherical aberration of lenses and optical systems is commonly characterized by a function $\Delta s'(h)$, where h is the radius of successive lens zones. Typically it is displayed as a graph using the rectangular x-, y-axes; the first axis is reserved for $\Delta s'$ and the other for h. Figure 2.23a shows an uncorrected spherical aberration of a single positive lens, whereas Figs. 2.23b–d refer to corrected systems. The plots such as shown in Figs. 2.23b and d represent, respectively, spherically

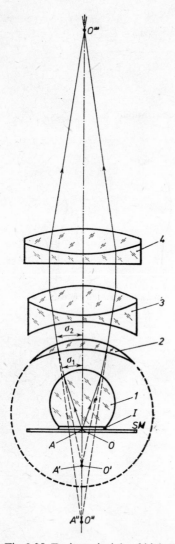

Fig. 2.25. Design principle of high-power oil-immersion objectives.

undercorrected and spherically overcorrected systems. Spherical aberration is, in fact, completely cancelled in the paraxial lens region and for a marginal zone, while for the remaining zones it is only reduced. This is a typical correction of spherical aberration.

It was said earlier that the spherical aberration of a positive lens can be reduced or eliminated by that of a negative lens made of a different glass; frequently such a combination is in the form of a *cemented doublet*. This is, in particular, a lens system of common low-power achromatic microscope objectives, which usually consist of a single doublet (objective 5×) or of two doublets (objective 10×). However, for middle and high magnifications more complicated systems are needed. They are universally based upon the *aplanatic points* of a ball-like lens (Fig. 2.24). These are the two conjugate points, A and A', for which spherical aberration does not exist. Let r and n be the radius and the refractive index of a glass sphere, respectively. It can easily be shown that if the object point A is distant r/n from the centre C of the sphere, then all rays emerging from A will, after refraction at the sphere surface, appear to be emerging from a point A' distant rn from C. The point A' (*outer aplanatic point*) is, of course, an aberration-free virtual image of A (*inner aplanatic point*). It is also important to note that after refraction the divergence of rays is much smaller than before refraction. If, for example, the ray AP forms an angle $\sigma = 65°$ with the optic axis AC of a glass sphere of refractive index $n = 1.5$, then behind the sphere the ray is inclined at an angle $\sigma' = 37°$.

These properties of a sphere are applied in the design of high-power microscope objectives, especially of oil-immersion objectives as shown in Fig. 2.25. To begin with, the sphere is replaced by a ball lens in the form of a truncated sphere with a plane or concave front surface passing through the inner aplanatic point A. Near this point the object under examination (O) is placed on a glass slide (SM). The gap between the slide and ball lens 1 is filled by a liquid (immersion oil, I) with a refractive index matching that of lens 1. Secondly, at an appropriate distance beyond the ball lens there is a positive meniscus 2, whose surface curvatures are so selected that the outer aplanatic point A' of lens 1 is simultaneously the centre of curvature of the first surface of lens 2 and the inner aplanatic point of the second surface of this lens. Extensions of rays emerging from the second lens (2) intersect each other at the outer aplanatic point A'' of the second surface of this lens where the secondary virtual image O'' of the object O is located. This image, as well as the primary one O', is free from spherical aberration and is transformed into a real image O''' by additional doublets 3 and 4. It is simple to correct the aberrations of these doublets by a combination of positive and negative lenses made of different glasses since they receive light beams of small divergence (σ_1 and σ_2 are, respectively, the aperture angles of the beam before and behind the meniscus 1, then $\sigma_2 \approx 0.65\sigma_1$).

From what has been said here, it is clear that attention should be paid to the principle of aplanatism not only by the designer of microscope objectives but also by the user. The latter must always remember to use a proper immersion liquid, a correct cover slip (in particular, for high-power dry objectives), a suitable ocular, and so on. Otherwise, the principle of aplanatism will be disturbed, leading to troublesome spherical aberration.

Coma. This aberration is closely related to spherical aberration, but is encountered for off-axis object points.[5] A light beam emerging from the off-axis point *B* (Fig. 2.26) is always asymmetric with respect to the lens axis; consequently

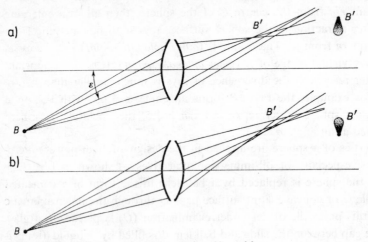

Fig. 2.26. Comatic aberration: a) negative, b) positive.

refraction of rays at successive lens zones and also within each zone is asymmetric as well. As a result, an asymmetrically blurred image *B'* of the object point *B* occurs. The blurring is comet-like, hence the name of this optical defect. The tail of the comet is turned either towards or away from the lens axis (compare Figs. 2.26a and b, respectively). In the first case we have an *inner* (or *negative*) *coma* and in the second an *outer* (or *positive*) *coma*.

As a rule, the larger the lens diameter and the greater the angle ε between the axis of the light beam and the optic axis of the lens, the more obvious is the coma. To reduce it, therefore, one may diaphragm the optical system, shift the position of the aperture diaphragm appropriately or select a proper lens shape. In general, for fixed positions of the object and its conjugate image it is

[5] If a multiple lens system is not correctly centred, a comatic aberration called *axial coma* also appears for axial object points.

possible to find a shape of the lens (lenses) that will effectively reduce comatic aberration; this shape will be found to be very similar to that required to reduce spherical aberration.

As regards microscope objectives, coma is corrected together with spherical aberration by applying the concept of aplanatism and *Abbe's sine condition*. The latter may be written as

$$\frac{n \sin \sigma}{n' \sin \sigma'} = \frac{y'}{y} = M = \text{const}, \tag{2.33}$$

where n and n' are, respectively, the refractive indices of the object-side space and the image-side space, whereas σ, σ', y, and y' are the conjugate quantities as shown in Fig. 2.17, and M is the transverse magnification. If a parallel beam strikes the optical system, the relation (2.33) takes the form

$$\frac{h_{en}}{\sin \sigma'} = f' = \text{const}, \tag{2.34a}$$

where h_{en} is the radius of the entrance zone, i.e., the distance of a ray from the optic axis of the lens in the object space, and σ' is the image-side aperture angle for this ray. If, however, a parallel beam must emerge from the optical system, Abbe's sine condition should be written as

$$\frac{h_{ex}}{\sin \sigma} = f = \text{const}, \tag{2.34b}$$

where h_{ex} is the radius of the exit zone of the end surface of the optical system through which exits the ray whose object-side aperture angle is σ.

The last condition is applied in the design of microscope objectives with infinite tube length. Remembering Eq. (2.6a), this condition may be written as

$$\frac{h_{ex}}{n \sin \sigma} = f' = \text{const}. \tag{2.34c}$$

The fulfilment of Abbe's sine condition may be interpreted as follows. If the image A' (Fig. 2.24) of an axial point A is free from spherical aberration, then the image B' of an off-axis point B, distant y from the optic axis of the lens, is without coma as long as y is small. Such a situation occurs when the points A and A' are aplanatic. But, as Fig. 2.24 shows, all points at distance r/n from the centre C of the ball lens are also aplanatic points. In particular, point B has its aplanatic point image B' distant y' from the optic axis of the lens. An inconvenient fact is that points A and B lie on a curved (spherical) surface. In microcopy, however, this curvature is without serious significance because the valuess of y to be taken into consideration are small (ca. 0.1 mm for an immersion objective of magnifying power $100 \times$). Thus one can say that a microscope objective (Fig. 2.25) will produce images free from spherical aberration and coma of all points of the object plane lying within the field of view of the objective.

Optical systems free from spherical aberration and coma are called *aplanatic systems*. Such systems cannot, however, be regarded as effective microscope objectives since they fail to correct other aberrations, especially chromatic aberration. Aplanatic systems are, however, widely used as microscope condensers and collectors of microscope illuminators.

Astigmatism. Causes of this aberration are almost the same as those of coma. It also occurs for oblique rays, but, unlike coma, astigmatism depends more strongly on the obliquity of the light beam and less strongly on the lens aperture. This aberration arises mainly because off-axis object rays are in different axial sections of the lens and are therefore focused at slightly different distances. To illustrate this, let us consider two specific sections: *tangential* (or *meridional*) *T–T* and *sagittal* (or *equatorial*) *S–S*, as shown in Fig. 2.27. The former is defined by the plane containing both the chief ray[6] and the optic axis of the lens, whereas the sagittal section is determined by the plane that is at right angles to the tangential section and contains the chief ray.

Leaving aside the comatic aberration, let us consider the refraction of rays within a lens zone. As can be seen in Fig. 2.27, the rays lying in the *tangential plane* and those lying in the *sagittal plane* are refracted differently. The tangential rays and the sagittal rays intersect the chief ray at different image points, B'_T and B'_S, respectively. The distance between these two points is called the *astigmatic difference* and is a measure of the astigmatism. As a result, a cone of rays emerging from the off-axial point object B produces no point image but a patch the shape of which is elliptical generally and linear in two particular positions where the intersections mentioned above, B'_T and B'_S, occur. These particular images are called the *sagittal line image*, denoted by b_S, and the tangential line image, denoted by b_T. Note that the sagittal line image b_S passes through the tangential point B'_T, and vice versa, the tangential line image b_T passes through the sagittal point B'_S. In another particular position the image patch is circular. This circle, known as the circle of least confusion, is between b_T and b_S, where the major and minor axes of the ellipse are equal.

[6] See Subsection 2.2.2 for definition. In Fig. 2.27 the chief ray is determined by the off-axis object point B and the centre of the aperture diaphragm opening marked by broken circle *TSTS*.

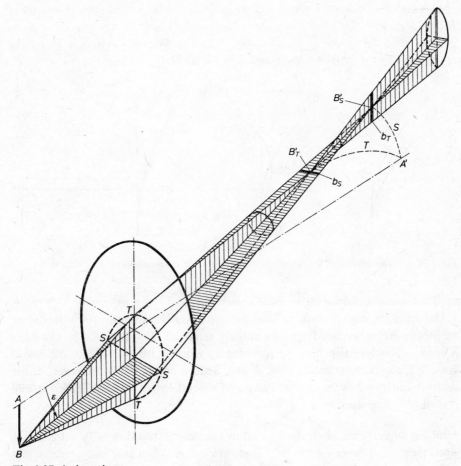

Fig. 2.27. Astigmatism.

Astigmatism is commonly characterized by a dependence $\Delta s'(\varepsilon)$, in which ε represents the angle between the chief ray and the optic axis of the lens (this angle is called the *field angle*), whereas $\Delta s' = s'_T - s'$ or $\Delta s' = s'_S - s'$, where s' is the vertex distance of the Gaussian image plane, s'_T is the vertex distance of the tangential image point B'_T, and s'_S is that of the sagittal image point B'_S. The dependence $\Delta s'(\varepsilon)$ is generally displayed as graphs plotted by using the rectangular x-, y-axes. Some examples are shown in Fig. 2.28. For an uncorrected optical system one obtains, in general, two parabolic curves, T and S, which indicate the positions of the tangential and sagittal image points, B'_T and B'_S, as a function of the field angle ε (Fig. 2.27). If $\varepsilon = 0$, the off-axis object point B coincides with the axial point A; consequently the image points B'_T and B'_S coincide with the Gaussian image A'. Note that the curves T and S in Fig. 2.28a correspond to the broken lines T and S in Fig. 2.27.

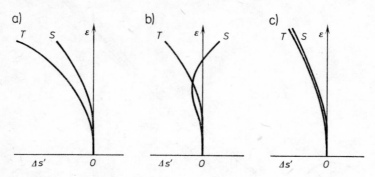

Fig. 2.28. Typical graphs of uncorrected (a) and corrected astigmatism (b, c).

To correct astigmatism it is necessary to reduce the astigmatic difference. If this quantity equals zero, at least for a particular value of $\varepsilon \neq 0$, the astigmatism is greatly reduced but not entirely removed (Fig. 2.28b). The complete removal of astigmatism can be regarded as causing the two astigmatic image curves, T and S, to coincide (Fig. 2.28c). Such a correction is, however, rather difficult and requires an optical system of several lenses made of different and special optical glasses.

Field curvature. This aberration is often considered together with astigmatism since their dependence on the field angle is similar. When field curvature is present, object points lying in a plane are imaged onto a curved surface; as a result, it is not possible to obtain an image that is both sharp in the axial region and simultaneously in a marginal zone of the field of view of an optical instrument. All microscopists are very familiar with this defect; when visual observations are performed, it is frequently insignificant, but it is particularly injurious in photomicrography where the image of a plate specimen should be focused on a flat photosensitive material.

Curvature of the field and its connection with astigmatism are shown in Fig. 2.29. When astigmatism is uncorrected (Fig. 2.29a), an object plane Π is imaged as two curved surfaces: *tangential image surface* Π'_T and *sagittal image surface* Π'_S. Note that these two surfaces result from Fig. 2.28 if the diagram is rotated round its horizontal axis; when rotated, the curves T and S form curved surfaces Π'_T and Π'_S, as shown in Fig. 2.29a. Between Π'_T and Π'_S there is an intermediate surface Π'_M, called the *medial image surface*, where the circles of least

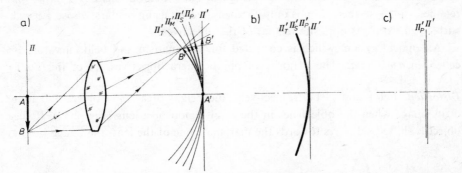

Fig. 2.29. Curvature of field and its connection with astigmatism: a) both the astigmatism and the curvature of field are present, b) no astigmatism but only curvature of field are present, c) both the astigmatism and the field curvature are corrected.

confusion are located. Within the scope of the third-order theory of aberrations, all these surfaces are paraboloids of revolution.

When astigmatism is corrected, the tangential and sagittal image surfaces are brought to coincidence to form a single curved surface, Π'_P, called the *Petzval surface*. In effect, there is no astigmatism and only field curvature is present (Fig. 2.29b). Its magnitude is generally expressed as deviations ($\Delta s'$) of an actual image surface from the Gaussian image surface Π'. The Petzval surface of an optical system depends only on the focal lenghts (f'_m). of the individual lens components and on their refractive indices (n_m). If the lenses are assumed to be very thin, the radius of curvature (r_P) of the Petzval surface results from

$$\frac{1}{r_P} = \sum_{m=1}^{k} \frac{1}{n_m f'_m}, \tag{2.35}$$

where k is the number of lens components.

Since field curvature is closely related to astigmatism, these two aberrations must be corrected together. This is a somewhat complicated procedure because an attempt to remove one of the aberrations generally has an adverse effect on the magnitude of the others. Thus a compromise is necessary. Field curvature is sometimes reduced by using field flattening lenses. Besides, in many optical systems the spacings between lens components can be varied to correct astigmatism. This operation, as can be seen from Eq. (2.35), does not change the Petzval surface curvature. Another procedure is generally used in the design of microscope objectives: astigmatism is removed, but Petzval field curvature is retained; this last aberration is only compensated by using oculars whose Petzval surface curvature is opposite to that of the objectives.

An optical system which is corrected for astigmatism and field curvature is called an *anastigmat*. The majority of photocamera objectives are of this kind.

Distortion. Let y and y' be, respectively, the height of a small object and that of its image when the object lies in the axial region of a lens (Fig. 2.30). If the object is shifted sideways towards the marginal zone of the lens, the image height

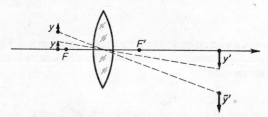

Fig. 2.30. Illustrating Eq. (2.36).

$\tilde{y}'$ will generally be greater or smaller than y'. Consequently, the lateral magnification will be changed by a magnitude

$$\Delta M = \frac{\tilde{M} - M}{M} = \frac{\tilde{y}' - y'}{y'}, \qquad (2.36)$$

where M and $\tilde{M}$ are the lateral magnifications related to the axial region and off-axis zones of the image plane, respectively. This defect is known as *distortion*. If the lateral magnification increases with the off-axis distance of the object, the distortion is said to be positive ($\Delta M > 0$). On the other hand, a decrease of magnification with the off-axis distance produces *negative* distortion ($\Delta M < 0$).

This aberration is easily observed when a test object in the form of a square grating is used (Fig. 2.31a). If $\Delta M < 0$, the grating image has a barrel shape (Fig. 2.31b), and when $\Delta M > 0$, a pincushion-like image occurs (Fig. 2.31c). Consequently, negative and positive distortions are also called *barrel* and *pincushion distortions*, respectively.[7]

[7] This nomenclature is not generally accepted. Some authors use the term "negative distortion" for "pincushion distortion" and "positive distortion" for "barrel distortion".

The type of distortion, as well as its magnitude, depends on the position of the aperture diaphragm. For example, if there is an aperture diaphragm (or entrance pupil) in front of a positive lens, barrel distortion occurs, whereas a diaphragm (or exit pupil) located behind this type of lens causes pincushion distortion. This property is used to design optical systems in which distortion has been removed. This is achieved by a symmetric arrangement of two (or more) lens components separated by a single diaphragm. The diaphragm has the effect of producing negative distortion by the first lens component and complementary positive distortion by the second component. An optical system which is corrected for distortion is said to be *orthoscopic*.

a)   b)   c)

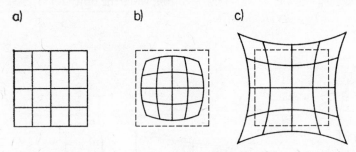

Fig. 2.31. Cross grating object (a) and its images deformed by barrel (b) and pincushion distortions (c).

In microscopical imaging distortion is mainly produced by oculars. When only visual and qualitative observations are performed, this aberration is not significant and $\Delta M = \pm 5\%$ can be admitted for typical oculars. If, however, an ocular is used as a micrometer eyepiece for measurements of length, width, and other linear quantities, the distortion is injurious, and ΔM must be as small as possible, typically less than $\pm 1\%$.

Chromatic aberrations. The foregoing optical defects of lenses have been discussed on the assumption of the presence of monochromatic light hence they are said to be monochromatic aberrations. However, all these aberrations are wavelength dependent. This dependence is known as chromatic aberration. It arises because of the refractive index variation with wavelength, or dispersion, of lenses and other elements in an optical system. The refraction index dispersion is a property of the material of which the optical elements are made hence this quantity does not depend on the divergence (or convergence) of light beams, and chromatic aberration occurs not only for off-axis rays but also in the paraxial region of the lens. Therefore, the concept of an ideal lens defined previously has no practical significance when polichromatic or white light is taken into consideration.

Off-axial theory of chromatic aberration is rather complicated; it is beyond the scope of this book, and only some chromatic variations in the paraxial image-forming properties of lenses will be discussed here.

In paraxial (Gaussian) optics the imaging properties of lenses depend mainly on the positions of the principal and focal planes of an optical system. If the position of any of these planes is wavelength dependent, chromatic aberration occurs. This dependence will be discussed in terms of a variation in image distance along the optic axis of the lens and a variation in lateral magnification. The first variation is known as *longitudinal* (or *axial*) *chromatism*, while the other is a consequence of *transversal* (or *lateral*) *chromatism*.

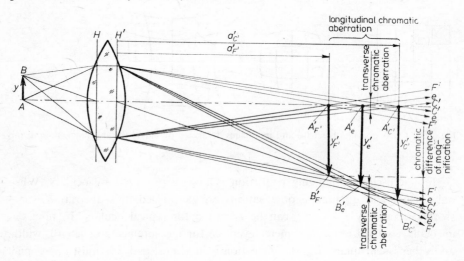

Fig. 2.32. Chromatic aberration.

To consider these two aberrations, let us consider a positive lens free from spherical aberration and other monochromatic aberrations (Fig. 2.32). Next, let us assume that the object AB emanates trichromatic light the wavelengths of which correspond, e.g., with those of F' (blue), e (green), and C' (red) spectral lines. The lens refracts blue light more strongly than green or red light therefore the image $A'_{F'}$ of the axial object point A is produced by blue rays at a smaller image distance $a'_{F'}$ than that $(a'_{C'})$ of the image $A'_{C'}$ produced by red rays, while the image A'_e is formed by green rays at an intermediate distance a'_e between $A'_{F'}$ and $A'_{C'}$. None of these images is a point of single colour, but a multichromatic patch occurs; the blue centre of $A'_{F'}$ is surrounded by red light, the red centre of $A'_{C'}$ is surrounded by blue light, while the green centre of A'_e is encircled by a purple border that results from an overlap of blue and red rays. A further variation of colours appears between $A'_{F'}$ and A'_e as well as between A'_e and $A'_{C'}$. The difference ($\Delta a'$) between image distances $a'_{F'}$ and $a'_{C'}$ is regarded as the *longitudinal* (or *axial*) *chromatic aberration*, i.e.,

$$\Delta a' = a'_{F'} - a'_{C'}. \tag{2.37a}$$

If a lens can be considered as thin, it may be shown that

$$\Delta a' = -\frac{f'_e}{V_e}(1 - M_e)^2, \tag{2.37b}$$

where f'_e is the focal length for e spectral line, V_e is the Abbe number, and M_e is the transverse magnification for the same e spectral line.

The size of the coloured image patch varies between $A'_{F'}$ and $A'_{C'}$ and is smallest near the image A'_e. This position fixes the plane of the sharpest image and the size of the spectrally dispersed image patch in this plane is regarded as the *transverse* (or *lateral*) *chromatic aberration*.

A similar spectrally dispersed formation of the image also occurs for any off-axial point B of the object AB.

A main consequence of transverse chromatic aberration is *chromatic difference of magnification* (CDM) also called *chromatic aberration of magnification* or *chromatism of magnification*. This quantity is expressed by the difference of the image heights, $y'_{F'}$ and $y'_{C'}$, for the F' and C' spectral lines referred to the image height, y'_e, for the e spectral line, i.e.,

$$\frac{\Delta y'}{y'_e} = \frac{y'_{F'} - y'_{C'}}{y'_e}. \tag{2.38a}$$

Frequently $\Delta y'/y'_e$ is given in percentages, and in that case

$$\frac{\Delta y'}{y'_e} = \frac{y'_{F'} - y'_{C'}}{y'_e} 100 \ [\text{in } \%]. \tag{2.38b}$$

Note that the image heights should be measured in the plane of the sharpest image, i.e., near the image $A'_e B'_e$. When traces of the external rays F', e, and C' are followed near this image, one easily find that $y'_{F'} > y'_e > y'_{C'}$ in the plane mentioned above. Thus for a convergent lens the chromatic difference of magnification is, in accordance with (2.38), positive. On the contrary, a divergent lens produces a reversed sequence of colours both for longitudinal chromatic aberration and for transverse chromatic aberration. Thus a positive lens can be achromatically corrected by a less-powerful negative lens (Fig. 2.33). A combination of this kind is called an *achromatic doublet*. Typically its positive component is made of crown glass while the negative component is made of flint glass. The two lenses have contact surfaces of the same curvature. They are cemented together with an optical cement the refractive index of which lies between those of crown and flint glass.

Fig. 2.33. Achromatic doublet.

In general, a system consisting of two lenses separated by a distance e (compare Fig. 2.19) may be free from transverse chromatic aberration if the following condition is fulfilled:

$$\frac{f_2'-e}{f_1'-e} = -\frac{V_1}{V_2}, \tag{2.39a}$$

where f_1' and f_2' are the focal lengths of component lenses for the wavelength $\lambda_e = 546.1$ nm (e spectral line), whereas V_1 and V_2 are the Abbe numbers of these lenses for the spectral lines e, F', and C' (compare Eq. (1.77)). If two lenses are in contact, the distance e is small in comparison with f_1 and f_2, and one can assume $e = 0$. In that case Eq. (2.39a) reduces simply to

$$\frac{f_2'}{f_1'} = -\frac{V_1}{V_2}. \tag{2.39b}$$

Because the numbers V_1 and V_2 are positive, then $f_2'/f_1' < 0$ and this inequality means that one of the lenses must be convergent and the other should be divergent. This is just the case of an achromatic doublet (Fig. 2.33). If, however, $V_1 = V_2$ (both lenses are made of the same glass), Eq. (2.39a) may be written as

$$e = \tfrac{1}{2}(f_1'+f_2'). \tag{2.39c}$$

Hence a two lens system will be free from transverse chromatic aberration if the component lenses made of the same glass are separated by a distance equal to half the sum of their focal lengths. Such a system is the basis of the Huygens eyepiece largely used in microscopes on account of its low cost (see Fig. 2.51).

A combination of two lenses is free from longitudinal chromatic aberration
if it satisfies the following relation, called *achromatization condition*:

$$\frac{f_2'}{f_1'} = -\frac{V_1}{V_2}\left(1 - \frac{e}{a_1'}\right)^2, \qquad (2.40)$$

where a_1' is the image distance of the first lens (to be achromatized), while the
remaining parameters are the same as in Eq. (2.39a). Note that a_1', of course,
pertains to the e spectral line. If two thin lenses are in contact ($e = 0$), Eq. (2.40)
takes the form of Eq. (2.39b). Thus the condition for correction of both the
transverse chromatic aberration and the longitudinal chromatic aberration is
the same.

With regard to the correction of chromatic aberration, microscope objectives
are divided into *monochromats, achromats, semiapochromats* (or *fluorits*), *apochromats, planachromats, planapochromats*, and *chromatic aberration-free systems*
(*CF objectives*).

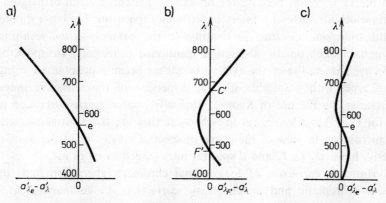

Fig. 2.34. Typical graphs of uncorrected chromatic aberration (a), and corrected achromatically
(b) and apochromatically (c) (λ in nm).

Monochromats, which may be useful only in some very special cases, are
optimally designed for one wavelength (λ_1) only, and chromatic aberration is
retained (Fig. 2.34a). Thus rays of different colours produce images of a single
object point at different image distances and monochromatic light is necessary
for sharp imagery. In white light, on the other hand, the image of a point object
is dispersed in colours and blurred. This colour dispersion is called the *primary
spectrum*. In the past, monochromats for $\lambda_1 = 275$ nm and $\lambda_1 = 257$ nm were

produced by C. Zeiss Jena. They were intended for ultraviolet microscopy, and their lenses were therefore made of quartz glass.

Achromatic correction means that the longitudinal chromatic aberration is removed for two colours: red and blue as shown in Fig. 2.34b; hence only images formed by red rays and blue rays are in coincidence (points A'_C, and A'_F, in Fig. 2.32), whereas those formed by rays of other colours are more or less deviated from the coinciding red and blue images. This deviation is observed as so-called *secondary spectrum*. Typically an achromatic correction is performed for wavelength $\lambda_{C'} = 644$ nm (C' spectral line) and $\lambda_{F'} = 490$ nm (F' spectral line) or $\lambda_C = 656$ nm (C spectral line) and $\lambda_F = 486$ nm (F spectral line).

For *apochromatically corrected* objectives the longitudinal chromatic aberration is removed for three colours: red, blue, and green, as shown in Fig. 2.34c; thus the positions of axial images produced by red, blue, and green rays are in coincidence (image points $A'_{C'}$, $A'_{F'}$, and A'_e in Fig. 2.32), whereas the image points for the remaining wavelengths of the visible spectrum deviate from the coinciding images only very slightly. This deviation is practically not perceived but only in the axial region because for off-axis points chromatism of magnification occurs which also gives a primary or secondary spectrum. The latter should be called the *transverse spectrum* in contrast to the *axial spectrum* resulting from the longitudinal chromatic aberration mentioned previously. In apochromats an effective elimination of the axial chromatism became possible by using special glass types with a suitable spectral dispersion of the refractive index and, in particular, by the use of fluorite. Typically apochromatic correction is performed for C', F', and e spectral lines because they are the most convenient for visual microscopy in view of the curve of spectral sensitivity of the human eye. Formerly, however, C, F, and d spectral lines tended to be in use.

Semiapochromatic correction of longitudinal chromatic aberration finds its level between achromatic and apochromatic correction. In comparison with achromats the secondary axial spectrum is considerably reduced, but is not minimized to being practically invisible as in apochromats.

All microscope objectives described above are, of course, also corrected for spherical aberration, coma, and astigmatism. In general, this correction is performed for one wavelength, but in apochromats spherical aberration is corrected for two wavelengths lying between those for which longitudinal chromatic aberration is removed.

Planachromat and *planapochromat* are terms for achromatic and apochromatic objectives in which a considerable correction of field curvature has been achieved. These objectives contain so-called *thick positive menisci*, the Petzval curvature of which corresponds to that of negative lenses; hence the field curvature of other positive lenses of the objective is balanced. Today planachromats are standard equipment of the majority of research and laboratory microscopes. On the other hand, in view of their cost planapochromats are on the whole used for more advanced photomicrography rather than visual microscopy.

Occasionally one may meet the name *new achromat* or *neoachromat*. This is simply an achromatic objective which is somewhere between the standard achromat and planachromat as regards correction of field curvature. It contains lenses made of special glass with a high refractive index and low V-number.

Compensation optics. No objectives described above are corrected completely for transverse chromatic aberration, which must be minimized by the use of compensating oculars. This compensation is, in general, obtained by including

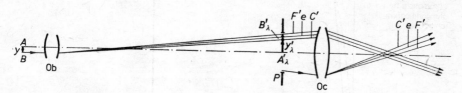

Fig. 2.35. Compensation of chromatic difference of magnification.

a chromatic difference of magnification in the oculars suitably matched to that of the objectives (Fig. 2.35). Matching is achieved by such an overcorrection of the ocular system (Oc) that its focal length increases as the wavelength of light decreases ($f_{F'} > f_{C'}$), in exact contrast to the objective system (Ob). In this case the rays C', e, and F' (for achromats of high power and planachromats the rays C' and F'), which form primary images $A'_\lambda B'_\lambda$ of different heights y'_λ for different wavelength λ, are so arranged that after passing through the ocular Oc they leave the microscope system parallel to each other and at infinity form a colour-free final image of the object AB. If the eye is accommodated to infinity (such an accommodation is typical for correct microscopical observation), it perceives the final image with the same angular magnification for each wavelength λ. In reality, however, a complete compensation of the chromatic difference of magnification across the entire angular field of view is extremely difficult so that a *residual chromatism of magnification* (*transverse secondary spectrum*) is left, which occurs especially in the marginal zones of the image field. This is due by the fact that the compensating ocular differs in transverse chromatic aberration (TCA) from the objective (Fig. 2.36a). In general, the TCA of compensating oculars increases noticeably with the primary image height (y'_λ, Fig. 2.35), whereas that of objectives changes almost linearly with the object height (y). Thus the chromatic difference of magnification (CDM), defined by Eqs. (2.38), is nearly constant for the objective and variable for the compensating ocular when the object height increases. Consequently, if a perfect compensation for the CDM is obtained

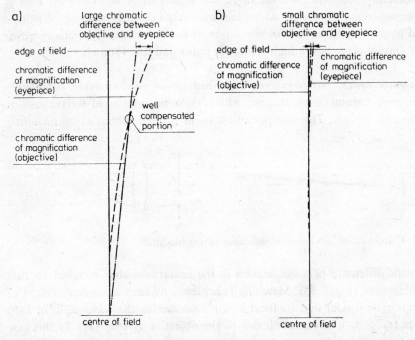

Fig. 2.36. Chromatic differences of magnification of microscope objective and ocular: a) compensating system, b) CF system.

in the central region of the field of view, an under-compensation (or over-compensation) results at the edges. It is therefore impossible to compensate completely the CDM over the entire image field, especially for microscope optics with a wide field of view.

The second disadvantage of the compensating oculars, especially those with a front field diaphragm, is the red-yellow or orange colour fringe around the field diaphragm edge. This phenomenon is shown in Fig. 2.35, which illustrates the path of colour rays C', e, and F', emerging from a point (P) of the field diaphragm edge. The angular dispersion of these rays is visible as a colouration of the image of the field diaphragm edge. Moreover, this dispersion means that the compensating ocular cannot be used as an entirely satisfactory measuring or counting eyepiece because marks, scales, graticules, etc. arranged in the front focal plane outside the centre of the image field are imaged with colour fringes.

CF optics. The aforementioned defects of the compensation microscope system do not apply to objectives and oculars developed in the last decade and called *chromatic-aberration-free optics*, in short *CF optics*. In a CF microscope system (Fig. 2.37) both longitudinal chromatic aberration and chromatic difference of

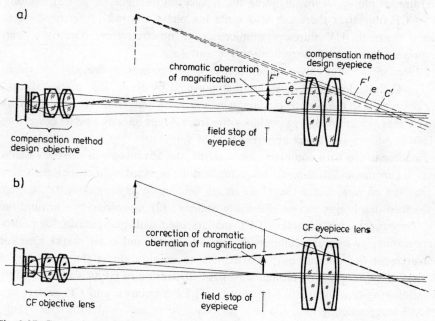

Fig. 2.37. Comparison between the compensation system (a) and CF system (b).

magnification are corrected in both the objectives and the oculars independently, without any compensation effects. The result is a dramatic elimination of transverse chromatic aberration across the entire field of view.

The first CF objectives and CF oculars for common microscopy both in transmitted and reflected light were developed in the years just before 1980 [2.2]–[2.5]. Now, several microscopes with CF optics, e.g., Biophot, Optiphot, and Labophot, are commercially offered by Nikon. CF microscope systems were made possible thanks to recent innovations in optical glass technology and advanced computer techniques in the design of optical systems. The extra-low dispersion glass developed by Nikon has similar optical properties to fluorite without the low primary fluorescence or other defects and processing difficulties to which fluorite is prone [2.4]. To ensure high-contrast imaging without flare and internal reflections, each lens of the CF system is coated with multiple layer antireflecting films.

A complete series of CF objectives manufactured by Nikon consists of *CF achromats* (chromatic aberration correction for *C* and *F* spectral lines), *CF planachromats*, and *CF planapochromats* (chromatic aberration correction across the entire visible spectrum including the *G* spectral line, see Table 1.3). Among Nikon CF objectives there are also some for phase contrast, differential interference contrast, UV flurescence microscopy, no-cover-glass objectives, and low-power objectives of magnification $1 \times$ and $2 \times$.

Objectives without chromatic difference of magnification are also manufactured by C. Zeiss Jena [2.5]. They are distinguished by an A. This mark indicates their correction to the A or P oculars, which are also free from chromatic difference of magnification. A-type oculars differ from P-type ones by a better correction of field curvature and larger apparent field of view.

In comparison with compensation systems, the advantages of the CF systems are: (1) chromatic difference of magnification is corrected right to the periphery of the field of view, hence the whole image field up to its limitation by the edge of the field diaphragm can be effectively utilized; (2) no colouration around the field diaphragm edge occurs; (3) CF oculars used with photomasks for photomicrography, as well as oculars with reticules, scales and other marks used for measurement or counting purposes exhibit no colouration at the periphery of the field of view; (4) longitudinal chromatic aberration of CF oculars (including CF photoeyepieces) is greatly improved; (5) CF objectives and CF oculars can be used independently with the best performance of each and with wider range of application; for instance, the intermediary image produced by the objective alone can be directly projected on a TV camera; (6) colour photomicrography is significantly improved. All these advantages mean that CF optics must be acclaimed as a very great advance in light microscopy.

It is worth noting here that the correction of aberrations of microscope objectives is a highly specialized procedure which in the past was a time-consuming operation, but has at present been greatly speeded up. This is mainly due to the use of computer-optimisation of design of optical systems.

The first period in the construction of modern microscope systems of high quality was ushered in by E. Abbe, who developed apochromatic objectives and compensation oculars (1886–1890). The second period was initiated in 1938 by

H. Boegehold, who constructed planachromatic and planapochromatic objectives. Other outstanding contributors to the improvement of plano-objectives were C.H. Claussen and C.G. Wynne (1960s). The third period has seen the development of CF optics, which has recently been undertaken by Z. Wahimoto for Nikon microscopes and H. Riesenberg for Zeiss microscopes (Jena).

It is customary for each renowned manufacturer of microscopes to employ one or two highly qualified specialists in the design and correction of optical systems. Abbe, Boegehold, and Riesenberg, for instance, developed successively more and more advanced optical systems for Zeiss microscopes (Jena); Claussen did the same for Leitz microscopes. In Poland the chief designes of highly qualified plano-objectives and PK oculars for PZO-microscopes are M. Gaj (Institute of Physics, Technical University of Wrocław) and T. Kryszczyński (Central Optical Laboratory, Warsaw). Almost the same highly qualified optical systems for Russian microscopes were mainly developed by D. Yu. Galpern, A. N. Grammatin, L. N. Andreev and V. P. Panov (see Ref. [4.38]).

2.3. The optical system of the microscope, its principal parameters and properties (bright-field microscopy)

The basic optical system of a compound microscope consists of an objective and an ocular arranged coaxially at opposite ends of a tube. Such an arrangement is a drastic simplification which in fact holds good only for the most common school microscopes; but for the general theoretical considerations covered by this section this simple arrangement is quite sufficient. It is, however, necessary right away to distinguish between a microscope with an objective designed for a *finite tube length* (or *finite image distance*) and a microscope with objectives designed for an *infinite tube length* (or *infinite image distance*). The latter has an additional lens, called the *tube lens*, situated between the objective and ocular; it will be discussed somewhat later.

It is also worth noting that though the objective and the ocular are always multiple lens systems, here they will frequently be represented by single positive lenses.

2.3.1. Magnifying power

Magnifying power is one of the most important optical parameters in microscopy, defined as the ability of a microscope to produce magnified images of micro-objects.[8] To evaluate this quantity, it is first useful to know how the *loupe* or *magnifier* works (Fig. 2.38). In general, the loupe is used together with the human eye of the observer to produce an enlarged virtual image $A'B'$ of a small object AB. The magnifying power or visual magnification of the loupe is definied as

$$M_L = \frac{\tan\omega'}{\tan\omega},\tag{2.41}$$

where ω and ω' are the angles at which the object AB and its image $A'B'$ are observed without and with a loupe, respectively.[9] Small objects are usually observed from a *distance of distinct vision*.[10] In this case (Fig. 2.38a) the *magnifying power of the loupe* is expressed by

$$M_L = -\frac{(f'-a')a_v}{(e-a')f'},\tag{2.42a}$$

where f' is the focal length of the loupe, a' is the image distance, e is the distance between the loupe and the eye, and a_v denotes the distance of distinct vision (Fig. 2.38b). Note that a' and a_v are negative while e and f' are positive quantities.

As can be seen from Eq. (2.42a), M_L is no constant value for a given focal length f' of the loupe; it depends on observation conditions, especially on the parameters e and a (or a'). In particular, we must note the very important situation when the object AB is in coincidence with the front focus (F) of the loupe (Fig. 2.38c). In this instance $a' = -\infty$, and Eq. (2.42a) reduces simply to

---

[8] There is no general agreement on the use of the terms "magnifying power" and "magnification". The latter is defined as the act or result of enlarging the apparent dimensions of objects by optical or other methods. According to Eqs. (2.14)–(2.17), the word "magnification" should be used for the ratio of conjugate quantities (see Figs. 2.11 and 2.17). On the other hand, the term "magnifying power" is defined as the ability of an optical (or other) system to produce magnified images of objects and adopted to express the ratios of both conjugate and non-conjugate quantities. There are, however, objections to such a nomenclature. Some authors prefer to use the term "magnifying power" for optical systems with a finite object and infinite image distances and reserve the word "magnification" for optical systems whose object and image distances are both finite. In Ref. [1.1] the term "magnifying power", however, refers to a situation where the image distance is equal to the reference visual distance (usually 250 mm). In practice, as well as in many books on optics or microscopy (and sometimes also in this book), the two terms are, however, used synonymously.

[9] This kind of magnification is also called *angular magnification* by some authors. Referring back to the definition of angular magnification (see Subsection 2.2.3 and Fig. 2.17), we can see that Eq. (2.15a) expresses another quantity than Eq. (2.41). A better term for M_L is then the *visual angular magnification*. Sometimes the latter is also defined as $M_L = \omega'/\omega$.

[10] In fact, this is the nearest distance of distinct vision, also called the *comfortable viewing distance*, *conventional visual distance*, or *reference viewing distance*.

$$M_L = -\frac{a_v}{f'}.$$ (2.42b)

This particular magnifying power is sometimes called *loupe magnification*.

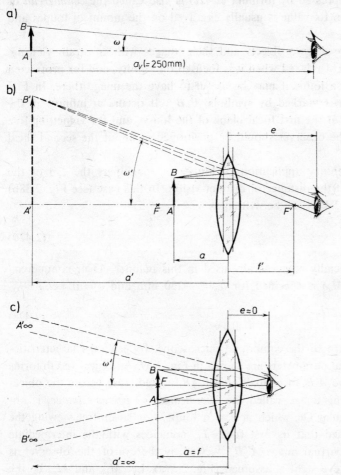

Fig. 2.38. Distance of distinct vision (a), and single convex lens used as a loupe (b, c).

Assuming normal eye sight, the internationally accepted value for the distance of distinct vision is 250 mm (strictly speaking, $a_v = -250$ mm), hence Eq. (2.42b) takes the form

$$M_L = \frac{250}{f'} \qquad [f' \text{ in mm}],$$ (2.42c)

which is conventionally taken as the definition of the magnifying power of a loupe, magnifier, and microscope ocular. The last device, as it will be shown later, works as a loupe with respect to the image formed by a microscope objective. The quantity expressed by formula (2.42c) is also called the *commercial* or *nominal magnification*. Its value is usually engraved on the mount of loupes and microscope eyepieces.

It is advisable to observe objects and their images without strain viewing. The normal eye is most relaxed when it is focused on infinity, and for protracted observation through a loupe it may be useful to have the image there. In Fig. 2.38c such an image is marked by symbols $A'_\infty B'_\infty$. It occurs at infinity if the object AB is located at the first focal plane of the loupe, and for vignetting-free viewing the eye of the observer should be positioned at about the second focal plane.

In another elementary application of loupes or magnifiers the eye of the observer is focused at the distance of distinct vision. In this case (see Fig. 2.38b) $a' - e = a_v$ and Eq. (2.42a) can be written as

$$M_L = 1 - \frac{a_v - e}{f'}.$$ (2.42d)

Reading glasses especially are generally used in this manner. Their commercial magnifying power (M_c) is specified for $a_v = -250$ mm and $e = 0$, i.e.,

$$M_c = 1 + \frac{250}{f'} \qquad [f' \text{ in mm}].$$ (2.42e)

We can now return to the compound microscope. Its general characteristics and some basic optical parameters are shown in Fig. 2.39. An objective Ob forms an inverted real image $A'B'$ of a small object AB located close before the objective first focus F_{ob}. This image, called the *intermediate* or *primary image*, is observed through an ocular Oc, which acts as a loupe. For strain-free viewing the ocular is so positioned that its first focus F_{oc} coincides with the intermediate image $A'B'$. A final virtual image $A''_\infty B''_\infty$ viewed by the eye of the observer is at infinity if normal eye sight is assumed (Fig. 2.39a). Let M_{ob} and M_{oc} be the magnifications due to the objective and ocular, respectively. Then the total magnification of the final image $A''_\infty B''_\infty$ due to the microscope is given by

$$M_m = M_{ob} M_{oc}.$$ (2.43)

Here M_{ob} is the transverse magnification, whereas M_{oc} and M_m are the visual magnifications. Taking into account Eq. (2.14c), M_{ob} may be expressed as $M_{ob} = -b'/f'_{ob}$, where f'_{ob} is the image-side focal length of the objective and b' is the

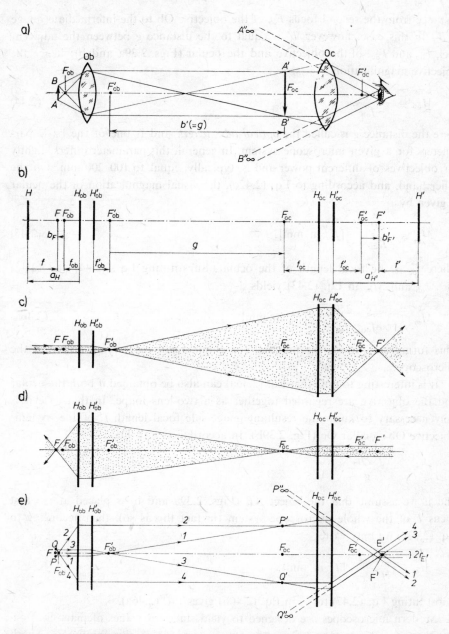

Fig. 2.39. Principle of the compound microscope (a), its basic parameters (b) and general characteristics (c, d, e).

distance from the second focus F'_{ob} of the objective Ob to the intermediate image $A'B'$. In this case, however, b' is equal to the distance g between the adjacent foci, F'_{ob} and F_{oc}, of the objective and the ocular (Figs. 2.39a and b); hence the objective magnification

$$M_{ob} = -\frac{g}{f'_{ob}}. \tag{2.44}$$

Here the distance g is called the *optical tube length* and is one of the basic parameters for a given microscope system. In general, this parameter differs slightly for objectives of different power and is, typically, equal to 100–200 mm. On the other hand, and according to Eq. (2.42c), the visual magnification of the ocular is given by

$$M_{oc} = \frac{250}{f'_{oc}} \qquad [f'_{oc} \text{ in mm}], \tag{2.45}$$

where f'_{oc} is the focal length of the ocular. Substituting Eqs. (2.44) and (2.45) for M_{ob} and M_{oc} in Eq. (2.43) yields

$$M_m = -\frac{g}{f'_{ob}}\frac{250}{f'_{oc}} \qquad [f'_{oc} \text{ in mm}]. \tag{2.46a}$$

This formula expresses the nominal (or commercial) magnifying power of the microscope.

It is interesting to note that Eq. (2.46a) can also be obtained if both the ocular and the objective are regarded together as a two-lens loupe. In that case it is only necessary to know the resultant image-side focal length f' of the system: objective Ob+ocular Oc (Fig. 2.39b). In accordance with Eq. (2.18),

$$f' = -\frac{f'_{ob}f'_{oc}}{g}, \tag{2.47}$$

and if we assume that the object AB (Figs. 2.39a and b) is placed at the first focus F of the whole microscope system (in fact this is so), then according to Eq. (2.42c) we may write

$$M_m = \frac{250}{f'} \qquad [f' \text{ in mm}]. \tag{2.46b}$$

Substituting Eq. (2.47) for f' in Eq. (2.46b) gives Eq. (2.46a).

Modern microscopes are designed to yield standard values of magnification, but there are also systems known as *zoom systems*, which give continuously variable magnification. The standardized values of magnification are based on the international R_a-10 standard series of ISO (Table 2.1). This is a decimal geometric series in which each successive number is greater than the preceding number by about 25%, and the product of any two numbers is again a standard value. When deciding on the magnifying power of objectives and oculars on the basis of this series, some practical circumstances must be taken into account. First of all, four to five main objectives should enable the user to cover the range of total magnification required in his work since only this number of objectives can be simultaneously screwed onto the conventional quadruple or quintuple revolving nosepiece. Secondly, the values of initial magnifying powers for objectives and oculars must be properly selected from the standard series of magnification. The most frequently used magnifying powers of objectives and oculars are listed in Table 2.1. Microscope objectives with ranges of magnifying power from $1 \times$ to $10 \times$, $10 \times$ to $40 \times$ and $40 \times$ to (or more than) $100 \times$, are classified as *low-*, *middle-* and *high-power objectives*, respectively.

At the present time it is only in rare cases or for some special purposes that the magnifying powers of objectives and oculars deviate to some extent from the standard values. For instance, the magnifying powers 1.5, 3, 6, 15 and 60 are also in use instead of the standard values 1.6, 3.2, 6.3, 16 and 63. Moreover, factory tolerances of $\pm 5\%$ for magnifying power of both the objectives and the oculars are generally accepted, while the magnification tolerances for certain additional optical components such as tube lenses and intermediate magnification changers are $\pm 2\%$. In the last instance the total visual magnification of the microscope is, of course, defined by the product

$$M_m = M_{ob} M_{oc} M_t, \tag{2.48}$$

where M_t denotes the magnifying power of the intermediate optical system situated between the objective and ocular. Frequently M_t is called the *tube factor*. It can be greater or smaller than 1.

Some manufacturers produce microscopes with binocular tubes having a tube factor greater than 1, e.g., equal to 1.25 or even 1.5. In general, these binoculars are less advantageous than those whose tube factor is unity. Some advantages are, however, provided by special intermediate magnification-changing systems such as the wide-field Optovar manufactured by C. Zeiss Oberkochen. This device provides different tube factors step by step as follows: $0.8 \times -1 \times - 1.25 \times -1.6 \times$. The first of them ($0.8 \times$) causes the field of view of the microscope to be enlarged. Another intermediate magnification changer is manufactured by E. Leitz Wetzlar. This makes it possible to obtain continuous variation of the tube factor within a range between $1 \times$ and $3.2 \times$. This zoom magnification

TABLE 2.1
Standard series of magnifications

For optical systems (in general)	0.5	0.63	0.8	1	1.25	1.6	2	2.5	3.2	4	
	5	6.3	8	10	12.5	16	20	25	32	40	
	50	63	80	100	125	160	200	250	320	400	
	500	630	800	1000	1250	1600	2000	2500	3200		
For light microscopes	1	1.25	1.6	2	2.5	3.2	4	5	6.3	8	
	10	12.5	16	20	25	32	40	50	63	80	
	100	125	160	200	250	320	400	500	630	800	
	1000	1250	1600	2000	2500						
For microscope objectives: basic series	1	2.5	6.3	10	16	25	40	63	100	(160)	
main series		2.5	6.3		16		40		100		
simplified series I		2.5		10			40		(100)		
simplified series II		(2.5)		10			40		100		
For oculars	3.2	4	5	6.3	8	10	12.5	16	20	25	32

changer is an attachment to the Orthoplan microscope.

It is worth noting that German literature especially uses different expressions for the transverse type of magnification when a real image is formed by a lens, and for the angular type of magnification when a virtual image is observed. For the former the same notation is used as for the image scale, 25 : 1 for instance, whereas for the latter the symbol " × " follows or precedes the value of magnifying power (e.g., $25\times$ or $\times 25$). Here such double specification is not observed and the second notation is used for both types of magnification through out the book.

2.3.2. Foci and exit pupil

In order to determine exactly the ray tracks in a compound microscope (Fig. 2.39), it is necessary to know the positions of the foci (F_{ob}, F'_{ob}; F_{oc}, F'_{oc}; F and F') and principal planes (H_{ob}, H'_{ob}; H_{oc}, H'_{oc}; H and H') of the objective Ob, the ocular Oc and the whole optical system (Ob+Oc) as indicated in Fig. 2.39b. The focal lengths f and f' of the compound microscope, as well as the distances a_H, $a'_{H'}$, b_F, and $b'_{F'}$ between the corresponding principal planes and foci can easily be determined from Eqs. (2.18)–(2.21) for a system with dry objectives; it is only necessary to replace the suffixes 1 and 2 by ob and oc, respectively.

Like a single lens, the compound microscope has its first (object-side) and second (image-side) foci, F and F'. If a parallel beam enters the microscope axially from its objective Ob (Fig. 2.39c), it is focused at the focus F', but was also earlier focused at the second focus F'_{ob} of the objective; thus both these foci are mutually conjugate points. If, on the other hand, such a beam enters the microscope from its ocular (Fig. 2.39d), it is focused at the focus F, but was also earlier focused at the first focus F_{oc} of the ocular; thus these foci are also two mutually conjugate points. According to the reciprocity theorem, a divergent beam which emerges from the focus F converges at the first focus F_{oc} of the ocular, and leaves the microscope as a parallel beam. It is clear that any off-axial parallel beam which enters the microscope from its objective focuses at the second focal plane of the objective and at the second focal plane of the microscope. Similarly, an off-axial parallel beam which enters the microscope from its ocular focuses at the first focal plane of the ocular and the first focal plane of the microscope.

Previously it was stated that the intermediate image $A'B'$ (Fig. 2.39a) is normally coincident with the first focus F_{oc} of the microscope ocular. This coincidence is possible only when the object AB is located at a particular point which is conjugate with respect to F_{oc}. From what was said above it is clear that this point is the first focus F of the microscope; thus the front focal plane and the object plane of the microscope are in coincidence. This situation is the same as for a loupe used according to the principle shown in Fig. 2.38c.

In order to have an idea of the location of the foci F, F_{ob}, F'_{oc}, and F' (Fig. 2.39), let $f'_{ob} = 4.5$ mm (such a focal length corresponds to objectives of magnifying power $40 \times$), $f'_{oc} = 20$ mm and $g = 180$ mm. For such a microscope system, according to Eqs. (2.21), $b_F = -f'^2_{ob}/g = -0.1$ mm and $b_{F'} = f'^2_{oc}/g = 2.4$ mm.

Now, let us take Fig. 2.39e and consider certain rays 1, 2, 3 and 4, emanated by any two off-axial points P and Q of the object plane; rays 1 and 2 emerge from P and 3 and 4 from Q. Initially rays 1 and 3 are parallel to the optic axis of the microscope, thus next pass through the foci F'_{ob} and F'. Rays 2 and 4, on the other hand, are oblique at the start and pass through the focus F_{ob}; they thus run parallel to the optic axis of the microscope between the objective and the ocular, and intersect rays 1 and 2 at points P' and Q', respectively. These intersection points belong to the first focal plane of the ocular and are the primary images of the object points P and Q, respectively. Next, rays 2 and 4 leave the ocular parallel to rays 1 and 3, and both parallel pairs produce the final

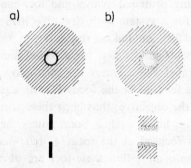

Fig. 2.40. Exit pupil of microscope: a) correct, b) aberrated.

images of the object points P and Q at infinity. The same situation occurs at the rear of the ocular if any other ray pairs emerging from the object points P and Q and acceptable to the microscope objective are taken into consideration. As a result, there is an infinite number of mutually parallel rays behind the ocular. By intersection, these rays form a cone with a distinct narrowing E', which constitutes the *exit pupil of the microscope* also called the *eye point* or the *Ramsden disc*. This pupil can easily be observed on a ground glass screen or a scrap of thin paper placed a few millimetres behind the ocular.

For a correct observation of the object image, the exit pupil of the microscope must be in coincidence with the pupil of the eye of the observer. Moreover, the diameter $2r_{E'}$ of the microscope exit pupil should be smaller than or at most equal to the diameter $2r_{eye}$ of the pupil of the eye, i.e., $2r_{E'} \leqslant 2r_{eye}$. Otherwise, the light from the outer zones of the microscope objective will be lost and an effect of vignetting will occur in the image plane. In fairly bright light (e.g., diffuse sunlight) $2r_{eye} \approx 3$ mm. The axial size of the exit pupil of the microscope must also be very small, i.e., the light emerging from the ocular must converge sharply to a thin circle (Fig. 2.40a). Otherwise, microscopical observation is imperfect because the eye of the observer looks into the microscope as through a small tube (Fig. 2.40b). This defect frequently occurs in practice and is mainly caused by an inadequate correction of spherochromatic aberration in the exit pupil of the microscope. Some wide field compensating oculars suffer from this defect which means that the eyes of the observer cannot be fixed precisely in position and small eye movements cause random partial obscuring of the outer zones of the field of view. Formation of an appropriate exit pupil of the microscope is therefore a troublesome problem in designing optical systems for advanced microscopy.[11]

2.3.3. Standard length parameters

Any compound microscope is characterized by many optical and mechanical parameters. Some of them are in use as international standards, while others are not universally accepted and are valid only as national or even factory standards. Some basic length parameters and their importance for the proper use of the microscope will be discussed in this subsection.

Mechanical tube length. Among standardized parameters of the microscope, mechanical tube length is the most important. Roughly speaking, this is the length t_m of a mechanical tube to the ends of which the objective Ob and the ocular Oc are attached (Fig. 2.41). At present, a straight tube is rarely used in more advanced visual microscopy. Instead, modern microscopes normally have a *revolving nosepiece* and a *binocular tube* coupled by the head of the limb (Fig. 2.42). In particular, the binocular body includes reflecting prisms, which, by comparison with an empty tube, increase the optical path of rays between the objective and ocular. Therefore, the concept of *reduced mechanical tube length* is used, i.e., as related to the air medium. Strictly speaking, this parameter is defined as the length of an empty tube which distances the objective and ocular

---

[11] This problem was extensively studied by T. Kryszczyński (Central Optical Laboratory, Warsaw).

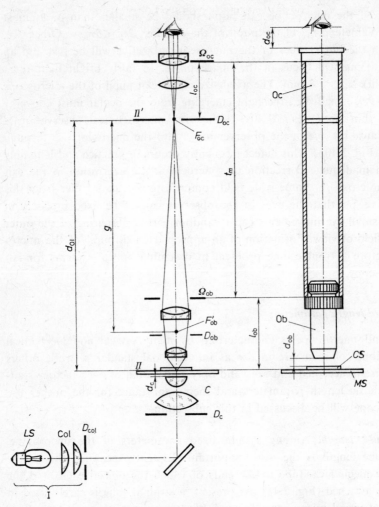

Fig. 2.41. Standard length parameters of microscope objectives and oculars.

in such a manner that the image plane Π' of the objective Ob is brought into coincidence with the first focus F_{oc} of the ocular Oc (Fig. 2.41).

Sometimes there is some confusion on the subject of mechanical tube length (t_m) and optical tube length (g). The latter is normally unknown to the user, but can be calculated from Eq. (2.44) if the magnification M_{ob} and focal length f'_{ob} of the objective are given by the manufacturer. Besides, the length g is a variable parameter for a set of different power objectives attached to a microscope whereas the length t_m is a constant normally engraved on the mount of the objective. Conventionally it is equal to 160 mm or 170 mm for transmitted light microscopes while for metallographic microscopes this parameter is generally greater than 170 mm (see Table 2.2).

Microscope objectives are normally designed so that their aberration, especially spherical aberration, in the image plane Π' (Fig. 2.41) exhibit minimum values when the objectives are used at the mechanical tube length for which they were optimally corrected. Some objectives are more sensitive to mechanical tube length variation than others and there are some limits within which this parameter can be changed without detriment to the performance of the microscope (see Table 2.3). Usually microscopes of high quality have their mechanical tube length fixed with an accuracy ± 1 mm. Such a tolerance results mainly from the need for the parfocality of objectives of different power. It may be found that a change of the mechanical tube length t_m by Δt_m displaces the object plane Π (Fig. 2.41) by

$$\Delta a = - \frac{f_{ob}\Delta t_m}{f'_{ob} M^2_{ob}}, \tag{2.49a}$$

where a denotes the object plane distance. For dry objectives, $f_{ob} = -f'_{ob}$, and Eq. (2.49a) reduces simply to

$$\Delta a = \frac{\Delta t_m}{M^2_{ob}}. \tag{2.49b}$$

This displacement can be regarded as defocusing which introduces spherical aberration into a microscope system well corrected for a standard mechanical tube length. A microscope tube that is too short introduces undercorrected spherical aberration, characteristic of a simple positive lens, whereas a tube that is too long introduces overcorrected spherical aberration. Table 2.3 shows tolerances to mechanical tube length change within which the *defocusing spherical aberration* is so small that it cannot be detected by the eye of the observer. As can be seen, the values given by different authors are similar for high numerical apertures but disparate for lower numerical apertures. This discrepancy is quite understandable because the optical (but not mechanical!) requirement for exactly the correct tube length diminishes very rapidly as the numerical aperture of the objective decreases.

Sometimes, however, a tube with variable length can be useful. Firstly, spherical aberration due to variations in thickness of the cover slips can be corrected by adjustment of the mechanical tube length. Secondly, small changes of the

TABLE 2.2

Basic parameters of microscopes produced by different manufacturers

Producer	Mechanical tube length t_m [mm]	Object-to-image distance d_{OI} [mm]	Parfocal lengths of		Examples of microscopes
			objectives l_{ob} [mm]	oculars l_{oc} [mm]	
1	2	3	4	5	6
VEB C. Zeiss Jena	160	192	45	13	Mikroval series Metallographic (Neophot); Peraval-interphako; 250-CF series (Jenaval,
	∞	∞	45	13	Jenamed) Metallographic with
	∞	∞	75	13	LWD objectives
E. Leitz Wetzlar	170	197	45	18	Orthoplan, Dialux
	∞	∞	45	18	Metallographic
	170	189	37	18	Some biological
Carl Zeiss Oberkochen	160	195	45	10	Standard series Axiomat; Axioplan,
	∞	∞	. 45		Axiophot Reflected-light (e.g.,
			33	10	with Epiplan objectives)
C. Reichert Wien	160	188	37	9	Transmitted-light
	190				Reflected-light
	250				Metallographic MeF 2
					UnivaR, Polyvar,
	∞	∞			Neovar, Biostar
Polish Optical Works (PZO) Warsaw	160	189	45	16	Biological (Studar series, Biolar series) Projection microscope
	170	189	35	16	MP3
LOMO Leningrad	160	179.5	32	12.5	Biological
	190				Reflected-light
	∞	∞			Metallographic

TABLE 2.2 continued

1	2	3	4	5	6
Meopta-Prague	170	195	36	11	Biological
Vickers	160	185	45	20	Transmitted-light (Photoplan, M17 series)
Wild Heerbrugg	160	188	37	9	Biological M20
Nachet (Sopelem) Paris	160 215			—	Biological Reflected-light
American Optical Corp. (Spencer)	160 180 ∞	∞		11.3 11.3	Biological Reflected-light Microstar series (10 and 11)
Bausch and Lomb	160 215 250 ∞			11 11 11	Balplan Reflected-light Metallographic Dynazoom
Nikon	160 210		45 45		Biological with CF optics (Optiphot, Biophot, Labophot) Reflected-light with CF optics (Metaphot)
Olympus	160 160	181 195	37 45	16 10	Biological Vanox with LB objectives, BHS/BHT series

TABLE 2.3

Mechanical tube length tolerances Δt_m within which defocusing spherical aberration is so small that may be neglected

Type of objective	Δt_m [mm]		
	acc. Ref. [2.6]	acc. Ref. [2.7]	acc. Ref. [2.8]
Achromats of magnifica- tion/numerical aperture:			
$10 \times /0.25$	27	≈ 180	
$50 \times /0.40$	16	35	
$40 \times /0.65$	7.6	4.3	± 7
$60 \times /0.85$	2.0	1.5	
$100 \times /1.25$ (oil immer- sion)	6.5	4.0	
Apochromats $(90 \times /1.3$ oil immersion)			± 2

objective magnification M_{ob} may be obtained when the mechanical tube length is varied. As it results from Eq. (2.44),

$$\Delta M_{ob} = \frac{\Delta g}{f'_{ob}} = \frac{\Delta t_m}{f'_{ob}}. \tag{2.50}$$

This procedure can be useful in microscopical measurements when it is desirable to match the divisions of an ocular micrometer with those of a stage micrometer.

Object-to-image distance. The second important standard parameter of the microscope is the distance d_{OI} between the object plane II and the primary image plane II' (Fig. 2.41), called the *object-to-image distance*. For a given microscope model this parameter must be kept constant. This requirement results from a practical consideration, namely the parfocality mentioned earlier. This term means that the image observed through the microscope must remain sharp (within some tolerances) when objectives and oculars are exchanged. To satisfy this requirement, the optical tube length (g) cannot be the same for low-, middle-, and high-power objectives; it is the mechanical tube length (t_m) and the object-to-image distance (d_{OI}) that must remain invariable. The latter is typically equal to about 190 mm (see Table 2.2).

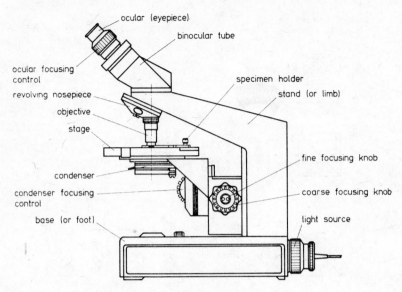

Fig. 2.42. Drawing showing the components of a typical microscope.

Parfocal length of objective. For a given mechanical tube length (t_m), the constant object-to-image distance d_{OI} can only be attained by a suitable design of the mounts of objectives and oculars. In order for the position of the image plane II' to remain invariable when the objectives are exchanged, the distance l_{ob} from the flange (Ω_{ob}) of the objective mount to the object plane II must also be constant (Fig. 2.41). This parameter is called the *parfocal length of objective* or *object distance of objective*. It is typically equal to 45 mm, and only some special objectives have another value of l_{ob} (see Table 2.2).

The parfocal length of objective is one of the most critical parameters in microscopy. For modern microscopes of high quality, l_{ob} should be kept constant within extremely small tolerences of the order of 1 μm. For testing l_{ob} within such a narrow range of tolerances, a special measuring instrument is necessary. That developed in the Central Optical Laboratory Warsaw [2.9] is shown in Figs. 2.43 and 2.44. A reflected-light microscopical system is equipped with a special epi-illuminator which includes a light source (LS), collector (Col), beam-splitter (BS_1), and rotating radial grating (RG). The latter element is situated in the image plane II' of the microscope objective Ob to be tested. The grating image given by the objective is focused on the mirror M, whose reflecting surface can be brought into coincidence with the object plane II of the objective

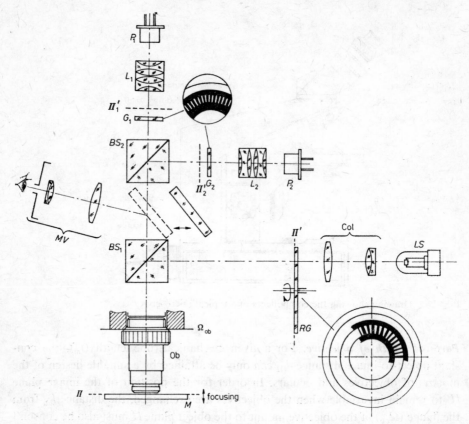

Fig. 2.43. Optical system of an instrument for testing the parfocal length of microscope objectives [2.9].

Ob. The light reflected by the mirror M is then divided by another beam-splitter BS_2 into two beams of equal intensity. Both beams pass through two other radial gratings (G_1, G_2) followed by auxiliary lenses (L_1, L_2) and photodetectors (P_1, P_2). The spatial frequency of the gratings G_1 and G_2 is the same as that of the input grating RG. Gratings G_1 and G_2 are positioned so that one of them lies before the image plane (Π_1') of the objective Ob while the other lies behind this plane (Π_2'). The distance between G_1 and Π_1' equals that between Π_2' and G_2. The output signals from the photodetectors P_1 and P_2 are amplified and then compared to each other. Differences in output signals are displayed on a digital meter. When the objective under examination is optimally focused, the difference in output signals is zero and a control lamp in the meter lights up. The vertical

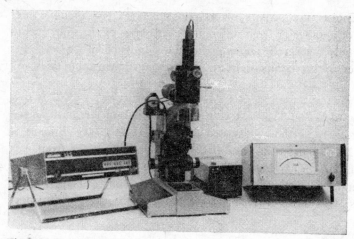

Fig. 2.44. Instrument for testing the parfocal length of microscope objectives, built in Central Optical Laboratory, Warsaw, according to the principle shown in Fig. 2.43 [2.9].

extra-fine focusing movement of the mirror M is electronically coupled with the digital meter; focusing accuracy is 0.1 μm.

This instrument is arranged so that it measures the deviations Δl_{ob} of the object distance l_{ob} of tested objectives with respect to the parfocal length of a master objective. The deviations Δl_{ob} can be measured with accuracies 0.5 μm, 1 μm, 2 μm, 10 μm, and 30 μm for objectives of magnification/numerical aperture $100 \times /1.30$, $40 \times /0.65$, $20 \times /0.40$, $10 \times /0.25$, $5 \times /0.15$, and $2.5 \times /0.08$, respectively.

The monocular viewer MV shown in Fig. 2.43 is an additional unit only used occasionally for preliminary adjustment of the instrument.

Parfocal length of ocular. In order for the image be remain sharp when the eyepieces are exchanged, the distance l_{oc} from the flange Ω_{oc} of the ocular mount to the primary image plane Π' must also be constant (Fig. 2.41). This length parameter is called the *parfocal length of ocular* or *intermediate image distance of ocular*. It is selected to be as short as possible so that oculars of high power can be as easily parfocalized as those of low power. Usually l_{oc} is between 10 and 20 mm (see Table 2.2).

Relation between mechanical tube length and parfocal lengths of objective and ocular. The combination of these three mechanical parameters gives an equation

$$d_{OI} = t_m + l_{ob} - l_{oc}. \tag{2.51}$$

Frequently microscopists use oculars or/and objectives of one manufacturer with a microscope of another or from different design series of the same manufacturer. This practice is generally unfortunate, but admissible if Eq. (2.51) is fulfilled and correction (or compensation) of chromatic aberration does not vary.

Working distance of objective. The next very important parameter, but one not standardized as precisely as those discussed above, is the *working distance of objective* d_{ob} (Fig. 2.41). This is the clear (free) distance between the front part of the objective Ob and the cover slip *CS* or object plane II for no-cover-glass objectives. In general, the distance d_{ob} decreases as the numerical aperture of objectives increases. This dependence is shown in Fig. 2.45. The diagram does

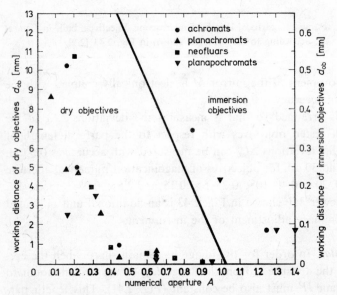

Fig. 2.45. Diagram showing the interdependence between numerical aperture and working distance of standard microscope objectives.

not refer to so-called *long working distance objectives* (in short, *LWD objectives*), but only to standard achromats, planachromats, neofluars, and planapochromats.

In general, it is useful to have dry objectives with working distances as long as possible. This demand is, however, limited by the need for a high numerical aperture and thus for high resolving power; thus a compromise is necessary between these parameters. In any case, however, the working distance cannot be less than a critical value (Table 2.4). Somewhat other requirements are for immersion objectives. If immersion liquid is to be contained between the front lens of the objective and the specimen under examination, the working distance cannot be too large. Objectives with close working distance are *spring loaded* and will retract if they touch the specimen. Such a resilient mount guarantees adequate protection against accidental damage to either the specimen or the front lens of the objective.

TABLE 2.4

Minimum working distance d_{ob} of microscope objectives according to their numerical aperture (after Polish Standard BN-74/5555-12)

Numerical aperture (A)		Minimum working distance d_{ob} [mm]	
for dry objectives	for immersion objectives	for achromats	for planachromats, apochromats and planapochromats
$A \leqslant 0.1$		10	5
$0.1 < A \leqslant 0.2$		8	2
$0.2 < A \leqslant 0.3$		4	0.5
$0.3 < A \leqslant 0.5$		0.35	0.15
$0.5 < A \leqslant 0.9$		0.10	0.10
	$0.5 < A \leqslant 1.0$	0.25	0.20
	$1.0 < A$	0.06	0.03

The LWD objectives mentioned above are required for biological microscopy *in vitro*, where specimens are observed through thick walls of glass, as well as for chemical and metallurgical microscopy, where the objective must be protected against the effects of heat, caustic vapors, etc., by means of a thick glass (or quartz glass) plate. The working distance of these objectives is several times as long as that of typical objectives with comparable or only slightly higher numerical apertures.

Eye relief. The nature of this parameter is similar to the working distance of objective but concerns the image space; it determines the distance d_{oc} between the ocular mount and the exit pupil of the microscope (Fig. 2.41). In order to avoid difficulties in microscopical observation, d_{oc} cannot be too small because the exit pupil of the microscope must be in coincidence with the eye pupil of the observer. In general, d_{oc} cannot be smaller than 7 to 5 mm. Otherwise, eye lashes or even eyelids touch the ocular and the microscopical observation is disturbed. However, eye relief which is too long is also inconvenient for the observer with normal eye sight. A range of d_{oc} between 7 and 13 mm is generally accepted. For spectacle wearers some manufacturers produce special eyepieces, so-called *highpoint oculars*, which can be used together with spectacle lenses (Fig. 2.46). The eye relief of these oculars is 15 to 20 mm. It should be noted,

a)   b)

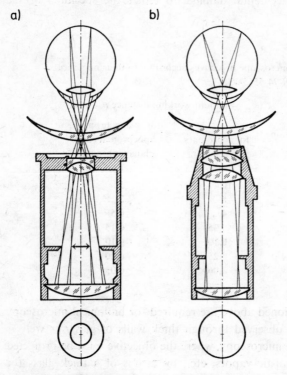

Fig. 2.46. Comparison between a typical ocular (a) and a highpoint ocular (b) when they are used by a spectacle wearer.

however, that spectacle lenses affect the distance of the microscope exit pupil: negative lenses increase, and positive lenses decrease eye relief according to their power. Someone who wears spectacles may therefore be unable to bring the pupil of his eye into coincidence with the exit pupil of the microscope even with the highpoint oculars, looking into the microscope would in this case be like looking through a keyhole into a room.

Strictly speaking, eye relief relates to two somewhat different situations: (1) an ocular is inserted into the microscope tube (the instance discussed so far),

(2) the ocular is considered independently of the microscope. In the second instance, eye relief is defined as a distance d'_{oc} from the front mount to the second focus F'_{oc} of the ocular. Now, this parameter is a constant and characterizes the ocular only, whereas in the first instance eye relief expresses, as defined previously, the distance d_{oc} between the front mount of the ocular and the second focus F' (or exit pupil) of the microscope. According to Eqs. (2.21), the distance between the foci F'_{oc} and F' (see Fig. 2.39) is expressed by

$$b'_{F'} = d_{oc} - d'_{oc} = \frac{f'^2_{oc}}{g}.$$ (2.52)

As can be seen, d'_{oc} is approximately equal to d_{oc} ($d'_{oc} \approx d_{oc}$), but only when the focal length f'_{oc} of the ocular is much smaller than the optical tube length g of the microscope ($f \ll g$). In this case the exit pupil of the microscope and the second focus of the ocular are practically in coincidence.

2.3.4. Role of diaphragms and modes of illumination

It is not enough to be able to determine the position and magnification of a microscopic image becuase the image brightness and the size of the field of view are also very important for every microscopist. The two latter problems require a discussion of the role of the aperture and field diaphragms which limit the spatial and angular sizes of light beams (see Subsection 2.2.2).

In a typical microscope there are four basic diaphragms. In Fig. 2.41 they are marked by D_{col}, D_c, D_{ob}, and D_{oc}. Two of them are the aperture diaphragms and the other two are the field diaphragms. One of the aperture diaphragms, D_{ob}, is part of the objective and the other, D_c, part of the condenser C, whereas one of the field diaphragms, D_{col}, belongs to the elements of the illuminator I and the other, D_{oc}, is part of the ocular Oc. Diaphragms D_{ob} and D_{oc} are usually inaccesible to the user because their axial positions and diameters are fixed by the manufacturer. Each microscopist should, however, be familiar with the use of diaphragms D_{col} and D_c; on their proper use depends the correct illumination of the specimen under examination and therefore a satisfactory image.

Numerical aperture and the exit pupil of the microscope objective. Apart from some low-power microscope objectives, other objectives have an aperture diaphragm as part of their optical system. Normally the role of this diaphragm is fulfilled by the rim of the lens which is close to the second focus F'_{ob} of the objective. However, there are some special objectives, usually of magnifying power not smaller than $60 \times$, which are equipped with an iris aperture diaphragm whose opening

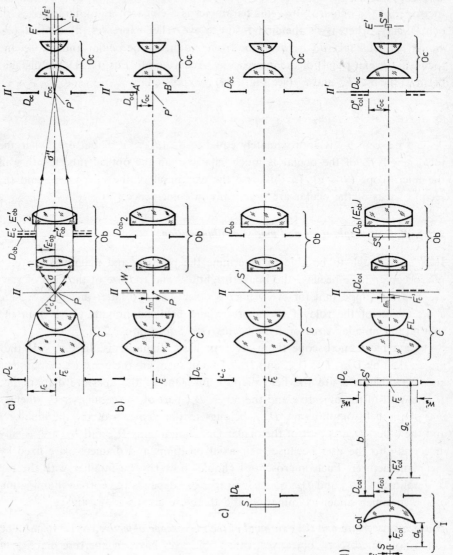

Fig. 2.47. Illustrating the role of the aperture and field diaphragms and modes of illumination in the microscope.

can be continuously varied by the user. Such objectives are frequently used for dark-field microscopy.

For the present, it will be assumed in what follows that the microscope objective Ob (Fig. 2.47) consists of two parts, 1 and 2, separated by the aperture diaphragm D_{ob}. The images of D_{ob} formed by the lens parts 1 and 2 are, respectively, the entrance pupil and the exit pupil of the objective. The first is practically localized at infinity, whereas the exit pupil E'_{ob} coincides (or is nearly in coincidence) with the second focus F'_{ob} of the objective Ob. The pupil E'_{ob} is then imaged by the ocular Oc as the exit pupil E' (Ramsden disc) of the microscope (compare the final text of Subsection 2.3.2). The last pupil is coincident with the second focus F' of microscope, and so the entrance pupil of the whole microscope system (Ob+Oc) is localized at infinity. All these pupils are optically conjugate because each is an image of the preceding one.

The objective aperture diaphragm D_{ob} determines a maximum angle σ, at which rays emerging from the axial point P can, after refraction, pass through the objective Ob. As defined previously (see Subsections 2.2.2 and 2.2.3), this angle is called the object-side aperture angle or the angular semi-aperture, while its conjugate counterpart σ' in the image space is said to be the image-side aperture angle. According to Eq. (2.33), we may write the relations

$$\frac{n\sin\sigma}{\sin\sigma'} = \frac{A}{A'} = M_{ob} \tag{2.53a}$$

for immersion objectives, and

$$\frac{\sin\sigma}{\sin\sigma'} = \frac{A}{A'} = M_{ob} \tag{2.53b}$$

for dry objectives, where n is the refractive index of the object space, while $A = n\sin\sigma$ (or $A = \sin\sigma$) and $A' = \sin\sigma'$ are, respectively, the object-side and image-side numerical apertures of the objective, and M_{ob} denotes the transverse magnification of the objective. For instance, a typical dry microscope objective of magnifying power $10\times$ has a numerical aperture $A = 0.25$, so $\sigma' = 1°25'$. On the other hand, the numerical aperture A of a high-power immersion objective of magnifying power $100\times$ is equal to 1.30; hence $A' = 0.013$ and $\sigma' = 45'$. From these examples it appears that the image-side aperture angles of microscope objectives are very small, and we can assume $\tan\sigma' = \sin\sigma'$. In this case, as can be seen from the geometry of Fig. 2.47a, the radius $r_{E'_{ob}}$ of the exit pupil of the objective is given by $r_{E'_{ob}} = g\sin\sigma'$, where g is the optical tube length. Then, just as $\sin\sigma' = A' = A/M_{ob}$ and $M_{ob} = g/f'_{ob}$, so also

$$r_{E'_{ob}} = f'_{ob}A. \tag{2.54}$$

The focal length f'_{ob} of the objectives $10 \times /0.25$ and $100 \times /1.30$, which was taken into consideration above, is approximately equal to 16 mm and 1.9 mm, respectively. Then the radii $r_{E'_{ob}}$ of the exit pupils of these objectives are equal to 4 mm and 2.5 mm.

It was stated earlier that the exit pupil E' of the microscope is the image of the exit pupil E'_{ob} of the objective Ob (Fig. 2.47a). This image is formed by the ocular Oc; hence there is a simple relation between the radii $r_{E'}$ and $r_{E'_{ob}}$ of both pupils, viz., $r_{E'}/r_{E'_{ob}} = M_{oct}$, where M_{oct} denotes the transverse magnification of the ocular (in this and the following formulae the negative sign is left out). Applying the first part of Eq. (2.14c) gives $M_{oct} = f'_{oc}/g$. Next, by combining Eqs. (2.44) and (2.54) we obtain

$$r_{E'} = A \frac{f'_{oc}}{M_{ob}}. \tag{2.55a}$$

Furthermore, by applying Eq. (2.45) we can write Eq. (2.55a) in the form

$$r_{E'} = A \frac{250}{M_{ob} M_{oc}} = A' \frac{250}{M_m} \text{ [mm]}, \tag{2.55b}$$

where M_{oc} and M_m are the visual magnifications of the ocular and the microscope, respectively. Returning to the exemplary objectives $10 \times /0.25$ and $100 \times /1.30$ and taking, e.g., $M_{oc} = 12.5\times$, yield $r_{E'} = 0.5$ mm and $r_{E'} = 0.26$ mm, respectively. Recalling what was said in the final paragraph of Subsection 2.3.2, it is now clear that the diameter $2r_{E'}$ of the exit pupil of the microscope can be smaller than the diameter $2r_{eye}$ of the pupil of the observer's eye; in fact this is so because under conditions of typical microscopical observation $2r_{eye} = 2$ to 3 mm.

It is generally known that the resolving power of a microscope depends primarily on the numerical aperture $A = n \sin \sigma$ and the quality of the objectives. However, within a given degree of aberration correction the numerical aperture cannot be greater than an optimum value prescribed for standard objective magnifications. This interdependence between the A and the magnifying power M_{ob} for different microscope objectives is shown in Figs. 2.48 and 2.49. Usually the numerical apertures of apochromats and semiapochromats are greater by 15 to 30% than those of achromats of the same magnifying power, but there is no significant difference between the numerical apertures of planachromats and achromats. This property relates also with planapochromats and apochromats. Both the numerical aperture and the magnifying power of microscope objectives are engraved on their mounts. The engraved quantities are only nominal values. In practice, the real values of A do not differ from the nominal values by more

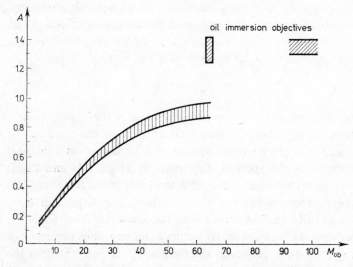

Fig. 2.48. Graph showing the interdependence between the numerical aperture (A) and magnifying power (M_{ob}) of achromatic and planachromatic objectives.

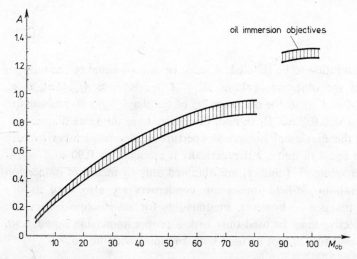

Fig. 2.49. Graph showing the interdependence between the numerical aperture and magnifying power of apochromatic and planapochromatic objectives.

than $\pm 10\%$, $\pm 5\%$, and $\pm 2\%$ for low-, middle-, and high-power objectives, respectively. The numerical aperture A is measured by means of a special device called the *apertometer*. But this parameter can also be determined from Eq. (2.54) if the exit pupil diameter $2r_{E'_{ob}}$ of the objective and its focal length f'_{ob} are known.

Numerical aperture of the condenser. One can take advantage of the large numerical aperture of the objective only when the numerical aperture of the condenser is also sufficiently large. For a given optical system of a condenser its numerical aperture A'_c is determined by the aperture diaphragm D_c (Figs. 2.41 and 2.47a) and defined as $A'_c = n' \sin \sigma'_c$, where n' is the refractive index of the image space[12] and σ'_c is the image-side aperture angle of the condenser. The diaphragm D_c is normally situated so as to be in coincidence with the condenser front focus F_c. Consequently, the exit pupil of the condenser occurs at infinity while the entrance pupil is the diaphragm D_c itself. However, the former, E'_c, is intercepted by the microscope objective Ob and imaged at its own exit pupil E'_{ob}. Notice that E'_c is also the image of the aperture diaphragm D_c. In a well assembled microscope the diameter $2r_{E'_c}$ of the image E'_c cannot be smaller than the diameter $2r_{E'_{ob}}$ of the objective exit pupil $E'_{ob}(2r_{E'_c} \geqslant 2r_{E'_{ob}})$. If $2r_c$ denotes the diameter of the diaphragm D_c and f'_c is the focal length of the condenser, then the condenser numerical aperture can be expressed as

$$A'_c \approx \frac{r_c}{f'_c}. \tag{2.56}$$

In order for this condition to be fulfilled, A'_c must be at least equal to the numerical aperture A of the objective. Taking $A'_c = A$ yields $r_c \approx A f'_c$. Let, e. g., $f'_c = 10$ mm and $A = 1.30$ so the diameter $2r_c$ of the diaphragm D_c amounts to 26 mm. The lens which follows D_c should, of course, have the same diameter.

Theoretically, the maximum numerical aperture of dry condensers (or dry objectives) can be equal to unity, but practically it amounts to 0.90 and sometimes 0.95. Greater values of A and A'_c are obtained only by means of immersion systems. In general, bright-field immersion condensers are also used as dry condensers. This practice is, however, inadmissible for microscope objectives: an immersion objective must be used only with a proper immersion liquid, and, vice versa, a dry objective cannot be used with any immersion liquid.

---

[12] Note that the object space of the microscope objective is the image space of the condenser.

Field of view. The field diaphragm of the ocular is the element which determines the field of view of the microscope. In conventional microscope oculars, this diaphragm precedes the optical system of the ocular or is found within it. Let us consider the first situation which is shown in Figs. 2.41 and 2.47. The field diaphragm D_{oc} is in the front focal plane of the ocular Oc. Since this plane is normally coincident with the primary image plane Π' of the objective Ob, the edge of the diaphragm D_{oc} is seen sharply together with the image of the specimen under examination. It is clear that this diaphragm makes it possible to observe a circular object area the radius r_m of which (Fig. 2.47b) is given by

$$r_m = \frac{r_{oc}}{M_{ob}},\qquad(2.57)$$

where M_{ob} is the transverse magnification of the objective Ob. The diameter $2r_{oc}$ of the field diaphragm D_{oc} determines the *linear field of view* of the ocular.

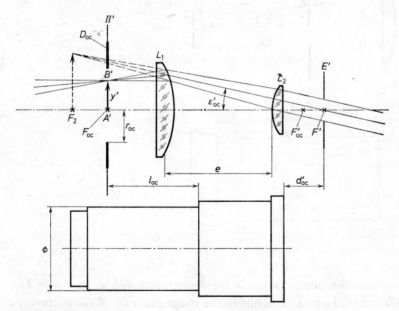

Fig. 2.50. Ramsden ocular.

This quantity is usually called the *field-of-view number* or, in short, *field number*. It will be denoted by FN.

Equation (2.57) is true only for oculars whose first focus F_{oc} is outside their optical system. A classical representative of such oculars is the *Ramsden eyepiece* shown in greater detail in Fig. 2.50. Its optical system consits of two planoconvex lenses L_1 and L_2, the separation e of which is suitably fixed according to their focal lengths f_1' and f_2'. The first lens, L_1, is called the *field-lens* because it is near the image plane Π', in which the primary image $A'B'$ of an object is form- ed by the microscope objective. The field-lens L_1 must, however, be adequately separated from plane Π', partly to prevent any dust or some surface defects on the lens from being visible together with the image of the specimen under examin- ation, and partly to render plane Π' accessible to cross-wires or a graticule. The second lens, L_2, is called the *eye-lens* because it is close to the eye of the observer who looks into the microscope.

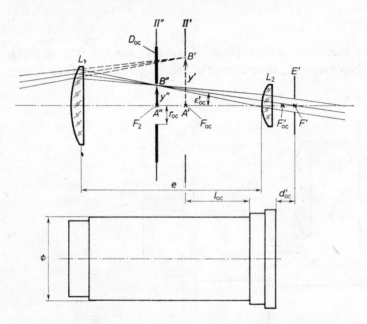

Fig. 2.51. Huygens ocular.

The other common ocular in use is the *Huygenian eyepiece*, shown in Fig. 2.51, which has a field-lens L_1 anterior to the image plane Π' of the microscope objective. Unlike to the Ramsden eyepiece, the Huygenian ocular has its first focus F_{oc} between the field-lens L_1 and the eye-lens L_2. In this case only the last lens acts as a loupe, whereas the former influences both the position and the size (y') of the primary image $A'B'$ formed by the microscope objective. This image is moved to the secondary image plane Π'', which is coincident with the first focus F_2 of the eye-lens L_2. Exactly in this plane the field diaphragm D_{oc} is situated. The relation between the sizes y' and y'' of the images $A'B'$ and $A''B''$ of an object under examination is as follows:

$$y'' = y' \frac{f_2'}{f_{oc}'}, \tag{2.58}$$

where f_{oc}' is the focal length of the whole ocular and f_2' is that of its eye-lens L_2. Consequently, Eq. (2.57) now takes the form

$$r_m = \frac{r_{oc} f_{oc}'}{M_{ob} f_2'} = \frac{r_{oc}'}{M_{ob}}, \tag{2.59}$$

where M_{ob} is the transverse magnification of the microscope objective Ob (Fig. 2.47a), r_{oc} the radius of the field diaphragm D_{oc} (Fig. 2.51), and

$$r_{oc}' = r_{oc} \frac{f_{oc}'}{f_2'}. \tag{2.60}$$

The quantity $2r_{oc}'$ is the diameter of the linear field-of-view in the image plane Π'. This is in fact the field number (FN) for Huygenian oculars.

The field number FN of typical oculars varies from 20 mm to 6 mm as the ocular magnification M_{oc} increases. For instance, an ocular of magnification $10\times$ has typically $FN = 18$ or 16 mm, while for $M_{oc} = 5\times$ the field number is usually equal to 20 mm. The diameter $2r_{oc}$ of the field diaphragm depends not only on the optical system of the ocular but also, or even primarily, on the correction of the off-axial aberrations of microscope objectives. In last two decades we have, however, seen systematic progress in the direction of larger fields of view. Finally, some highly corrected microscope systems with *wide-field oculars* have been developed. The first of them, the Orthoplan, was done by E. Leitz Wetzlar. The maximum field number of this microscope is equal to 28 mm (with GW Periplan oculars of magnification $6.3\times$ and $8\times$). Next, C. Reichert Wien designed a series of highly advanced microscopes (Univar, Polyvar) equipped with PK wide-field oculars ($FN = 28$ mm, 24 mm, and 15.5 mm for $M_{oc} = 18\times$, $10\times$, and $16\times$, respectively). Wide-field microscope systems have also been developed by other firms (see Subsection 2.4.5).

An eyepiece can be considered as a wide-field ocular if the product of its field number FN and focal length f_{oc}' is not smaller then 175 mm, i.e.,

$$FN \times f_{oc}' \geqslant 175 \text{ mm.} \tag{2.61}$$

Apart from the linear field of view, the *angle of view* or the *image field angle* is also important for visual microscopy. This is determined by the angle $2\varepsilon_{oc}'$

subtended at the centre of the microscope exit pupil by the diameter of the linear field of view (Figs. 2.50 and 2.51). It can easily be shown that

$$\tan \varepsilon'_{oc} = \frac{r'_{oc}}{f'_{oc}}, \tag{2.62a}$$

then, just as $f'_{oc} = 250/M_{oc}$ [mm], so also

$$\tan \varepsilon'_{oc} = \frac{r'_{oc} M_{oc}}{250}, \tag{2.62b}$$

where r'_{oc} is expressed in mm (for Ramsden-type oculars $r'_{oc} = r_{oc}$). As can be seen, the image field angle $2\varepsilon'_{oc}$ is the greater the higher is the ocular magnifying power M_{oc}. A microscope ocular of a given M_{oc} is, of course, the better the greater its field angle $2\varepsilon'_{oc}$. Typical values of this parameter are from 20° to 40°, but for *wide-angle oculars* $2\varepsilon'_{oc} = 40°$ to 55° (or more).

Huygenian eyepieces are advantageous especially for the ability to correct effectively their chromatic aberration and coma. They are therefore commonly used with achromatic objectives, particulary with those of low- and middle-power. Among the disadvantages of these eyepieces are the rather short eye relief d'_{oc} (Fig. 2.51), which diminishes dramatically as the magnification M_{oc} increases; hence Huygenian oculars of magnification $M_{oc} > 10\times$ are rarely used. Moreover, these oculars are inconvenient for measuring and counting purposes because they cannot be used with a graticule external to their optical system. Apart from this, they are not able to give such a good correction for spherical aberration as do the Ramsden eyepieces. Chromatic aberration of the latter may, in any case, be greatly reduced by using an achromatic doublet as eye-lens. This modification is known as the *Kellner ocular*, which is especially suitable for stereoscopic microscopes. Another, more complicated modification constitutes the *orthoscopic* and *symmetrical oculars*. In general, they are similar to the Ramsden eyepiece, but are composed of four lenses and have a higher degree of correction for aberrations, especially distortion; hence they are frequently used as measuring eyepieces. Moreover, their long eye reliefs are readily available for high magnifications (15× and more). On the other hand, compensating oculars for using together with planachromatic, apochromatic, semiapochromatic and planapochromatic objectives (and possibly high-power achromats) are similar to the Huygenian eyepiece but composed of a larger number of lenses. Compensating or plancompensating oculars are manufactured by various firms under different trade names, e.g., "periplan" (E. Leitz Wetzlar) or "hiperplan" (Bausch and Lomb).

It is worth mentioning that the external diameter Ø (Figs. 2.50 and 2.51) of an ocular mount is an international standard and is equal to 23.2 mm or 30 mm. The latter value pertains to wide-field oculars and those used in stereoscopic microscopes which, in general, have a larger linear field of view than do typical biological or metallographic microscopes. It is clear that in the case of the smaller external diameter (Ø = 23.2 mm) the diameter $2r_{oc}$ of the field diaphragm cannot be greater than about 20 mm. Consequently, in the case of the larger Ø (= 30 mm) the diameter $2r_{oc}$ cannot exceed 28 mm.

The attentive reader may ask where the windows conjugated with the ocular field diaphragm D_{oc} are located (Fig. 2.47a). The answer is readily arrived at: the entrance window W is coincident with the object plane Π and its diameter $2r_W$ is consistent with that of the object area seen through the microscope ($2r_W = 2r_m$); the exit window, on the other hand, like the final image of the specimen under examination, appears at infinity.

Illumination. High quality microscopic imaging can only be obtained when the specimen under examination is correctly illuminated in accordance with certain important requirements. First, the specimen must be illuminated only over that area under observation whose diameter ($2r_m$) results from Eq. (2.57) or (2.59); this ensures that stray light and glare are considerably reduced. Secondly, illumination should be uniform with controllable intensity. Thirdly, the angular aperture of illumination must be variable to fill a desired portion of the whole aperture of the objective. These requirements can only be fulfilled when an appropriate illumination system is installed.

There are two main principles of illumination in microscopy. One was devised by Nelson [2.10] and the other by Köhler [2.11]. The *Nelsonian method of illumination* is generally called *critical illumination* although the term is rather unfortunate. In this method (Fig. 2.47c) the condenser C is axially adjusted (focused) so that the image S' of a uniform light source S is produced in the object plane Π. A desirable cone of rays incident upon the specimen can be obtained if the condenser has an adequate numerical aperture and the iris D_c (aperture diaphragm) is properly opened. This method of illumination requires an extended light source S with uniform brightness. This is, of course, a disadvantage. Satisfactory results are obtained if a pearl bulb, opal bulb, or sometimes an illuminated ground glass screen are used. Focusing a source of this kind in the object plane Π may give an injurious grainy or speckled background. This defect can, however, be removed by slightly defocusing the condenser. Frequently, close behind the light source there is a field diaphragm (D_s) in the form of an iris close behind the light source. In the preliminary stage of the adjustment of illumination the iris D_s is contracted to a small opening and the condenser axially adjusted so as to image this diaphragm rather than the light source surface in the object plane Π. The iris D_s is then opened until the field of view is only just covered by the image of this iris opening. A possible excentricity of the field diaphragm D_s with respect to the field of view is corrected by a transverse adjustment of the condenser.

At present time Nelsonian illumination is likely to be used in simple rather than in more advanced microscopes. As a rule, each modern research or laboratory microscope has a built-in illuminator which makes use of the *Köhler principle of illumination*. The latter is sketched in Fig. 2.47d. Close behind the light source S is the *collector (lamp condenser)* Col followed by an iris (*illumination field diaphragm*) D_{col}. The collector and its iris are usually combined with the light source into a *microscope lamp* or *illuminator I*, which is fitted into the microscope or installed in its base. The collector Col forms an image S' of the light source S in the front focal plane of the condenser C, which also has an aperture diaphragm D_c. A secondary image S'' of S is formed by the condenser and the objective Ob in the back focal plane of the latter; thus S'' is coincident with the exit pupil of the objective. The third image S''' of S is produced by the ocular Oc in its exit pupil E'. The source S and all its images (S', S''. S''') lie in optically conjugate planes because each is an image of the preceding one. Another sequence of conjugate planes originates from the iris D_{col}. The first and second its images, D'_{col} and D''_{col}, are produced by the condenser C and the objective Ob in the object plane Π and the image plane Π', respectively. The image D''_{col} is coincident with the ocular field diaphragm D_{oc} and both can be observed together with the image of the specimen situated in the object plane Π. The iris D_{col} must never be opened more than is necessary to illuminate an object field whose radius is expressed by Eq. (2.57) or (2.59).

An image D'_{col}, which is as sharp as possible in the object plane Π, is obtained by focusing the condenser C. Initially the iris D_{col} is contracted to a small hole and the condenser C is focused and centred to obtain a sharp image D''_{col} at the centre of the image field. The illumination iris D_{col} is then opened so as just to fill the field of view with light. It is important to know the conditions under which the entire field of view of the microscope can be illuminated by the iris D_{col}, whose diameter is $2r_{col}$. In general, these conditions result from the relation

$$r_{col} = r_m \frac{b}{f'_c} = r_m M_c, \tag{2.63}$$

where r_m is defined by Eq. (2.57) or (2.59), b is the distance of the iris D_{col} from the first focus F_c of the condenser C, f_c' the focal length of the latter, and M_c denotes the transverse magnification of the condenser ($M_c = b/f_c'$), when the image D_{col}' of D_{col} is projected into the object plane II. A typical distance b is equal to about 300 mm. Let, e.g., $f_c' = 15$ mm, which corresponds to a typical high-power condenser. In order to cover the object field $2r_m$, the iris D_{col} must be opened until $2r_{col} = 20 \times 2r_m$. If a microscope consists of an objective and an ocular with magnifying powers equal to $5\times$ and $10\times$, respectively, then the object field of this system is given by $2r_m = 2r_{oc}/5 \approx 18/5 = 3.6$ mm; hence $2r_{col}$ should be equal to $20 \times 3.6 = 72$ mm, and it is clear that the collector lens must have a diameter of the same magnitude. Such a big collector is, of course, unhandy. Low-power objectives therefore require low-power condensers. Normally, a high-power can be converted into a low-power condenser by removing or exchanging its front lens.[13] A better solution is a *pancratic* (*zoom*) *condenser* which has a focal length that can be varied continuously. A condenser of this kind is manufactured by VEB C. Zeiss Jena. Another satisfactory solution is also in use, viz., an additional lens is inserted between the condenser and collector, which makes it possible to change the size of the illumination iris image D_{col}' in the object plane II (Fig. 2.47d).

Frequently it is necessary to have a very sharp image D_{col}'' of D_{col} in the image plane II'. This does not present any serious difficulty when low- and middle-power objectives are used. If, however, a high-power objective is applied, the edge of the iris D_{col} is seen as a blurry image. The degree of blurring primarily depends on the correction of condenser aberrations. Several types of condensers are available for bright-field microscopy. The most popular of them being the *Abbe condenser*. This consists of two lenses and suffers from excessive spherical and chromatic aberrations, so that it produces an unsatisfactory image of the iris D_{col} in the object plane II and, of course, in the image plane II'. A better condenser than the Abbe is the *aplanatic*, which consists of three lenses and is corrected for spherical aberration and field curvature. Although this too suffers from chromatic aberration, it gives a satisfactory image of the illumination iris when used with monochromatic light. However, for critical work with high-power objectives an *achromatic-aplanatic condenser* is necessary. Usually this consists of five lenses, among which there are two achromatized doublets and a single top lens. The latter is frequently exchangeable for another of different numerical

---

[13] This particular condenser lens, which is on the side of the specimen, is, in general, called the *front lens* or *top lens*. These names are, however, unfortunable.

aperture (compare, e.g., achromatic-aplanatic condenser Z of numerical aperture 1.4, 0.9 and 0.6 manufactured by Carl Zeiss Oberkochen). This condenser is corrected for field curvature, chromatic aberration at two wavelengths (in the red and blue spectral regions), and spherical aberration. The latter is completely removed only for one particular wavelength from the middle of the visible spectrum ($\lambda \approx 550$ nm). Condensers of the highest quality are *apochromatic-aplanatic*. These consist of a large number of lenses, are therefore expensive and used mainly together with highly-corrected oil immersion objectives.

The light source S is sharply imaged in the entrance pupil of the condenser by varying the distance d_s between the source and the collector (Fig. 2.47d). Usually either the source S or the collector Col is moved slightly along the optic axis. This operation is performed by the user thus he should be familiar with the correct adjustment of the light source.

Any built-in illuminator must be designed and arranged so as to light up the whole aperture of the condenser and that of all objectives which are attached to a given microscope. It is therefore important to know the conditions under which the entrance pupil of the condenser and the exit pupils of the objectives are completely filled with the images S' and S'' of the light source S (Fig. 2.47d). It may be stated that these conditions result from the relation

$$w_s = \frac{2Af_c'f_{\text{col}}'}{g_c}, \tag{2.64}$$

where w_s is the width (transverse size) of the light source S, A denotes the numerical aperture of the objectives, f_c' and f_{col}' are the focal lenghts of the condenser C and collector Col, while g_c is the distance between the first focus F_c of the condenser and the second focus F_{col}' of the collector. Equation (2.64) shows that for a given numerical aperture of objective and fixed focal lenghts of condenser and collector, the desired size w_s of a light source is a function of the distance between the condenser C and the illuminator I. Let, e.g., $A = 1.30$ and $f_c' = 10$ mm (both the objective and the condenser are oil immersion systems), then $w_s/f_{\text{col}}' = 0.1$ for $g_c = 300$ mm. If. e.g., $f_{\text{col}}' = 20$ mm, the size w_s of a light source cannot be smaller than 2 mm.

Equation (2.64) also shows that the desirable transverse size of a light source diminishes as the numerical aperture of the objective decreases. For economy of light it will be useful to have an illumination system which permits to "pack" the entire image S'' into the exit pupil E_{ob}' of both high-power objectives and low-power objectives, without a significant cut-off of light by the rims of objective lenses. As a matter of fact, some peak performance microscopes (e.g., UnivaR from C. Reichert Wien and Apophot from Nikon) are equipped with a versatile illumination system of this kind. It contains a zoom collector which enables the image sizes of both the light source and the illumination iris to be matched to objectives of any power (compare Fig. 2.86).

If a microscope has a built-in illuminator, the user cannot arbitrarily change the distance between the adjacent foci F_c and F'_{col} of the condenser and collector (Fig. 2.47d). The axial shift, if any, of the collector is normally limited and the distance g_c is strictly defined for illumination by the Köhler principle. The situation differs when the illuminator or microscope lamp is separate from the microscope. In that case the user can adjust the Köhler illumination for different distances between the condenser and illuminator, and according to Eq. (2.64), choose the most suitable distance g_c for a light source of given size.

It is important to note that the intensity of illumination should never be controlled either by means of the aperture diaphragm of the condenser or by shifting the condenser axially. This is a bad practice of those who are not familiar with the elementary principles of illumination in microscopy. Illumination intensity can only be controlled by changing the lamp voltage or by inserting neutral filters in the light path. The neutral filters for decreasing the light intensity, colour temperature correction filters, heat absorbing filters, a green filter for improving the performance of chromatic objectives and condensers, etc., etc., are some of the very useful accessories in illumination techniques in microscopy (see Refs. [2.8] and [2.12]).

Working distance of the condenser. This parameter is directly related to the Köhler method of illumination because the collector field diaphragm D_{col} of the microscope illuminator I (Fig. 2.41) must be imaged in the object plane II when between this plane and the top lens of condenser C there is a free space for a glass slide or another substratum on which the object under examination is mounted. On the analogy of the working distance of the objective, that of the condenser is defined as the clear distance d_c between the condenser front lens (or its mount) and the microscope (object) slide MS. The latter, strictly speaking, is one of the elements of the illuminating system. It is worth noting that the slide MS acts with regard to the condenser as does the cover slip CS to the objective (Fig. 2.41). The microscope slide is therefore itself an optical component and should be suitable for correct use with the condenser and the whole illuminating system. First of all, the refractive index and thickness of microscope slides should be kept constant because these two parameters primarily influence the quality of the image of the collector field diaphragm and the working distance of the condenser.

The value of the refractive index $n_e = 1.53 \pm 0.02$ (or $n_D = 1.52$) and the thickness $t = 1.1^{+0.1}_{-0.2}$ have recently been fixed by ISO for standard glass slides.

Frequently the working distance d_c (Fig. 2.41) is confused with the *object distance of the condenser* (d_o). The latter indicates the position of the image D'_{col} (Fig. 2.47d) of an infinitely distant illuminator field diaphragm, on the assumption that the top surface of the condenser is coincident with the surface of a microscope stage on which there is a glass medium of refractive index $n_e = 1.520$. The parameter d_o indicates the thickness of the glass plate on which a mounted specimen can still be properly illuminated with a condenser adjusted according to the Köhler principle. For the majority of typical bright-field condensers, the object distance d_o is 1.6 to 2.2 mm. If instead of glass there is air between the condenser and the specimen, the object distance d_{oa} of the condenser is smaller than d_o, viz., $d_{oa} = d_o/n_e$.

Some special research, e.g., microscopy *in vitro*, needs condensers with d_o greater than 2.2 mm, sometimes several times as large. Such condensers are known as *long working distance (LWD) condensers*. However, the greater the d_o the smaller is the numerical aperture of the condenser (A_c). For instance, for a dry achromatic-aplanatic condenser, whose object distance is equal to about 5 mm, A_c cannot be greater than 0.8. To obtain a LWD condenser it is frequently enough to remove the top lens of a normal condenser. It will readily be seen that removing this lens approximately doubles the focal length, halves the numerical aperture, and considerably increases the working distance d_c. The correction of aberrations is, however, disturbed.

LWD condensers are especially necessary for biological microscopy *in vitro*, where illumination requirements are more complicated than in common transmitted-light microscopy. A frequent system for microscopy *in vitro* is shown in Fig. 2.52, which illustrates an inverted microscope used for observation of a transparent specimen, cell culture for instance, placed on a special slide. The most useful microscope slides for microscopy of this kind are cavity slides, cell-slides, troughs, and shallow glass dishes, but often small round aquaria, jars, tubes, bottles, and other glass or plastic containers are also in use.

Let a glass cell GC (Fig. 2.52) contains a thin specimen O. The cell walls, W_1 and W_2, on both the condenser side and the objective side are plane parallel plates. Above the specimen O there is a layer of liquid L followed by an air layer A. Next, let d_{oa} be the object distance of the condenser related to the air medium, while the illuminating system ($I+C$) is adjusted in accordance with the Köhler principle as in the case where specimen O was preceded only by an air medium. If this medium is then replaced by liquid, e.g., water, the thickness t_w of the water layer acceptable under these conditions is given by $t_w = d_{oa}n_w$, where n_w is the refractive index of water. If, however, the specimen O is preceded by a glass plate W_1 of thickness t_g, an air layer A of thickness t_a and a liquid layer L of thickness t_l, as shown in Fig. 2.52, then the global thickness of all layers must satisfy (still in accordance with the Köhler principle) the following relation:

$$\frac{t_g}{n_g} + \frac{t_l}{n_l} + t_a = d_{oa}, \qquad\qquad (2.65)$$

where n_g and n_l are the refractive indices of glass and liquid. If d_{oa} is known (from factory data on the parameters of condensers), one can calculate the total thickness t_{GC} of a microscope cell which can be used to observe a transparent specimen illuminated according to the Köhler principle.

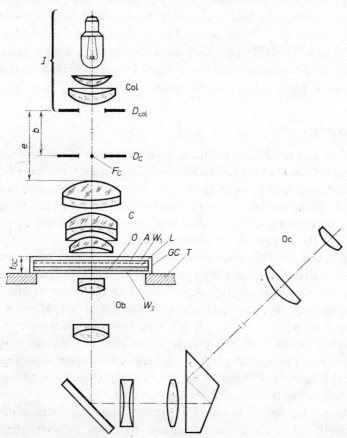

Fig. 2.52. Optical system of an inverted microscope for transmitted light.

It is worth noting that d_{oa} indicates the position of the second focus F'_c of the condenser (Fig. 2.47d) because this parameter is defined similarly as d_o, i.e., the field diaphragm of the illuminator is assumed to be infinitely distant. Therefore, d_{oa} is approximately equal to the back focal length of the condenser. In reality, however, the collector diaphragm D_{col} (Figs. 2.47d and 2.52) is at a finite distance, hence the *real (actual) object distance of the condenser*, d_{or} or d_{oar}, is longer than d_o or d_{oa}, and can be calculated from Eqs. (2.13) for any distance b between D_{col} and the first focus F_c of the condenser if its focal length f'_c is known. Bringing the illuminator I closer to the condenser C increases the actual object distance of the condenser and it makes possible to use longer (thicker) cell-slides. But this procedure also causes the size w'_s of the image S' (Fig. 2.47d) of the light source S to diminish. Thus the distance b (Fig. 2.52) must be a compromise between these two parameters (w'_s and d_{oar} or d_{oa}). Modern biological inverted microscopes (Biovert from C. Reichert Wien, Diavert from E. Leitz Wetzlar, Invertoscopes IDO2 and IDO2MT from Carl Zeiss Oberkochen, Biostar from AOCo., and others) are designed so that the distance between illuminator and condenser can be varied within a considerable range.

2.3.5. Immersion technique

Microscope objectives and condensers requiring a liquid instead of air between their front lenses and the specimen are said to be *immersion systems*. Consequently, systems which have an air gap between their front lens and the specimen are called *dry systems*. In biological microscopy the specimen (object) to be examined is usually mounted between a microscope slide and cover slip; the latter is on the objective side and the former on the condenser side.

One of the basic functions of the immersion liquid is to increase the numerical aperture A. Since $A = n \sin \sigma$, and since the aperture angle σ cannot exceed 90°, the maximum possible numerical aperture A_{max} is equal to the refractive index n of the immersion medium. For dry systems $n = 1$, hence $A_{max} = 1$. However, this is only a theoretical value. In practice, the maximum numerical aperture of dry systems cannot exceed 0.95, and greater values may be obtained only with immersion systems. For special purposes, however, immersion objectives of smaller numerical aperture and of middle or even low magnifying power are in use (e.g., Antiflex objectives from Carl Zeiss Oberkochen).

For most ordinary purposes the standard immersion liquid is immersion oil, i.e., a special synthetic oil or cedar oil. For special purposes, however, some other fluids (e.g., water, glycerine, paraffin oil) are also in use (see Table 2.5).

TABLE 2.5

Immersion liquids used in microscopy

Liquid	Refractive index n_D or n_d (at 20°C)	Abbe number V_D	Available upper limit of the numerical aperture of objective A_{max}
Methyl alcohol	1.3288		
Distilled water**	1.3330	56	1.20
Acetone	1.3592		
Ethyl alcohol	1.3617		
Ethyl acetate	1.3727		
n-Heptane	1.3872		
n-Butyl chloride	1.4022		
Methyl cyclohexane	1.4235		
Ethylene glycol	1.4318		
Ethyl citrate	1.4434		
Butyl stearate	1.4446		
Trimethylene chloride	1.4476		
Kerosene	1.4500		
Cyclohexanol	1.4678		
Glycerine**	1.4695	60	1.30
Paraffin oil**	1.480		1.33
Triethanolamine	1.4853		
s-Tetrachlorethane	1.4943		
Benzene	1.5014		
Ethyl iodide	1.5138		
Cedar oil*	1.515	50	1.40
Synthetic oil*	1.515	43	1.40
Anisole**	1.5178	30	1.40
Chlorobenzene	1.5250		
Ethylene bromide	1.5383		
o-Nitrotoluene	1.5466		
Tri-o-cresyl phosphate	1.5586		
o-Toluidine	1.5725		
Aniline	1.5864		
Bromoform	1.5973		
o-Iodotoluene	1.6095		
Quinaldine	1.6120		
Carbon disulphide	1.6277		
α-Chloronaphthalene	1.6317		
α-Monobromonaphthalene**	1.6585	20	1.53
Methylene iodide**	1.740		1.60

 * Standard immersion oil for microscope objectives and condensers.
** Frequently used as objective immersion liquid for special purposes.

Apart from increasing the numerical aperture, oil immersion has the important advantage of reducing the common and internal reflection of light at adjacent surfaces of the cover slip and objective front lens. These advantages are illustrated in greater detail in Fig. 2.53, where only the cover slip CS and the front lens FL followed by a meniscus of a high-power objective are shown.

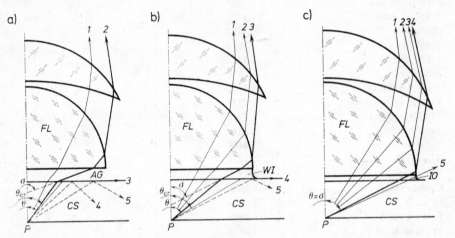

Fig. 2.53. Drawings showing the advantages of immersion objectives (b, c) with respect to dry objectives (a).

Consider a number of skew rays emerging from an axial object P being in contact with the under surface of the cover slip CS. Rays 1 and 2 (Fig. 2.53a) pass through the cover slip and on emerging into the air gap AG are refracted according to Snell's law (see Subsection 1.5.1) and accepted by the objective. The more oblique ray 3 emerges parallel to the cover-slip surface. Any ray such as 4 or 5, which is more oblique than ray 3, does not emerge at all, but is internally reflected again into the cover slip. Let θ denotes the maximum angle of incidence, at which rays strike the top surface of the cover slip, pass through it and enter the objective. Next, let θ_{cr} be the critical angle of incidence and σ the maximum aperture angle. If the refractive index n_{CS} of the cover slip CS is equal to 1.515 (for D spectral line), the critical angle, according to Eq. (1.67), is equal to θ_{cr} = arcsin(1/1.515) = 41°. For this angle of incidence, the ray propagates along the interface of the cover slip and makes an aperture angle σ = 90°. Theoretically it can be accepted by the objective provided its working distance and the thickness of the cover slip are zero. This demand is, however, unrealistic and σ cannot practically be greater than 72°. Consequently, the maximum numerical aperture $A_{max} = 0.95$, to which corresponds an angle of incidence $\theta = 39°$.

Let us now assume there is water immersion WI between the cover slip CS and the objective front lens FL (Fig. 2.53b). Since the refractive index of water is equal to 1.333, the critical angle θ_{cr} is given by arc $\sin(1.333/1.515) = 61.5°$. In this case therefore θ_{cr} is greater by 20° than in the previous system (Fig. 2.53a), hence the portion of light reflected internally into the cover slip is, of course, smaller. For the practical reasons mentioned above, a water immersion objective cannot accept rays which form an angle σ greater than 65°, hence its maximum numerical aperture $A_{max} = 1.333 \sin 65° = 1.20$, to which corresponds an angle of incidence $\theta = 53°$. This means an increase of 14° over the previous value (Fig. 2.53a).

Finally, consider the effect of filling the gap between the cover slip CS and the objective front lens FL with an immersion oil the refraction index of which is equal to that of the cover slip and front lens, viz., $n = n_{CS} = n_{FL} = 1.515$. Now, there is no change in refractive index, none of the rays $1-5$ will be refracted between CS and FL, but will propagate along straight lines in their original directions, and there is no internal reflection. For practical reasons, an oil immersion objective cannot, however, accept rays which form an aperture angle σ greater than 68°. Consequently, the maximum numerical aperture is given by $A_{max} = 1.515 \sin 68° = 1.40$. For this value, the corresponding angle of incidence is $\theta = 68°$ (in this case $\theta = \sigma$). As can be seen, θ is now greater by 29° and 14°, respectively, than in the first and second cases (Fig. 2.53a and b), although the aperture angle σ is approximately the same.

The immersion technique also allows an increasing flux of light, F, to be picked up by the objective. This flux is proportional to the solid angle 2θ subtended at P. In other words, $F \propto \sin^2 \sigma$ (Fig. 2.53a). If an immersion liquid of refractive index n is employed between the cover slip CS and the objective front lens FL, as in Fig. 2.53b or c, the cone of rays picked up by the objective will come from a much wider solid angle 2θ so that $\sin\theta = n\sin\sigma$; the flux F will therefore increases to

$$F \propto n^2\sin^2\sigma. \tag{2.66}$$

Consequently, the image of the specimen produced by an immersion objective is much brighter than that given by a dry objective which accepts the same aperture angle σ.

An additional advantage of the immersion method occurs when the space between the top condenser lens and microscope slide is also filled with an immersion liquid (Fig. 2.54). If the refractive index (n) of this liquid matches that of the top condenser lens *TL*, the microscope slide *MS*, the cover slip *CS*, and the objective front lens *FL* ($n = n_{TL} = n_{MS} = n_{CS} = n_{FL}$), then we refer to the system as *homogeneous*, and the immersion liquid is said to be *homogeneous immersion*. Oil immersion (cedar wood oil or synthetic oil) has approximately the same refractive index as the glass used for slides, cover slips, top lenses of immersion condensers, and front lenses of immersion objectives. Strictly speaking, the homogeneous immersion system also requires the medium used to mount the specimen between the microscope slide and cover slip to have a refractive index equal to that of the immersion oil.

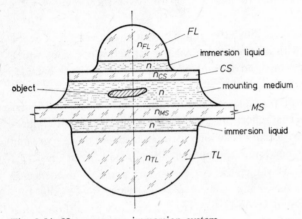

Fig. 2.54. Homogeneous immersion system.

Like the cover slip, the immersion oil is an optical part of the immersion objective and must satisfy certain requirements. First of all, its refractive index and Abbe number should be exactly defined and constant. Each manufacturer specifies the immersion oil to be used for its own microscopes. Many countries have their own industrial standards in this field (see Table 2.6) and the ISO 8036/1 standard for general purpose immersion oil was published in 1986. This lays down the following optical properties: $n_e = 1.5180 \pm 0.0005$ ($n_D = 1.515$), $V_e = 44 \pm 3$, valid for a temperature of $23 \pm 0.1°C$.

Apart from its controlled optical properties, immersion oil must also exhibit suitable physicochemical features, that is it should be non-drying, safe for lenses and specimens, non-fluorescing, chemically inert, easily removed, etc., etc.

The best known immersion oil for microscopy is that supplied by R.C. Cargille Laboratories, Inc., Cedar Grove, New Yersey, USA. Its basic optical properties

TABLE 2.6

Specifications of some national industrial standards related to immersion oils and those supplied to microscope users by different manufacturers

National standard or manufacturer of microscopes	Basic optical data on immersion oil
British Standard BS 3836: Part 1	$n_e = 1.5180 \pm 0.0004$ (at $t = 23°C$) $n_d = 1.515$ $V_e = 44 \pm 5$
German Standard DIN 58884 (oil immersion of type 518C)	$n_D = 1.515$ (at $t = 23°C$) $n_e = 1.518$
Japanese Standard JIS K-1980	$n_d = 1.516 \pm 0.001$ (at $t = 23°C$) $V_d = 43 \pm 3$
Polish Standard BN-79/5523-07	$n_D = 1.5150 \pm 0.0005$ (at $t = 25°C$) $V_D = 45 \pm 5$
E. Leitz Wetzlar	$n_D = 1.515$ $V_D = 49$
C. Zeiss Oberkochen	$n_D = 1.515$ (at $t = 23°C$) $n_e = 1.518$
Vickers, England	$n_D = 1.524$ (at $t = 20°C$) (oil of type ALPI)
VEB Carl Zeiss Jena	$n_D = 1.515$ (at $t = 20°C$)
Polish Optical Works (PZO) Warsaw	$n_D = 1.5150 \pm 0.0005$ (at $t = 25°C$) $V_D = 45 \pm 5$
Bausch and Lomb, USA	$n_D = 1.5150 \pm 0.0002$ (at $t = 25°C$)

TABLE 2.7
Basic properties of the Cargille immersion oil (at 23°C)

Type	A	B	VH
Refractive indices:			
n_F	1.5236	1.5236	1.5227
n_e	1.5180	1.5180	1.5176
n_D	1.5150 ± 0.0002	1.5150 ± 0.0002	1.5150 ± 0.0002
n_C	1.5115	1.5115	1.5118
Mean dispersion $n_F - n_C$	0.0121	0.0121	0.0109
Abbe number V_D	42.6	42.6	47.2
Temperature coefficient $(-dn/dt)$ within 15°−35°C	0.00033	0.00031	0.00031
Fluorescence (ultraviolet)	very low	low	low
Viscosity [cSt or mm²/s]	150 (low)	1250 (high)	120,000 (very high)

are shown in Table 2.7. This oil is available in three types, A, B, and VH, which differ in viscosity. Type A, of the lowest viscosity, is particularly suitable for objectives of short working distance, as well as for dark-field fluorescence microscopy. Type B, of high viscosity, is generally accepted for biomedical research. It is supplied by several manufacturers of microscopes as a standard accessory. Type VH, of very high viscosity, is especially intended for inverted, horizontal, and inclined microscopes, as well as for objectives and condensers with a relatively long working distance.

A drawback of immersion oil is its high absorption of ultraviolet light (below 360 nm). The most suitable immersion liquid for UV microscopy therefore is glycerine and water or a composition of glycerine (ca. 90%) and water (ca. 10%). The most restricted in its choice of an immersion liquid is IR microscopy because many substances that are transparent in ultraviolet and visible light, are, as a rule, opaque in infrared light, and vice versa. Paraffin oil is an exception and only this oil is successfully used as an immersion liquid for IR microscopy. Paraffin oil is frequently employed for fluorescence microscopy as well because it does not fluoresce when illuminated with blue-violet or ultraviolet light. Water immersion objectives are especially used in biological microscopy for the study of living cells and tissues surrounded by an aqueous medium. Water, however, has a low refractive index compared with glass. This is, of course, one of the disadvantages of water immersion. On the other hand, high refraction liquids, like methylene iodide, are frequently used in reflected-light microscopy for the study of minerals whose reflectance is low or strongly varying (see Subsection 2.4.3).

Finally, it is important to note that a series of versatile *multi-immersion objectives* (*Plan-Neofluars*) has recently been developed by Carl Zeiss Oberkochen. This series not only includes high-power but also middle- and even low-power immersion objectives, that are designed not only for use with standard immersion oil but for other immersion liquids as well, such as water, glycerine, paraffin oil, silicon oil, etc. These new Plan-Neofluars with a magnification range from $16 \times$ to $63 \times$ are partly fitted with a correction collar which makes it possible to adjust the optical system of the objective to the different optical properties of the immersion liquids mentioned above. These objectives have a considerably higher numerical aperture, consequently their resolving power and general image brightness are much better than is found with typical objectives of equal magnification. They are, therefore, particularly suitable for fluorescence microscopy, especially as care has been taken to ensure that the optical material of the lenses is free from autofluorescence. In addition, these objectives are flat-field corrected and thus well suited for photomicrography. The new series of multi-immersion Plan-Neofluars is available for phase contrast and differential interference contrast microscopy as well.[14] This series represents significant progress in immersion microscopy.

2.3.6. *The cover slip and its importance for the quality of microscopical imaging*

The *cover slip* (or *cover glass*) is an optical part of the image-forming system of the microscope and, in fact, suffers from the same aberrations as lenses (see Subsection 2.2.4) which must be corrected as part of the design of microscope objectives. In this case, however, correction is carried out for a plane parallel glass plate whose thickness and optical properties are strictly defined. It is therefore worthwhile looking at the cover slip in conjunction with aberrations of a plane parallel plate.

Spherical aberration of a plane parallel plate. Let us consider a plane parallel plate of glass (Fig. 2.55). At a distance from the plate there is an object AB. Ray *1* emerging from the object point B strikes the plate normally, therefore passes

[14] Data after the technical bulletin *Opton Newsletter*, 1/1983, p. 5 (see also *Proc. Roy. Micr. Soc.*, **19** (1984), 50).

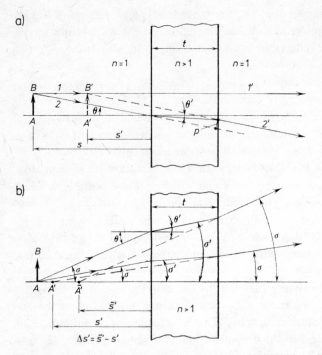

Fig. 2.55. Plane parallel plate as an optical element (a) and its spherical aberration (b).

through it without any refraction, and as conjugate ray $1'$ propagates in its original direction. Another ray, 2, strikes the plate obliquely, and is refracted on both plate surfaces, and shifted in such a way as to emerge from the plate as conjugate ray $2'$ parallel to the original ray. Reference to Fig. 2.55a shows that a simple application of Snell's law and elementary trigonometric identities yield the following value for this shift:

$$p = t\,\frac{\sin(\theta - \theta')}{\cos\theta'} = t\sin\theta\left(1 - \frac{\cos\theta}{\sqrt{n^2 - \sin^2\theta}}\right), \qquad (2.67a)$$

where t is the thickness of the plate, θ is the angle of incidence, θ' is the angle of refraction, and n is the refractive index of the plate relative to the surrounding medium (compare with Eq. (1.65)). If the angle of incidence θ is small ($\theta \ll \pi/2$), then $\cos\theta' \approx 1$, $\sin(\theta - \theta') \approx \theta - \theta' \approx \theta - \theta/n$ and Eq. (2.67a) reduces simply to

$$p = t\theta\,\frac{n-1}{n}. \qquad (2.67b)$$

The extensions of the rays $1'$ and $2'$ intersect at the image point B' distant by s' from the first surface of the plate. At that distance the image $A'B'$ of the object AB appears. This image is virtual and equal in size to the object (transverse magnification of a plane parallel plate is equal to $+1$). From the geometry of Fig. 2.55a it results that the image distance s' is given by

$$s' = s - s_{OI} = s - \frac{p}{\sin\theta}, \tag{2.68a}$$

where s is the vertex object distance, and s_{OI} is the distance between the object AB and its image $A'B'$. The distance s' therefore depends on the angle of incidence θ. If, however, $\theta \ll \pi/2$, Eq. (2.68a) reduces to

$$s' = s - \frac{p}{\theta} = s - t\frac{n-1}{n}, \tag{2.68b}$$

and s' is now independent of θ; in this case a plane parallel plate can be considered as an ideal afocal optical element. Otherwise, if θ is large, the plate suffer from spherical aberration $\Delta s'$. This defect is shown in Fig. 2.55b, where A' and $\tilde{A}'$ are, respectively, the paraxial and marginal images of the axial point object A. The image A' is formed by paraxial rays for which the aperture angle σ (or σ') is small, whereas the image $\tilde{A}'$ is produced by marginal rays whose aperture angle σ is large.

The spherical aberration mentioned above is of the longitudinal type. It is enough well expressed by

$$\Delta s' = s' - \tilde{s}' = \frac{t}{n}\left(1 - \frac{\cos\theta}{\cos\theta'}\right) = \frac{t}{n}\left(1 - \frac{\cos\sigma}{\cos\sigma'}\right), \tag{2.69a}$$

where s' and $\tilde{s}'$ are, respectively, the distances of the images A' and $\tilde{A}'$ from the plate. By substituting $n\sin\sigma' = \sin\sigma$ and by using fundamental trigonometric identities, we can express Eq. (2.69a) in the form

$$\Delta s' = t\left(n - \frac{1}{n}\right)\frac{2\sin^2(\frac{1}{2}\sigma)}{n^2 - 2\sin^2(\frac{1}{2}\sigma)}, \tag{2.69b}$$

which is more convenient for further consideration of the cover slip and its importance for the quality of microscopical imaging.

Chromatic aberration of the plane parallel plate. The refractive index n of the plate under consideration depends on the light wavelength λ. In white light, therefore, the plate suffer from longitudinal chromatic aberration as shown in Fig. 2.56. Because of the refractive index dispersion, a polichromatic ray emerging

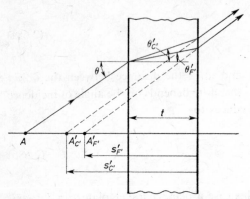

Fig. 2.56. Chromatic aberration of a plane parallel plate.

from the axial point object A is dispersed by the plate into rays of different colours from which two, F' (blue) and C' (red), are taken into consideration. They produce images $A'_{F'}$ and $A'_{C'}$ of the object A at slightly different distances $s'_{F'}$ and $s'_{C'}$ from the plate. The difference $s'_{F'} - s'_{C'}$ is a measure of the chromatic aberration. It is enough well characterized by

$$ s'_{F'} - s'_{C'} = t\,\frac{n_{F'} - n_{C'}}{n_{F'}n_{C'}}, \tag{2.70} $$

where $n_{F'}$ and $n_{C'}$ are the refractive indices of the plate for wavelengths $\lambda_{F'}$ (F' spectral line) and $\lambda_{C'}$ (C' spectral line). As can be seen, the right hand part of Eq. (2.70) is positive. Hence, like a negative lens the plane parallel plate suffers from longitudinal chromatic aberration of the positive type.

Astigmatism and coma of the plane parallel plate. When a divergent light beam is oblique to the plate under consideration, astigmatism and coma also occur. These aberrations, especially astigmatism, may be significant when the numerical aperture of a microscope system is large or the plate is oblique to the optical axis of the system. This situation is shown in Fig. 2.57. The divergent beam emerges from the object point B and strikes the plane parallel plate obliquely at an angle of incidence θ. The beam forms a cone the solid angle of which is

$\Delta\sigma$. The refraction of this cone must be considered in the tangential and sagittal planes (for definition, see Subsection 2.2.4). Extensions of the tangential rays, shown in Fig. 2.57, intersect at the image point B_T', whereas those of the sagittal rays (not shown in Fig. 2.57) intersect at image point B_S'. The distance $\overline{B_T'B_S'}$

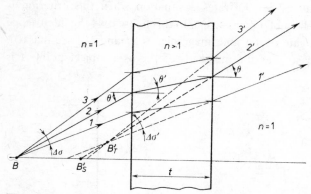

Fig. 2.57. Astigmatism of a plane parallel plate.

between B_T' and B_S' is the magnitude of the astigmatism. By using trigonometric analysis [2.14], it can be shown that

$$\overline{B_T'B_S'} = \frac{t}{n\cos\theta'}\left(1 - \frac{\cos^2\theta}{\cos^2\theta'}\right) = \frac{t}{n}\left(\frac{1}{\cos\theta'} - \frac{\cos^2\theta}{\cos^3\theta'}\right). \tag{2.71}$$

If, e.g., the thickness t and refractive index n are equal to 0.2 mm and 1.52, respectively, and the angle of incidence $\theta = 65°$, then $\overline{B_T'B_S'} = 0.11$ mm. With the same conditions, the spherical aberration $\Delta s'$ calculated from Eq. (2.69a) amounts to 0.06 mm.

Tolerances for cover slip thickness. Microscope objectives intended for use with a cover slip are normally corrected for a cover glass thickness $t = 0.17$ mm. If the cover slips actually used deviate from this conventional thickness, they will produce an undesirable over- or undercorrection of the objective, especially for spherical aberration. The larger the objective numerical aperture the smaller the permissible deviations Δt. Objectives of a numerical aperture below 0.40 and those designed for homogeneous immersion may, however, be used with or without cover slip.

What are the tolerances Δt of the cover slip thickness for dry objectives? To give a satisfactory answer, the *Rayleigh quarter-wave criterion* in respect of the *depth of field* will be used. According to this criterion, there will be no appreciable deterioration of the image of an object point provided the maximum optical path difference between aberrated light waves arriving at the centre of the image does not exceed $\lambda/4$. Figure 2.58 illustrates a situation when this criterion is used to define the depth of field of a microscope. In this figure only the microscope objective Ob and its conjugate object and image planes, Π and Π', are marked. The objective is corrected so that the image A' of the axial object point A is free from aberrations.

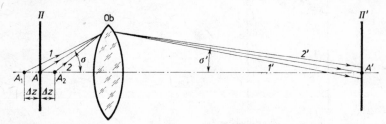

Fig. 2.58. Depth of field according the the Rayleigh quarter-wave criterion.

free from aberrations. Consider two other axial object points, A_1 and A_2, symmetrically separated by Δz from the object plane Π. Rays *1* and *2* emerging from A_1 and A_2 strike the objective Ob at the same zonal height as a ray emerging from A at an aperture angle σ. In the image space the angle σ has its conjugate counterpart σ'. Similarly, rays *1* and *2* have their image-side conjugate counterparts *1'* and *2'*. If under these conditions the maximum optical path difference between rays *1'* and *2'* does not exceed $\lambda/4$ in the image plane Π', then the images of the object points A_1 and A_2 are observed at point A' without an appreciable deterioration in sharpness.

A simple trigonometric analysis (see, e.g., Ref. [1.4], p. 307) shows that the criterion formulated above is fulfilled if

$$\Delta z = \frac{\lambda}{8\sin^2(\tfrac{1}{2}\sigma)}. \qquad (2.72)$$

The axial distance $2\Delta z$ between A_1 and A_2 is said to be the depth of field of the microscope, and, referring back to Fig. 2.55b, can also be considered as a permissible value $\Delta s'$ of the spherical aberration introduced by deviations Δt of the cover slip thickness. In other words, there will be no appreciable deterioration of the image provided the following demand is satisfied:

$$\Delta s' \leqslant 2\Delta z = \frac{\lambda}{4\sin^2(\frac{1}{2}\sigma)} \,. \tag{2.73a}$$

The spherical aberration $\Delta s'$ caused by Δt is expressed by Eq. (2.69b) except that Δt should be inserted instead of t. The relation (2.73a) therefore takes the form

$$\Delta t \left(n - \frac{1}{n} \right) \frac{2\sin^2(\frac{1}{2}\sigma)}{n^2 - 2\sin^2(\frac{1}{2}\sigma)} \leqslant \frac{\lambda}{4\sin^2(\frac{1}{2}\sigma)} \,, \tag{2.73b}$$

and

$$\Delta t \leqslant \frac{\lambda}{8\left(n - \dfrac{1}{n}\right)} \left(\frac{n^2}{\sin^4(\frac{1}{2}\sigma)} - \frac{2}{\sin^2(\frac{1}{2}\sigma)} \right). \tag{2.74a}$$

For standard cover slips, the following values of the refractive index n_e and Abbe number V_e were recently fixed by the ISO Standard No. 8255/1: $n_e = 1.5255 \pm 0.0015$, and $V_e = 56 \pm 2$. By substituting $n = n_e = 1.5255$ and $\lambda = \lambda_e = 0.5461$ μm in expression (2.74a), we obtain

$$\Delta t \leqslant 0.078 \left(\frac{2.33}{\sin^4(\frac{1}{2}\sigma)} - \frac{2}{\sin^2(\frac{1}{2}\sigma)} \right) \ [\mu\text{m}]. \tag{2.74b}$$

The maximum values of permissible deviations Δt, calculated from the formula (2.74b) for different numerical apertures of dry microscope objectives, are listed in Table 2.8. As can be seen, with objective of numerical aperture $A > 0.65$ the required accuracy of the cover slip thickness is extremely high (Δt should be smaller than 0.01 mm) and impossible to achieve in routine practice. For this reason, dry objectives of a numerical aperture $A \geqslant 0.75$ are fitted with correction mounts which allow one lens element to be axially adjusted so as to compensate for the over- or undercorrection due to a deviation Δt of the cover slip thickness from the nominal value. These so-called *correction collar objectives* allow their optical system to be adjusted to the effective cover slip thickness t within a range from 0.12 to 0.22 mm.[15] The "effective cover slip thickness" is the actual cover slip plus the layer of the mounting (embedding) medium between the object under examination and the cover slip. For this reason, cover slips of a thickness slightly below 0.17 mm are preferable in practice. The above-mentioned ISO standard has established two classes of cover slip thickness: No. 1 (general purpose) $t = 0.17 - 0.04$ mm and No.1-H (high performance) $t = 0.17 - 0.02$ mm.

---

[15] There are also objectives whose range of the cover slip thickness correction is from 0 to 2 mm (see Vol. 2, Chapter 5, Sec. 5.2).

TABLE 2.8

Thickness of cover slips for dry objectives

Numerical aperture of objective ($A = \sin\sigma$)	Admissible deviations $\pm\Delta t$ from nominal thickness $t = 0.17$ mm		Ranges of admissible cover slip thickness [mm]
	relating to spherical aberration correction	relating to chromatic aberration correction	
0.1	27.2	13.6	0–1
0.2	1.7	3.4	0–0.5
0.3	0.34	1.47	0–0.3
0.4	0.11	0.85	0.1–0.27
0.50	0.036	0.51	0.13–0.20
0.60	0.016	0.28	0.15–0.19
0.65	0.011	0.25	0.16–0.18
0.70	0.007	0.24	0.16–0.175
0.75	0.0054	0.20	0.165–0.175
0.80	0.0041	0.17	0.166–0.174
0.85	0.0025	0.14	0.167–0.173
0.90	0.0017	0.12	0.168–0.172
0.95	0.0011	0.09	0.169–0.171

Consider now the effect of cover slip thickness deviations on the correction of longitudinal chromatic aberration. For an analysis of this effect, it will be better to write Eq. (2.70) in the form

$$s'_{F'} - s'_{C'} = \Delta t \, \frac{n_e - 1}{n_e^2 V_e}, \tag{2.75}$$

where $n_e^2 \approx n_F . n_{C'}$ and $V_e = (n_e - 1)/(n_{F'} - n_{C'})$. From what was said above on depth of field and spherical aberration caused by cover slip thickness deviations, it is clear that $s'_{F'} - s'_{C'}$ cannot be larger than $2\Delta z$ defined by Eq. (2.72); hence this demand may be expressed as

$$\Delta t \, \frac{n_e - 1}{n_e^2 V_e} \leqslant \frac{\lambda_e}{4\sin^2(\frac{1}{2}\sigma)}, \tag{2.76}$$

and

$$\Delta t \leqslant \frac{n_e^2 V_e}{n_e - 1} \, \frac{\lambda_e}{4\sin^2(\frac{1}{2}\sigma)}. \tag{2.77a}$$

By substituting $n_e = 1.5255$, $V_e = 56$, and $\lambda_e = 0.5461$ μm in the above expression we obtain

$$\Delta t \leqslant \frac{33.9}{\sin^2(\frac{1}{2}\sigma)} \quad [\mu m].$$

(2.77b)

Table 2.8 shows the maximum values of permissible deviations Δt calculated from the above formula. As can be seen, they are much larger than those resulting from formula (2.74b) for numerical apertures greater than 0.2. It appears therefore that some deterioration of chromatic aberration correction by cover slip thickness deviations is negligible in comparison with the deterioration of the correction of spherical aberration.

In conclusion, tolerances for cover slip thickness t are primarily derived from the need to maintain optimum correction of spherical aberration. However, this relates only to middle- and high-power dry objectives. For low-power objectives ($A < 0.3$), the tolerances Δt resulting from Eqs. (2.74) are large and their actual values are therefore dependent on the mechanical rather than the imaging parameters of a microscope. The values for the permissible ranges of cover slip thickness listed in the last column of Table 2.8 have been derived from this approach. Almost identical data are specified by Carl Zeiss Oberkochen (Table 2.9) and E. Leitz Wetzlar (Table 2.10). Experimental confirmation of the theoretical results presented above can be found in Chapter 4.

From what has been said until now, it is clear that microscope objectives designed for use with cover slips cannot be employed for the study of uncovered specimens, unless the desired numerical apertures are low ($A < 0.3$). Such specimens require specially corrected objectives designed for use without a cover slip. In general, reflected-light microscopes are equipped with objectives of this kind.

As was mentioned previously, immersion objectives designed for use with cover

TABLE 2.9

Thickness of cover slips for dry objectives specified by Carl Zeiss Oberkochen [2.15]

Numerical aperture of objective	Admissible deviations $\pm\Delta t$ from nominal thickness $t = 0.17$ mm	Ranges of admissible cover slip thickness [mm]
0.08–0.30	—	0–0.3
0.30–0.45	0.07	0.1–0.24
0.45–0.55	0.05	0.12–0.22
0.55–0.65	0.03	0.14–0.20
0.65–0.75	0.02	0.15–0.19
0.75–0.85	0.01	0.16–0.18
0.85–0.95	0.005	0.165–0.175

TABLE 2.10

Thickness of cover slips for dry objectives specified by E. Leitz Wetzlar [2.16]

Numerical aperture of objective	Admissible deviations $\pm\Delta t$ from nominal thickness $t = 0.17$ mm	Ranges of admissible cover slip thickness [mm]
0.40	0.09	0.08–0.28
0.60	0.013	0.157–0.183
0.75	0.004	0.166–0.174

slips are less sensitive to variations of cover slip thickness than dry objectives, and those designed for homogeneous immersion may even be used with or without a cover slip. If, however, the principle of homogeneity is abandoned, the use of a cover slip is necessary. This requirement should be remembered when an immersion objective designed for use with a cover slip is employed for the study of uncovered smears. This is typical practice in the microscopy of blood and bacteria. It is true that a slight deterioration of image quality may be permissible for some routine research, but a nominal cover slip is definitely necessary in critical studies, where the high performance potential of the objective must be fully exploited; otherwise, a specially corrected immersion objective for use without cover slip should be employed. It is worth noting, however, that there are other ways of balancing the lack of a cover slip, such as increasing tube length [2.12] or using an immersion oil of higher refractive index, but this practice is now obsolete and unsuitable for present-day microscopes.

2.3.7. Microscope with objectives corrected for infinite tube length

So far we have confined ourselves to microscopes with objectives corrected for a finite tube length, but most of the text preceding and following this subsection is also valid for microscopical systems equipped with objectives whose image plane is infinitely distant. This latter system has certain optical and mechanical advantages over the former; hence present-day microscopes are more and more often designed for use with infinity corrected objectives (see Table 2.2). Their principle is illustrated in Fig. 2.59.

The basic image-forming system of such a microscope consists of an objective Ob and an ocular Oc between which there is a tube lens TL. The object plane II coincides with the first focus F_{ob} of the objective; hence an object AB situated in plane II is imaged by the objective Ob at infinity. Due to this property, the name "objective with infinite image distance" or "objective corrected for infinite tube length" (in short, *infinity corrected objective*) has been adopted. After the light has passed through Ob, the rays emerging from any object point form parallel beams. They are focused by the tube lens *TL* in the image plane Π', which is coincident with the second focus F'_{TL} of this lens. The image $A'B'$ of the object AB is enlarged by the ocular Oc adjusted so that its first focus F_{oc} coincides with the second focus F'_{TL} of the tube lens *TL*. The final image $A''_\infty B''_\infty$ is observed at infinity (if the observer's eye is normal); the ocular-side situation is then identical with that in a typical microscope (compare Fig. 2.39a).

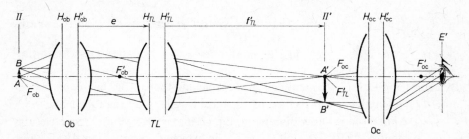

Fig. 2.59. Principle of the microscope with objectives corrected for infinite tube length.

It is worth noting that the tube lens *TL* and ocular Oc form a telescopic system, whereas the objective Ob and lens *TL* behave together as a compound objective which produces a primary image $A'B'$ of the object AB at a finite distance. In order to determine the size of the image $A'B'$, let us consider Fig. 2.60. It shows only the principal planes of the objective Ob, those of the tube lens *TL*, two rays (*1* and *2*) emerging from the object point B, and the general geometry of the imaging process between the object AB and its image $A'B'$. From the triangles $A'B'H'_{TL}$ and $F'_{ob}H'_{ob}H'_B$ it results that $\tan\sigma'_2 = y'/f'_{TL}$ and $\tan\sigma'_1 = -H'_{ob}H'_B/f'_{ob} = -H_{ob}H_B/f'_{ob} = -y/f'_{ob}$, where y is the height of the object AB, y' is that of the image $A'B'$, whereas f'_{ob} and f'_{TL} are the focal lengths of the objective Ob and tube lens *TL*. Then, just as $\sigma'_2 = \sigma_2 = \sigma'_1$ (rays *1* and *2* are parallel to each other in the space between Ob and *TL*) so also $\tan\sigma'_2 = \tan\sigma'_1$. Hence it appears that the transverse magnification of the system Ob + *TL*, defined as $M'_{ob} = y'/y$, is given by

$$M'_{ob} = -\frac{f'_{TL}}{f'_{ob}}. \tag{2.78}$$

The image $A'B'$ is observed through the ocular Oc as shown in Fig. 2.59. The visual magnification M_{oc} of this ocular is defined by Eq. (2.45), consequently

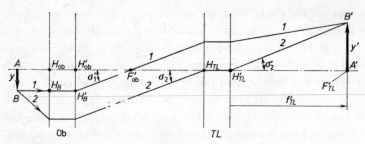

Fig. 2.60. Illustrating the discussion of Eqs. (2.78)–(2.84).

the magnifying power of the microscope system under consideration is given by

$$M_m = -\frac{250}{f'_{oc}}\frac{f'_{TL}}{f'_{ob}} \qquad [f'_{oc} \text{ in mm}]. \tag{2.79}$$

By comparing Eqs. (2.46a) and (2.79), we conclude that in the last equation the optical tube length g is represented by the focal length f'_{TL} of the tube lens.

For the microscope under discussion, Eq. (2.46b) is also valid; hence we may write an identity $250/f' = -250 f'_{TL}/f'_{ob}f'_{oc}$, in which f' denotes the focal length of the microscope. Here

$$f' = -\frac{f'_{ob}f'_{oc}}{f'_{TL}}. \tag{2.80}$$

This formula is similar to Eq. (2.47), which only gives the optical tube length g instead of the focal length f'_{TL} of the tube lens. Thus f'_{TL} is directly correlated with g and, therefore, called the *related focal length* ; consequently the lens *TL* is said to be the *related tube lens*.

Strictly speaking, the magnifying ability of an infinity corrected objective cannot be characterized similarly to an objective corrected for finite tube length because the former produces a real image at a finite distance only in combination with a properly corrected tube lens. The correct definition of the magnifying power of infinity corrected objectives is identical with that given for microscope oculars normally used in visual microscopy, i.e.,

$$M_{ob} = \frac{250}{f'_{ob}} \qquad [f'_{ob} \text{ in mm}]. \tag{2.81}$$

By combining Eqs. (2.78) and (2.81) we obtain

$$M'_{ob} = -\frac{M_{ob}f'_{TL}}{250} = -M_{ob}q \qquad [f'_{TL} \text{ in mm}], \tag{2.82}$$

where q denotes the *tube lens factor* or *tube factor*. It is given by

$$q = \frac{f'_{TL}}{250} \qquad [f'_{TL} \text{ in mm}]. \tag{2.83}$$

By combining Eqs. (2.81) and (2.83) we obtain $f'_{TL}/f'_{ob} = qM_{ob}$, and Eq. (2.79) may also be written in the form

$$M_m = -qM_{ob}M_{oc}. \tag{2.84a}$$

The focal length f'_{TL} of the tube lens is usually equal to 250 mm. In that case $q = 1$, $M'_{ob} = -M_{ob}$, and

$$M_m = -M_{ob}M_{oc}. \tag{2.84b}$$

This formula is identical with Eq. (2.43), and there is no difference between the magnifying properties of microscopes equipped with traditional and infinity corrected objectives. (Note that M_{ob} is formally negative in Eq. (2.43) but positive in Eqs. (2.84).)

Figure 2.61 shows more in detail the path of illuminating and image-forming rays in a microscope with infinity corrected objectives and an illuminator adjusted according to the Köhler principle of illumination. The reader is now referred back to Subsection 2.3.4, and no further discussion of this figure need detain us here. It is only worth noting that the diameter $2r_m$ of the object field is, in general, given by

$$2r_m = \frac{2r'_{oc}}{M_{ob}q}, \tag{2.85}$$

where $2r'_{oc}$ is the diameter of the linear image field. When a Ramsden-type ocular is used, as shown in Fig. 2.61, $2r'_{oc}$ is simply the diameter $2r_{oc}$ of the ocular field diaphragm, whereas for a Huygens-type ocular $2r'_{oc}$ is defined by Eq. (2.60). If $q = 1$, Eq. (2.85) reduces simply to Eq. (2.59). Apart from that, it should be mentioned that the image-side numerical aperture of infinity corrected objectives is equal to zero, so that Eq. (2.53a) now takes the form

$$\frac{n\sin\sigma}{\sin\sigma'} = \frac{A}{A'} = M'_{ob}, \tag{2.86}$$

where σ' and A' are, respectively, the image-side aperture angle and the image-side numerical aperture of the optical system which consists of the objective Ob and tube lens TL (Figs. 2.59 and 2.61).

As was previously mentioned, a microscope with infinity corrected objectives has some advantages over one equipped with objectives corrected for a finite tube length. First of all, the distance e of the objective Ob from the tube lens TL

can vary within a range $\Delta e = e_{max} - e_{min}$ because image-forming beams emerging from individual object points are parallel in the space between Ob and *TL*. This feature offers new scope for focusing the microscope. Normally this is done by adjusting two knobs, one for *coarse focusing* and the other for *fine focusing*. By means of a *rack and pinion mechanism* the two milled knobs move the microscope stage or/and body tube along its axis. Both coarse and fine focusing operate so that the distance between the objective and the ocular is kept constant. How-

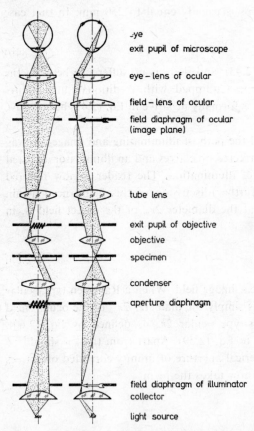

eye
exit pupil of microscope

eye – lens of ocular

field – lens of ocular

field diaphragm of ocular
(image plane)

tube lens

exit pupil of objective
objective

specimen

condenser

aperture diaphragm

field diaphragm of illuminator
collector

light source

Fig. 2.61. Optical system of the microscope with infinity corrected objectives.

ever, this distance needs not to be constant for a microscope with infinity corrected objectives. Here the focusing mechanism can therefore be coupled with the lower part of the body tube to which the objective is attached; the nosepiece, in particular, may be axially moved to focus the objective on the specimen by means of the coarse and/or fine adjustment knobs. In general, the mechanical performance of such a focusing arrangement is better than that of these mentioned earlier. It is employed by the American Optical Corporation among others in its microscopes of the MicroStar series, as well as by Reichert–Jung in the Biostar inverted microscope.

Different additional optical elements may also be situated in the space between the infinity corrected objective and the tube lens, such as a polarizer (analyser), filters, retarders, compensators, etc., etc., without causing a disturbance in the path of the image-forming beams. In short, infinity corrected objectives make it possible to design more versatile microscopes by using simpler means than where objectives corrected for finite tube length are employed.

2.4. A bird's eye view of modern microscopes for general purposes

The development of light microscopes themselves and their applications has become greatly accelerated during the last two decades. This is due both to the gradual improvement in industrial technology and the use of computers for optimization of design of optical systems. At the present time, high quality images can be provided for large fields of view (up to 32 mm). Most present-day microscopes are much easier to handle than their predecessors of twenty years ago. Due to rapid progress in mechanical engineering and electronics, all leading manufacturers offer a great number of different types of microscopes for both everyday and specialized work.

A bird's eye view of modern microscopes will given in this section, starting with some routine types and ending with the latest (1986) most sophisticated and commercially available instruments.

2.4.1. Binocular observation

There are several basic methods of presenting or displaying microscopical images. The most popular discussed so far depends on observation of the image through a monocular or binocular tube. The latter is, in general, part of the standard equipment of more advanced microscopes.

There are two main types of binocular tube. The first and more popular comprises a system of reflecting prisms and a beam-splitter which divides the light into two parts and directs them into two oculars. Depending on the *interpupillary distance* of the observer, different separations of both oculars are obtained by the transverse movement of the ocular tubes. But this movement means that the optical path (the effective tube length of the microscope) is varied and decoincidence occurs between the image plane of the objective and the first focal planes of the oculars. This is, of course, a defect which must be compensated by an appropriate variation of the length of both ocular tubes. The necessary compensation can be carried out in different ways, but today automatic compensation within an interpupillary range from 55 to 75 mm is generally accepted. An example of such compensation is shown in Fig. 2.62. The two oculars Oc1 and Oc2 are shown both at their narrowest and at their widest separation. In the first position let y be the smallest distance between the right-angle prisms P_1 and P_2 and x the distance between the prism P_3 and P_4. When the oculars are transversally separated into the widest position, the prisms P_1 and P_2 are moved apart to the greatest distance y', and the system of prisms P_1, P_2 and P_3 is lowered in the direction of prism P_4. The new distance between P_3 and P_4 will then be x'. This distance is defined so that the optical path of the light from P_4 to the

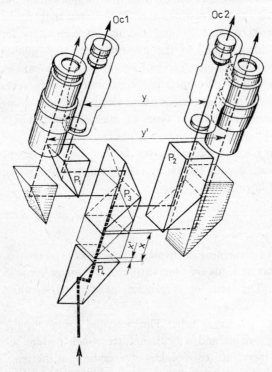

Fig. 2.62. Optical system of a binocular tube with automatic compensation of interpupillary distances of individual observers [2.16].

primary image plane of the objective is identical to that in the narrowest separation of both oculars. This is true of any interpupillary distance from 55 to 75 mm, and the parfocal length of the oculars is always correct independently of their separation.

The interpupillary compensation described above is a feature of the FSA binocular for the Orthoplan microscope manufactured by E. Leitz Wetzlar. A similar compensation, but obtained by means of a different arrangement of prisms, is also applied in the MND8 binocular for the Biolar microscope produced by the Polish Optical Works (PZO).

The second type of binocular tube comprises a somewhat different system of prisms which are arranged so that different separations of the oculars are obtained by rotating the viewing tubes, together with the prisms. This rotation does not change the optical path of the light beams and no interpupillary compensation of the tube length is needed. However, the mechanical construction of this binocular is less convenient for the user than that of the former type (Fig. 2.62).

Binocular observation gives maximum possible image resolution and brightness, but suffers from several disadvantages: (a) keeping the head and eyes in a fixed position is tiring, (b) the fact that the exit pupil (Ramsden disc) of the microscope contracts with increasing magnification gives rise to so-called *entoptic disturbances* making observation troublesome (compare Fig. 3.41), (c) certain individuals experience difficulties in obtaining true binocular vision.

2.4.2. Projection microscopy and photomicrography

The other very popular method of presenting microscopical images is that of *screen projection*. The advantage here is that the user of a projection microscope can sit in a relaxed position without close viewing contact with the microscope. Some serious disadvantages of this method, however, are rather poor resolution and insufficient image brightness. Moreover, the required intensity of illumination of the specimen to be examined is high, making observation with high-power-objective unsatisfactory.[16]

At present, screen projection is being more and more widely replaced by *closed circuit television* (CCTV). As in screen viewing the CCTV technique has the advantage of freedom of movement for the observer, but for reasonable viewing comfort the TV monitor must be placed some distance away. Currently, this

[16] An attempt to overcome this defect by using laser technique will be presented in Vol. 2, Chapter 12.

technique is extensively used in conjuction with microscopes for routine industrial inspection. Quite recently, the TV technique has been applied to *video-enhanced microscopy* (VEM). When combined with the differential interference contrast or polarizing microscope, the VEM method shows up linear elements and particles of biological specimens that are normally not visible in the intermediate optical image (see Vol. 2, Chapter 7, and Vol. 3, Chapter 15).

Like intermediate visual microscopy, photomicrography is the most widespread method of recording and presenting microscope images. Generally it may be stated that present-day photomicrographic equipment has reached the highest standards and is in routine use.

Projection microscopy and photomicrography, as well as CCTV microscopy, have a common feature, namely the microscope ocular is adjusted so as to produce a real image at a finite distance behind it in the plane of a screen or a photosensitive material.

Principle of screen projection microscopy. Microscopical screen projection can be accomplished by means of either a positive or a negative eyepiece[17] which are normally designed as special *projection* (or *photographic*) *oculars*. Occasionally, however, visual eyepieces can also be used for this purpose. First, we shall consider a projection microscope with a positive ocular (Fig. 2.63a).

The image $A''B''$ projected on the screen PS is upright with respect to the object AB but inverted in relation to the primary image $A'B'$. According to Eq. (2.14c), the size (height) of the image $A''B''$ is given by

$$M_{mp} = \frac{y''}{y} = \frac{b'}{f'}, \tag{2.87a}$$

where M_{mp} denotes the actual transverse magnification of the projection microscope, b' is the distance between the screen PS and the second focus F' of the microscope, and f' is the focal length of the microscope. If the objective Ob is corrected for a finite tube length, the magnification M_{mp} may, however, be expressed as

$$M_{mp} = M_{ob} M_{oct}, \tag{2.87b}$$

where M_{ob} is the transverse magnification of the objective given by Eq. (2.44) and M_{oct} is that of the ocular. According to Eq. (2.14c), $M_{oct} = y''/y' = -b'_{oc}/f'_{oc}$, where b'_{oc} is the distance of the screen PS from the second focus F'_{oc} of the ocular

---

[17] In some books on optics and microscopy Huygenian eyepieces are called *negative oculars*. This is misleading because, strictly speaking, these eyepieces cannot be considered as negative lens systems.

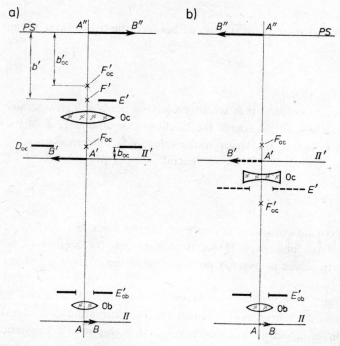

Fig. 2.63. Rear microprojection systems: a) with positive ocular, b) with negative ocular.

Oc. As can be seen, the image size y'' increases as the focal length f'_{oc} of the projection ocular decreases.

If, however, the objective Ob is corrected for infinite tube length,

$$M_{mp} = M'_{ob} M_{oct},$$ (2.87c)

where M'_{ob} is given by Eq. (2.78).

The diameter $2r_p$ of the projection image field is determined by the diameter $2r_{oc}$ of the ocular field diaphragm D_{oc}. Note that now this diaphragm is no longer coincident with the first focus F_{oc} of the ocular but precedes it because the projection ocular Oc is axially adjusted so that its first focus F_{oc} follows the image plane II' of the objective Ob. The space b_{oc} between II' and F_{oc} increases as the distance b'_{oc} between the ocular second focus F'_{oc} and the screen PS decreases. In practice, $b_{oc} \ll b'_{oc}$ and no serious disturbance occurs if the diaphragm D_{oc} is coincident with F_{oc} as in a visual eyepiece.

When the field diaphragm D_{oc} precedes the lenses (Ramsden-type ocular), the diameter of the projection image field is given by

$$2r_p = 2r_{oc} M_{oct}. \tag{2.88a}$$

On the other hand, if this diaphragm is situated between the lenses (Huygenian ocular),

$$2r_p = 2r'_{oc} M_{oct}, \tag{2.88b}$$

where r'_{oc} is given by Eq. (2.60).

It has already been mentioned that screen projection needs a high intensity of illumination. With a given light source the brightness or luminance L of the image on the screen is higher the larger is the image-side numerical aperture A' of the microscope, viz., $L \propto A'^2$. As, in general, $A' = A/M_{mp}$, hence

$$L \propto \left(\frac{A}{M_{mp}}\right)^2, \tag{2.89}$$

where A is the object-side numerical aperture of the microscope (or effective numerical aperture of the objective). Hence it appears that for a given A the image brightness on the screen is inversely proportional to the squared transverse magnification of the microscope.

Now let us consider a projection microscope with a negative ocular (Fig. 2.63b). The advantage of such an ocular is that the field curvature and lateral chromatic aberration of apochromatic objectives and high-power achromatic objectives may be compensated better than by using a positive projection ocular. The negative ocular cannot, however, be used for visual observation because the exit pupil E' of the microscope is in this case inside the optical system of the ocular Oc, which is normally adjusted so that its first focus F_{oc} follows the image plane II' of the objective Ob. Note that E' is the image of the exit pupil E'_{ob} of the objective. This image is now virtual.

In contrast to the previous situation (Fig. 2.63a), the image $A''B''$ projected on the screen PS is inverted with respect to the object AB but upright in relation to the primary image $A'B'$. Apart from this, there are no differences between the two types of screen projection systems.

Screen projection microscopes facilitate observation by small groups and permit rapid direct measurement using a graduated screen. They are especially useful in the examination of textile fibres and yarns, a purpose for which some manufacturers produce special instruments. These are among others the Lanameter (VEB Carl Zeiss Jena), Visopan (C. Reichert Wien) and MP-3 (Polish Optical Works, Warsaw). These microscopes are specially designed for measuring the thickness of fibres and grain sizes, for counting small particles, and exploring thin sections, as well as for comparing specimens with reference material.

Usually each more advanced laboratory or research microscope is easily converted into a projection microscope by attaching a screen head to or instead of the binocular (trinocular) tube. Recently projection screens for microscopy have been improved, resolution of the projected image is finer, and they are free from the glare typical of most standard screens.[18] In addition, some special screens suitable for *stereoscopic rear-projection microscopy* have been developed [2.17] and employed by VEB Carl Zeiss in the design of their unique stereoscopic projection microscope Plastival. The stereoscopic screen of this microscope is a *Fresnel screen* adequately modified in comparison with the screens of standard projection microscopes. However, Astvatsaturov and his co-workers [2.18] have recently developed a *holographic screen* which appears to be particularly suitable for stereoscopic rear-projection microscopy.

The microscope mentioned above are all rear-projection instruments. i.e., the observer sees the image in the light which is diffused through the screen. There are, however, other projection microscopes (*microprojectors*) which produce the image on a large screen some distance away, the projected image being observed in the light diffusely reflected from the screen. For example, Pictoval (VEB Carl Zeiss Jena) and Tele-Promar (E. Leitz Wetzlar) are instruments of this kind. The former comprises four semiplanachromatic or planachromatic objectives ($3.2 \times /0.10$, $6.3 \times /0.16$, $16 \times /0.32$ and $40 \times /0.65$). With screen image diameters ranging from 0.8 to 3.2 m the projection distance may be varied within a range from 2.5 to 25 m. A decisive factor for the performance of this instrument is the fact that a highly intensive condenser is correlated to each objective. The light source is exchangeable; the basic outfit includes a highly efficient gas discharge lamp. The specimen is protected against heat by a cold mirror and filters.

The Tele-Promar microprojector is equipped with plano-objectives up to the highest magnification ($100 \times$ oil immersion). With the $2.5 \times$ and $4 \times$ wide-field projection oculars a screen image of up to 3.5 m diameter can be obtained at projection distances of up to about 15 m. Like the Pictoval, it contains four condensers which are correlated to individual objectives for optimum illumination of the specimen; they are mounted on a permanently centred turret.

These instruments are especially intended for group viewing; hence they may be classified as *didactoscopes*. Today, their role in the classroom is, however, being largely superceded by CCTV microscopy.

[18] For the interested reader, more details regarding diffusing screens may be found in Ref. [1.19].

Dynascopic projection system. It was pointed out earlier (see Subsection 2.3.2) that for the correct observation of microscopical images the diameter $2r_E$ of the exit pupil of the microscope should be smaller than the diameter $2r_{eye}$ of the pupil of the observer's eye. If the former could, however, be artificially spread over an area E'' (Fig. 2.63), much larger than $2r_{eye}$ then the observer would also be able to view the entire image produced by the microscope, and additionally he would now be able to move his head laterally over a range of area E'' without loss of image. This possibility was known for many decades, but was not implemented in practice. The first satisfactory solution to the problem of spreading the exit pupil of the microscope was arrived at by Vision Engineering Ltd. (England) in 1972 and used in the design of their aerial image projection system (Dynascope). The principle of a reflective version of this system is illustrated in Fig 2.64.

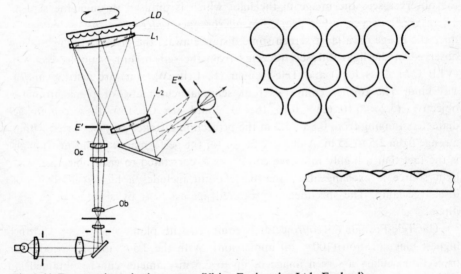

Fig. 2.64. Dynascopic projection system (Vision Engineering Ltd., England).

The projection ocular of a microscope projects the image of the specimen on the *lenticulated disc LD*. This disc consists of an aluminized plastic material embossed with very high optical precision with a myriad of lenticular forms, as shown on the right of Fig. 2.64. The lenticulated relief spreads the microscope exit pupil E' over an area E'' at a distance from the disc *LD*. Additional field lenses, L_1 and L_2, are provided, but are not always required, depending on the design of the system. When the lenticulated disc is static, the observer will see a series of small dots of light, one to each lenticular, and these light dots will in fact constitute parts of the image of the specimen under examination. If, however, the disc is rotated, the lenticulars scan the entire image area and therefore a composite image is formed. The rotational speed of the disc depends on the field lens magnification and the type of microscope on which the dynascopic system is employed, but is usually a maximum of 1500 rpm. Exit pupils E'' for individual or group display with included angles up to 160° are possible by varying the radius of curvature of the lenticular forms which can either be positively or negatively shaped.[19] The dynascopic system of a transmission version includes a lenticulated transparent disc rotated about its axis and the image is observed from the back of the disc as with a conventional rear-projection screen (Fig. 2.63).

When using the dynascopic system, the observer can sit in a relaxed viewing position with the freedom of head and eye movement as shown in Fig. 2.64 by arrows. In addition, because an aerial image is produced rather than a diffused image, as in the case of a conventional rear-projection screen, the high image brightness and contrast only associated with intermediate eyepiece observation is retained. However, a higher level of specimen illumination is necessary than would be required for use with eyepiece observation although substantially less light than would be needed for conventional screen projection. The ability to present binocular images is also retained but without the binocular tube whose interpupillary and/or dioptre adjustments frequently lead to optical misalignment. Despite its advantages, the dynascopic system has not hitherto been widely adopted.

Photomicrography. During the past three decades photomicrography has become a very large field of light microscopy, too extensive to be discussed exhaustively; only some basic data relating mainly to geometrical optics will therefore be given here. For further details one of several standard monographs on photomicrography, e.g., [2.7], [2.19], or [2.20], may be consulted.

Photomicrography can be defined as the photography of magnified images formed by a microscope. There are also other definitions which fix an approximative limit of minimum magnification for photomicrography. With an objective of magnification 1 × or 2.5 × and a 3.2 × ocular, the minimum final magnification would be 3.2 × or 8 ×, respectively. With a conventional microscope, which is

---

[19] Data from a Vision Engineering leaflet of 1972.

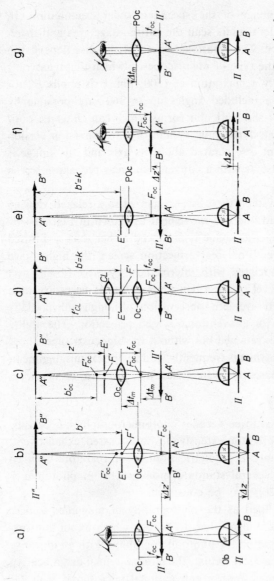

Fig. 2.65. Drawings showing different possibilities of making the photomicrographs. Ob—objective, Oc—visual ocular, POc—projection ocular, CL—camera lens, II—object plane, II′—primary image plane, II″—secondary image plane (plane of a photographic plate).

a combination of two magnifying systems, i.e., objective and ocular, photomicrography of magnifications lower than $3 \times$ is impractical. Sharp, clear photographs, when attainable, require prodigious skill due to the uneven illumination and the inherent limitations of the lenses and equipment available. Photography at low magnification is therefore undertaken by means of an apparatus called the *macroscope* which contains only one magnifying optical system. Photography at magnification lower than about $10 \times$ is therefore known as *photomacrography*. Special equipment for high performance photomacrography is available, e.g., from Wild Heerbrugg (M420 Zoom Makroskop or M400 Photomakroskop [2.21]).

In general, the optical principles of photomicrography do not differ from those of screen projection microscopy (Fig. 2.63), except in the use of different materials for displaying microscopical images. First of all, a photographic material (roll-film, sheet-film, photo-plate) replaces the screen. The photosensitive material must be protected against external light and correctly exposed with image-forming rays. Optical requirements for high performance photomicrography are stricter than for screen projection microscopy. Therefore, some additional theoretical discussion is now necessary. This will centre on Fig. 2.65, which shows different possible ways of making photomicrographs.

The first diagram (a) shows a typical system for visual microscopy. This system can be directly used for photomicrography by defocusing the microscope (diagram b) or by increasing the mechanical tube length (diagram c). In the first case the object AB is lowered by Δz with respect to the object plane II of the objective Ob; consequently, the primary image $A'B'$ is to be found below the ocular focus F_{oc}, which is normally coincident with the image plane II' of the objective. It may be shown that

$$\Delta z = -n \frac{250^2}{M_m^2 b'} = -n \frac{250^2}{M_{ob}^2 M_{oc}^2 b'} = -n \frac{f_{oc}'^2}{M_{ob}^2 b'}, \tag{2.90}$$

where n is the refractive index of the space between the object AB and the objective Ob, M_m and M_{oc} are the visual magnifications of the microscope and ocular, M_{ob} is the transverse magnification of the objective, f_{oc}' the focal length of the ocular, b' the distance between the exit pupil of the microscope and the plane II'', in which the photographic plate is positioned (f_{oc}' and b' are expressed in mm). The above formula shows that Δz decreases as the distance b' and the ocular magnifying power M_{oc} increase. To maintain the objective at a high performance level, Δz cannot be larger, roughly speaking, than tolerances for the cover slip thickness (see Tables 2.8–2.10). This means that the method outlined above is suitable for photomicrography with large distances b', as in the case of bellows cameras. These are mounted separately from the microscope on a rigid stand and used for large image sizes (photo-plates or sheet films up to 13×18 cm). Taking, for example, a dry objective of magnifying power $M_{ob} = 40 \times$ (numerical aperture $A = 0.65$) and an ocular of magnifying power $5 \times$, and assuming b' $= 250$ mm, the defocusing Δz will be found to be -0.006 mm. This value is quite tolerable although the objective will suffer from a small defocusing spherical aberration, especially if its numerical aperture is larger than 0.65. For this reason, the method of using a visual ocular for photomicrography, as shown in Fig. 2.65c, is preferable. The object AB is now coincident with the object plane Π of the objective Ob, but the ocular Oc is raised by Δt_m with respect to the primary image plane Π' of the objective. From Eq. (2.13b) it follows that

$$\Delta t_m = \frac{f_{oc}'^2}{b_{oc}'} \approx \frac{f_{oc}'^2}{z'}, \tag{2.91}$$

where b_{oc}' is the distance between the second focus F_{oc}' of the ocular Oc and plane Π'' ($b_{oc}' \approx z'$). The ocular movement Δt_m can be considered as an elongation of the mechanical tube length. Taking $b_{oc}' = 250$ mm and $f_{oc}' = 50$ ($M_{oc} = 5 \times$), the elongation $\Delta t_m = 10$ mm and is greater than the tolerance of mechanical tube length for objectives whose numerical aperture is equal to or larger than 0.65 (see Table 2.3). However, for a $10 \times$-ocular ($f_{oc}' = 25$ mm) the comparable value Δt_m is 2.5 mm and is tolerable. Hence, this method requires either long photographic distances z' and/or short focal length oculars.

In any case, a microscope with visual oculars used for photomicrography is not longer working strictly in accordance with the requirements for which it was designed since the light beams forming the real image of an object point are not now parallel to each other when they emerge from the ocular. In order to maintain the parallelism of these beams, an additional lens (*camera lens*) CL should be used behind the ocular Oc as shown in Fig. 2.65d. It is, however, most desirable that the entrance pupil of the lens CL should be coincident with the exit pupil of the microscope. The real image $A''B''$ occurs, of course, in the rear focal plane of the camera lens and is magnified as follows

$$M_{mp} = -M_{ob} \frac{f_{CL}'}{f_{oc}'}, \tag{2.92a}$$

where f_{CL}' is the focal length of the camera lens CL. The above formula relates to microscopes equipped with objectives corrected at a finite image distance. For infinity corrected objectives Eq. (2.92a) takes the form

$$M_{mp} = -M'_{ob} \frac{f'_{CL}}{f'_{oc}}, \qquad\qquad (2.92b)$$

where M'_{ob} is defined by Eq. (2.78).

According to this principle some photomicrographic adapters, as well as cine-cameras, are attached to microscopes equipped with visual oculars. They are mounted directly, or through a supporting arm, on the microscope tube. These photomicrographic adapters are especially designed for 35 mm film (strictly 24×36 mm), as well as for medium film sizes (6.5×9 cm) and for film sheets 9×12 cm.

The most popular attachments for photomicrography are, however, designed according to the principle shown in Fig. 2.65e. They contain positive *photo-oculars* or projection oculars (POc) which form a real image $A''B''$ at a finite distance b'. This system of photomicrography has already been discussed and no further comment is required here. In fact, there is no difference between Figs. 2.65e and 2.63a. One of the best known items of photomicrographic equipment designed in accordance with the principle shown in the figures mentioned above is the "mf" attachment camera available from VEB Carl Zeiss Jena (Fig. 2.66).

Some microscope manufacturers, e.g., the Polish Optical Works (Warsaw), produce photomicrographic cameras with negative projection oculars [2.22]. These work according to the principle shown in Fig. 2.62b. In the past, negative photo-oculars, called *homals* [2.23], were very popular. They were manufactured by C. Zeiss Jena and intended for use with apochromatic and high-power achromatic objectives to compensate the field curvature and transverse chromatic aberration of these objectives. At present, planachromatic or planapochromatic objectives and projection oculars specially designed for use with them are generally available from almost all microscope manufacturers and interest in homals has completely diminished. It is worth mentioning that Boegehold [2.24] and Claussen [2.25] were the first designers who developed microscope objectives with plano-correction. Such objectives have become a matter of routine in modern photomicrography.

To return to our main subject, it is worth noting that the distance b' of the photographic material from the exit pupil of the microscope is called the *optical length of the camera*, usually denoted by k. Since the visual magnification of the ocular is defined by Eq. (2.45) and the transverse (projection or photographic) magnification is expressed by $M_{oct} = -b'_{oc}/f'_{oc} \approx -b'/f'_{oc} = -k/f'_{oc}$, so also (Fig. 2.65e)

$$M_{oct} = -M_{oc} \frac{k}{250} \qquad [k \text{ in mm}]. \qquad\qquad (2.93)$$

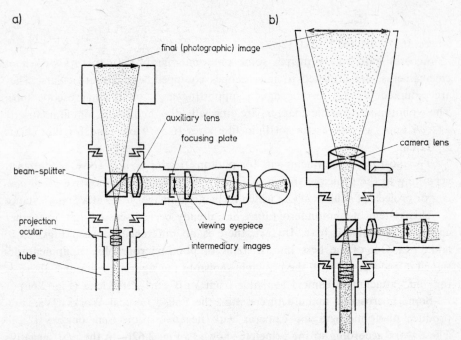

Fig. 2.66. Attachment cameras for photomicrography: a) small standard format (24×36 mm), b) large format (6×6 or 6×9 cm).

From this formula it follows that the photographic magnification of the ocular (M_{oct}) is larger or smaller than the visual magnification (M_{oc}) by a factor

$$p = \frac{k}{250} \qquad [k \text{ in mm}], \tag{2.94}$$

called the *camera-* or projection-factor. By substituting $M_{oct} = -M_{oc}\,k/250 = -M_{oc}\,p$ into Eqs. (2.87b) and (2.87c), we obtain

$$M_{mp} = -M_{ob}\,M_{oc}\,p \tag{2.95a}$$

for the photographic (or projection) magnification of a microscope equipped with objectives corrected at a finite tube length, and

$$M_{mp} = -M'_{ob}\,M_{oc}\,p \tag{2.95b}$$

for the same magnification of a microscope equipped with infinity corrected objectives, where M'_{ob} is given by Eq. (2.78).

   If in the situation characterized by Fig. 2.65d the camera lens *CL* is as close as possible to the exit pupil of the microscope, then the photographic distance

b' is approximately equal to the focal length f'_{CL} of the camera lens, and Eqs. (2.92) take the form of Eqs. (2.95).[20]

It is worth noting that a positive photo-ocular (Fig. 2.65e) can on occasion be used for visual microscopy as well (Fig. 2.65f). In this case the microscope must, however, be defocused by a value Δz, which enables the primary image $A'B'$ of the object AB to be brought into coincidence with the first focus F_{oc} of the ocular POc. Consequently, the object AB follows the object plane Π and the objective Ob suffers from a small defocusing spherical aberration. For this reason, another way which is shown in Fig. 2.65g may be found more acceptable. This consists in shortening the mechanical tube length by a value Δt_m, which is defined by the distance between the primary image plane Π' and the ocular focus F_{oc} in Fig. 2.65e. The relation between Fig. 2.65e and Figs. 2.65f, g is the converse of the relation between Fig. 2.65a and Figs. 2.65b, c.

It is generally true to say that photomicrographic instruments today are of very high quality. During the past two decades there has been a rapid improvement in the variety, quality, and versatility of the equipment available. Fully automatic or semi-automatic photomicrographic cameras have become routine equipment. A device for photomicrography is classed as "fully automatic" if the exposure time is assessed (and sometimes also indicated), the measured information is processed, exposure is controlled, and the film is transported after exposure. Instruments which have all the features mentioned above with the exception of automatic film transport are said to be "semi-automatic". This classification means that only cameras for roll-films can be regarded as fully automatic.

For about 25 years the Leitz Orthomat camera was synonymous with automatic photomicrography in the field of the 35 mm film. Recently this attachment has been replaced by the Leitz Vario-Orthomat (Fig. 2.67), which is now available in all conventional photographic formats: 24×36 mm, 3.25×4.25 in (Polaroid) and 9×12 cm (or 4×5 in). This latter instrument contains a zoom photographic ocular, which allows the overall microscope magnification to be continuously changed by a factor $2.5 \times$ in order to match the desired image area to the selected format of the photomicrographs. Similar equipment (with or without zoom photographic oculars) is also available from almost all microscope manufacturers (VEB Carl Zeiss Jena, C. Reichert Wien, Carl Zeiss Oberkochen, Vickers, Nikon, Olympus, and others). All these instruments for photomicrography have reached

---

[20] In practice, the magnification M_{mp} is usually determined by photographing a stage micrometer (see Vol. 3, Chapter 13).

a high ("push-button") level of operating comfort, and only some sophisticated improvements in the selection of correct exposure times are suggested by some users. For further details on automatic photomicrography Refs. [2.27] and [2.28] may be consulted.

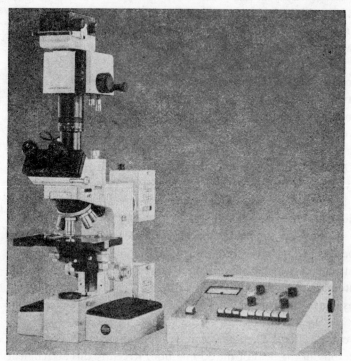

Fig. 2.67. Leitz Vario-Orthomat attached to the universal microscope Orthoplan [2.26] (by courtesy of E. Leitz Wetzlar).

Finally, it is necessary to mention *camera microscopes*. These instruments feature either a built-on or built-in photomicrographic system which forms a functionally integrated unit for both visual microscopy and photomicrography with 35 mm film and other photographic materials of conventional image formats. For example, Docuval manufactured by VEB Carl Zeiss Jena (Fig. 2.68) and Photomicroscope III available from Carl Zeiss Oberkochen are of this type of instruments. Some universal microscopes, e.g., Axiophot (Carl Zeiss Oberkochen), UnivaR (C. Reichert Wien), and Photoplan (Vickers) can also be regarded as photographic (camera) microscopes.

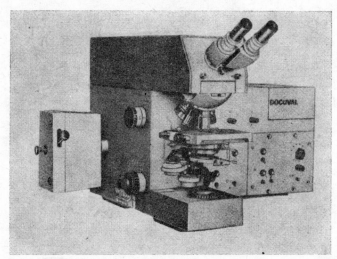

Fig. 2.68. Photomicroscopę Docuval (C. Zeiss Jena).

2.4.3. Reflected-light microscopy

So far we have confined ourselves to *transmitted-light microscopy*, which is normaly used for the examination of thin transparent specimens. Now we shall consider *reflected-light microscopy*, which is intended for the study of opaque specimens in reflected light. This kind of microscopy is also called *incident-light microscopy*. In the past, transmittted-light microscopes were mainly used in biological and medical research, and for this reason are also called *biological microscopes*, while reflected-light microscopes, which were mostly employed in metallography and metallurgical research, are frequently called *metallographic* or *metallurgical microscopes*. At present, this nomenclature is no longer applicable as the uses of both microscope systems are hard to delimit.

Similarities and differences between transmitted-light and reflected-light microscopes. A reflected-light microscope with bright-field illumination may be considered to be equivalent to a typical transmitted-light microscope since only its illumination system appears to be folded at the object plane Π of the objective Ob and partly coupled with the image-forming system (Fig. 2.69). Therefore, the objective of the reflected-light microscope also functions as the condenser in the transmitted-light microscope. This similarity means that our previous discussion concerning the imaging properties of transmitted-light microscopy applies almost in its entirety to reflected-light microscopy. Only the illuminating system needs some additional discussion.

Several different—more or less complicated—illumination systems exist for reflected-light microscopy. A detailed description may be found in Refs. [2.8] and [2.29]–[2.31]. The most popular bright-field reflected-light illuminator based on the Köhler principle of illumination is shown in Fig. 2.69. Its principal com-

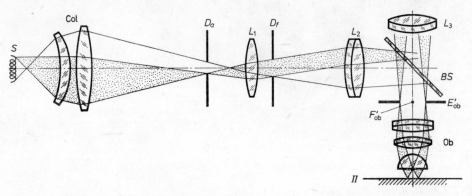

Fig. 2.69. Bright-field epi-illuminator.

ponents are: the light source S, the collector Col, the aperture diaphragm D_a, the field diaphragm D_f, two auxiliary lenses L_1 and L_2, and the semitransparent mirror (beam-splitter) BS. The common optic axis of the collector and succeeding lenses is at right angles to the optic axis of the objective Ob, hence the plate BS must be fixed at an angle of $45°$ with respect to these axes. The collector Col projects the image S' of the light source S onto the aperture diaphragm D_a. The lenses L_1, L_2 and the mirror BS image D_a and S' on the exit pupil E'_{ob} of the objective Ob. Furthermore, the lenses L_1, L_2 and the objective Ob image the field diaphragm D_f on the surface of the specimen under examination. The light reflected back by the specimen enters the objective, passes partly through the beam-splitter BS, and forms the image of the specimen surface in the primary image plane of the objective. The primary image is observed (projected, photographed) by means of appropriate oculars using the already described procedures of transmitted-light microscopy. The whole illuminating system ($S + \text{Col} + D_a + L_1 + D_f + L_2 + BS$) usually forms either a unit integral with a reflected-light microscope or constitutes a replaceable attachment to a transmitted-light microscope. In the latter case the reflected-light illuminator is additionally equipped with a correction lens L_3. This lens corrects the primary image distance of objec-

tives corrected for finite tube length, which is distorted by the height of the illuminator.

Reflected-light microscopes are mainly used to observe surface structures
of objects that reflect light, and for this reason no cover slips are used. In general,
all objectives for reflected-light microscopy are designed for uncovered specimens.
In view of the height of the illuminator to be inserted between the ocular tube
and the nosepiece, objectives in reflected-light microscopes frequently have a relatively short parfocal length. For instance, the parfocal length of the Epiplan
objectives available from Carl Zeiss Oberkochen is equal to 33 mm (see Table
2.2). According to ASTM[21] requirements, the standard final magnifications
used in metallurgical work are $50 \times$, $100 \times$, $200 \times$, $500 \times$, $1000 \times$ and $1600 \times$.
For this reason, the initial magnifications of objectives and those of oculars for
reflected-light microscopy should be selected so as to fulfill the requirements
mentioned above.

a)   b)

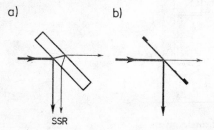

Fig. 2.70. Plane parallel plate (a) and pellicle (b) used as beam-splitters.

The semireflecting plate *BS* (Fig. 2.69) does not allow more than 25% of
the luminous flux incident on this plate to reach the ocular. This is, of course,
a disadvantage. Another, even more serious drawback is the *second surface
reflection* (SSR) as shown in Fig. 2.70a. This undesirable reflection is responsible
for the ghost images, which disturb the right image of the object under examination. To overcome this second surface reflection, the rear surface of the semitransparent mirror *BS* (Fig. 2.69) must be coated with a special antireflecting
film. Another solution is to use a very thin beam-splitter as shown in Fig. 2.70b.
Such beam-splitters are made of NPC pellicles [2.32]. They are so thin (2 to
8 µm) that the displacement between rays reflected from the first pellicle surface
and those reflected from the second surface is negligible. The desired reflection/
transmission ratio is achieved by coating the pellicle with metallic and/or dielec-

[21] American Society for Testing of Materials

tric materials. However, the use of pellicle beam-splitters in microscopical devices has not been reported so far.

Figure 2.71 shows another method of illumination used in reflected-light microscopy. Instead of a semireflecting plate, a mirror or a right-angle prism is employed which occludes one half of the objective aperture. The occluded part of the objective then functions as condenser and the non-occluded part as objective. This method is free from the drawbacks of the semitransparent plate, but reduces the numerical aperture of the objective and consequently the resolving

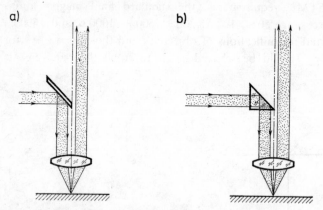

Fig. 2.71. Epi-illumination with a mirror or right angle prism.

power of the microscope. Moreover, it produces an oblique illumination which changes the actual appearance of specimen images although this kind of illumination may sometimes be advantageous. The reflecting prism arrangement (Fig. 2.71b) was first used by Nachet (Paris) and is, with greater or lesser modifications, still in use today, especially in polarizing reflected-light microscopes (see Vol. 3, Chapter 15).

Reduction of stray light. In bright-field reflected-light microscopes the illuminating beam when passing through the objective lenses towards the specimen (Fig. 2.69) is partly reflected back to the image plane from glass–air surfaces. The reflection appears in conformity with the Fresnel formulae (see Subsection 1.5.4) and is, in particular, the source of stray light which considerably diminishes the image contrast of objects under examination. The amount of such light depends on the type of objective, the curvature of objective lenses, the quality of antireflecting coatings, as well as on the design and adjustment of the illuminating system [2.33–2.35]. In general, the beam-splitting plate system (Fig. 2.69) produces the greatest amount of stray light in the image plane of the microscope. Of course, stray light also occurs in transmitted-light microscopes, but in reflected-light microscopes it causes much more trouble.

It has been stated that the greatest amount of stray light is caused by the reflection in the paraxial area of the objective lenses when a reflected-light illuminator is used as shown in Fig. 2.69. An annular type of illumination, developed by E. Leitz Wetzlar would therefore appear to be better than one using a circular aperture. Moreover, annular illumination increases the sensitivity of the reflected-light microscope when observing a fine micro-relief.

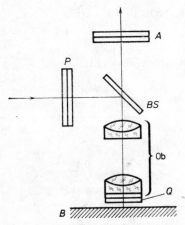

Fig. 2.72. Drawing showing the principle of Antiflex objectives.

Another, much more effective way of reducing stray light has been developed by Carl Zeiss Oberkochen [2.36]. This depends on the use of polarized light, as shown in Fig. 2.72. A polarizer (P) is placed in the illuminating beam ahead of the beam-splitting plate BS, while an analyser (A) is placed in the image-forming beams, behind BS. The polarizer is adjusted so that its axis of light vibrations is either at right angles or parallel to the incidence plane related to the plate BS. If, at the same time, the analyser A is adjusted so that its axis of light vibrations is at right angles to the vibration direction of the light specularly reflected from the lens surfaces of the objective Ob, this light will be extinguished. On the other hand, the light diffusely reflected from the specimen surface will largely pass through the analyser because it is depolarized in the process of diffuse reflection.

If, however, the specimen surface reflects the light specularly, as do polished surfaces, this method functions only when a quarter-wave plate (Q) is placed between the specimen B and the objective Ob. For rays being passed twice, the plate Q becomes a half-wave plate. If this plate is adjusted so that its fast (or slow) axis[22] forms an angle of 45° with the vibration direction of the light incident on it, then the light vibrations, which are reflected from the specimen and pass back through Q, become rotated by 90° and are transmitted by the analyser A. The result is a well contrasted image of the specimen surface under examination since stray light is almost completely extinguished in the image plane of the microscope. The reduction of stray light is additionally increased if the space between the quarter-wave plate Q and the specimen B is filled with an immersion liquid.

Some bright-field objectives of magnification 2.5× to 40× from Carl Zeiss

TABLE 2.11

Antiflex immersion objectives available from Carl Zeiss Oberkochen [2.15]

Designation	Magnifying power	Numerical aperture	Focal length [mm]	Working distance [mm]	Cover-slip thickness [mm]
Epiplan–Antiflex, 8× /0.20, oil	7.9×	0.2	18.75	0.4	—
Antiflex-Epi-Achromat, 16× /0.40, oil	16.0×	0.4	10.0	0.45	—
Antiflex-Epi-Achromat, 40× /0.65, oil	40.0×	0.65	4.6	0.5	0
Epiplan-Antiflex 2.5× /0.08, methylene iodide	2.9×	0.08	54.8	0.3	0
Epiplan-Antiflex, 4× /0.1, methylene iodide	4.1×	0.1	36.3	0.4	—
Epiplan-Antiflex, 8× /0.2, methylene iodide	7.9	0.2	18.75	0.4	—
Antiflex-Epi-Achromat, 16× /0.40, methylene iodide	16.2	0.4	10.0	0.35	0
Antiflex-Epi-Achromat 40× /0.65, methylene iodide	39.6	0.65	4.6	0.25	0

[22] For definition see Subsection 1.4.5.

Oberkochen are provided with rotatable quarter-wave plates [2.15]. They are called *Antiflex objectives* and are available as Epiplan or achromatic systems for reflected light microscopy with oil and methylene iodide immersion (Table 2.11). Almost identical objectives termed *immersion-contrast planachromats* are also manufactured by E. Leitz Wetzlar [2.36].

It should be pointed out here that a typical quarter-wave plate is chromatic. Nomarski therefore suggested using instead a birefringent (quartz) plate the thickness of which is adequate to produce a channelled spectrum between crossed polarizers. Five or six channels are sufficient for white light. As a matter of fact, this quartz plate is half-wave (for rays passing twice through it) but not for a single wavelength only; it applies as a high-order $\lambda/2$-plate to a series of wavelengths distributed within the visible spectrum, and can therefore be considered as truly achromatic (see Ref. [1.31], Chapter 4).

Antiflex objectives are especially suitable for the examination of materials of low to medium reflectance (coal, rocks, ceramics, plastics) and specimens of varying reflectivity caused by weak to strong absorbance or refraction. In general, differences of reflectivity and intrinsic colour of certain materials are more pronounced and more clearly visible in oil immersion than in air. Methylene iodide immersion generally produces the same result as oil immersion, but the former is better suited for objects of extremely low reflectivity (coal petrography). Recently the antiflex technique has successfully been applied in reflection-contrast microscopy to study cells growing in glass substrates (see Vol. 2, Chapter 8). However, the quarter-wave plates placed in front of the objective makes this technique unsuitable for certain measuring procedures in polarized-light microscopy, e.g., for the quantitative analysis of polarized reflected light.

Types of reflected-light microscopes. There are many different designs of reflected-light microscopes on the market today. Among them four main types are best known in practice. The first includes so-called *upright metallurgical microscopes*, which are similar in design to typical transmitted-light microscopes. The Nikon metallurgical Optiphot, Metavar (C. Reichert Wien), Standard 08 and Standard WL (C. Zeiss Oberkochen),[23] NS200 (Nachet, Paris), Laborlux 12 (E. Leitz Wetzlar [2.38]), and BHM (Olympus) are all examples of upright reflected-light microscopes. There is usually a corresponding version for transmitted light (such as that shown in Fig. 2.73); in fact, the reflected-light version is

---

[23] It is worth noting that the Standard WL microscope includes an epi-system for providing Köhler illumination using an 150 watt Schott fibre-guide light source.

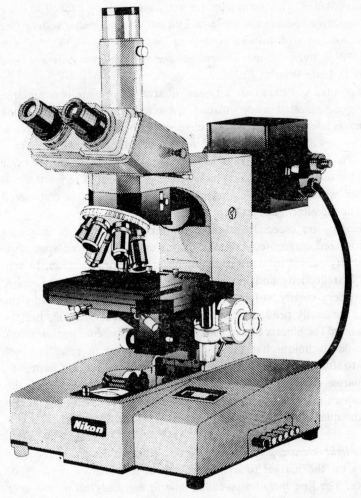

Fig. 2.73. Laboratory and research microscope Optiphot (by courtesy of Nikon).

normally obtained by attaching an appropriate reflected-light illuminator (in this case also called *vertical* or *epi-illuminator*) to the transmitted-light microscope. This design is convenient only for the examination of plane parallel specimens because the upper and lower specimen surfaces must be parallel to the microscope stage to avoid readjusting the illumination and refocusing the image each time the specimen is moved transversely. Frequently microscopes of this type are arranged so that they function either in reflected or transmitted light with little or no changeover necessary. For example, Standard T (C. Zeiss Ober-kochen), Ergolux (E. Leitz Wetzlar [2.40]) and BHMJL (Olympus) are instruments of this kind. They are especially useful in the materials sciences for the examination of polished thin sections of rocks and other materials containing both transparent and opaque phases, as well as in the electronic industry for the inspection and measurement of photomasks, integrated circuits and other objects.

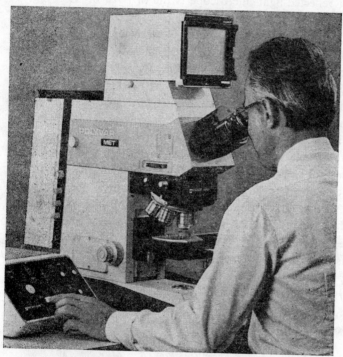

Fig. 2.74. Metallurgical microscope Polyvar-Met (by courtesy of C. Reichert Wien).

Reflected-light microscopes belonging to the second group are similar to the instruments described above except that they have a built-in illuminating system for reflected light instead of the built-on epi-illuminator. The metallurgical microscope Polyvar-Met available from Reichert-Jung is the best example here (Fig. 2.74). It retains the same general design as the Polyvar for transmitted light and can even be set up for operating in either reflected and/or transmitted light.[24] Such a versatile arrangement for metallurgical use is termed Polyvar-Met 66. It is especially intended for semiconductor technology and fitted for several microscopical techniques (bright-field, dark-field, DIC, polarization, fluorescence, microinterferometry, microhardness testing). Almost the same capacities for research and measurement are guaranteed for the Vickers M17 (metallurgical), Metalloplan (E. Leitz Wetzlar [2.40]), Vertival or Epival and Jenavert (VEB Carl Zeiss Jena), and some other large universal microscopes (see Subsection 2.4.5).

The third kind of reflected-light microscope is the *inverted* or *Le Chatelier type*. During the last two decades many changes have been introduced in the latter. Modern versions were initiated by E. Leitz Wetzlar and C. Reichert Wien when, respectively, they made available their Epivert and Metavert microscopes. Next, inverted microscopes IM35 and ICM405 (Carl Zeiss Oberkochen [2.41]), Metaval (VEB Carl Zeiss Jena [2.42]), and Metallovert (E. Leitz Wetzlar [2.43]) appeared on the market. These microscopes are in general devised for use in either reflected or transmitted light with only minor conversions necessary as shown in Figs. 2.75 and 2.76. The version for transmitted light is normally used as an inverted biological microscope, whereas that for reflected light is mainly employed in metallurgical work. With the last type of instrument (Figs. 2.75b and 2.76b), the upper and lower surfaces of the specimen need not be parallel to each other since the specimen surface under examination is coincident with the stage surface. Moreover, the large open space above the stage allows objects of any height to be examined.

An advantageous feature of this type of instrument is a relay system, which projects the exit pupil of objectives on a secondary, easily accessible plane. This feature improves versatility of the microscope, for instance, when a fundamental change of technique, say from bright-field to phase contrast, is involved. Phase plates normally present in the rear focal plane of the objective can now be inserted

---

[24] Recently Reichert–Jung U.K. has announced (see *Proc. Roy. Micr. Soc.*, **19** (1984), 52) that an automated version for biological uses—the Polyvar Telematic—is commercially available. This version incorporates a motorized nosepiece with push-button selection of objective plus automatic condenser selection. The motor control and push-button objective selector are built into the microscope base. All the operator has to do is depress the key for the objective required and the nosepiece will then rotate to position the desired objective above the specimen under examination. For a $100\times$ objective, the multimotor stops halfway between settings so that immersion oil can be applied. Depressing the key a second time moves the objective into position. As well as positioning the desired objective, the motorized control system will select the correct condenser depending on the objective power. This instrument is especially advantageous to anyone who has to rapidly examine specimens using a variety of objectives.

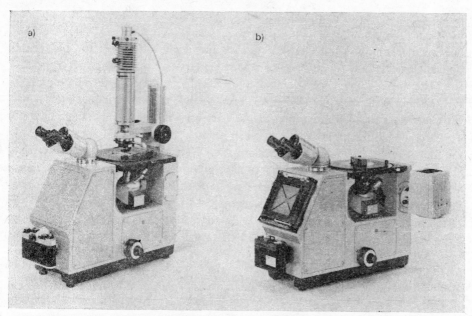

Fig. 2.75. Inverted microscope for both reflected and transmitted light available from C. Zeiss Oberkochen: a) model IM35, b) model ICM405 (with built-in camera).

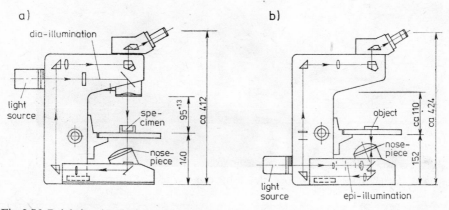

Fig. 2.76. Reinheimer's principle of the Leitz inverted microscope: a) transmitted-light version—Labovert, b) reflected-light version—Metallovert [2.43].

in the secondary exit pupil plane without any changeover necessary for the objectives and illuminating system. Moreover, scales, reticules, stops for central dark-field technique or dispersion staining (see Vol. 2, Chapter 6), and other optical elements may be inserted into the secondary exit pupil plane of the objective. The microscopes under discussion are usually equipped with the Nomarski DIC system. With the relay system it is now possible to change immediately from phase contrast to differential interference contrast without loss of time.

Large metallurgical microscopes, such as that shown in Fig. 2.77, belong to the fourth group of reflected-light microscopes. Their basic unit is also inverted (Le Chatelier system), but the light source, photographic camera or projection screen, and other facilities are arranged on an optical bench or very rigid support. Since these instruments are normally equipped for photomicrography, they are often called *metallographs*. The best known metallographs are: MeF-2 (C. Reichert Wien), Neophot 2 or Neophot 2I (VEB Carl Zeiss Jena), MM-5[25] or MM-6

Fig. 2.77. Metallographic microscope MeF-2 (by courtesy of C. Reichert Wien).

[25] A special version of this metallograph—the model MM5RT—is devoted for microscopy in radioactive contaminated hot cells.

(E. Leitz Wetzlar), Unimet or Versamet (Union), and Epiphot (Nikon).[26] Their massive construction makes them suitable for use with different accessories such as a stage furnace, which permits observation of specimens at high and even very high temperature. The MeF-2 metallograph, for instance, is equipped with a vacuum furnace (Vacutherm 2) capable of heating to 1800°C. A furnace of this kind requires special long working distance objectives (see footnote referring to Epiphot and Subsection 2.4.4).

Metallurgical microscopy is a very wide field and the interested reader may find more details on it in Refs. [2.30] and [2.44]–[2.49].

2.4.4. Catoptric and catadioptric systems

So far only *dioptric* (refractive) image-forming and illumination systems have been discussed while *catoptric* (reflecting) systems were referred to in only a few passing remarks. This omission was deliberate since dioptric optics predominates in microscopy. The time has come now, however, to say a few words about curved mirrors and their combinations which are in use as special microscope objectives.

Imaging properties of curved mirrors. Mirrors may be treated as lenses if the refractive index ratio n_L/n_M is taken to be -1 (n_L is the index of refraction of the lens material, n_M that of the surrounding medium). This substitution is valid for all the lens equations discussed previously (see Subsection 2.2.3). Mirror equations may be derived from them if it is assumed that a lens of zero thickness is reduced to a single refracting surface. In particular, substituting $n_L = -n_M$, $t = 0$, $r_2 = \infty$, and $r_1 = r$ in Eq. (2.8a) gives

$$f' = \tfrac{1}{2}r, \tag{2.96}$$

where f' is the focal length of the mirror and r is its radius of curvature (Fig. 2.78). These quantities are positive or negative according to the sign convention defined previously for lenses. Next, substituting $f' = r/2$ in Eqs. (2.12b) and (2.13b) yields the catoptric imaging equations in the form

$$\frac{1}{a'} - \frac{1}{a} = \frac{2}{r}, \tag{2.97}$$

---

[26] This instrument is equipped with metallurgical CF optics. Four of CF objectives are LWD systems. Two bright-field objectives—CF ELWD planachromats 20×/0.40 and 40×/0.50—have working distances of 10.5 mm and 10.1 mm, respectively. The other two objectives—the CF BD ELWD planachromates 20×/0.40 and 40×/0.50—have working distances of 8.5 mm and 9.8 mm, respectively, and can be used both for bright-field and dark-field microscopy.

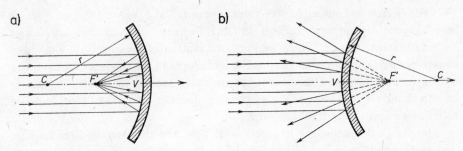

Fig. 2.78. Spherical mirrors: a) concave, b) convex (*F'*—focal point, *C*—centre of curvature, *V*—vertex of mirror).

and

$$bb' = -\frac{4}{r^2},\tag{2.98}$$

where a, b and a', b' are, respectively, the object and image distances as marked in Fig. 2.79.

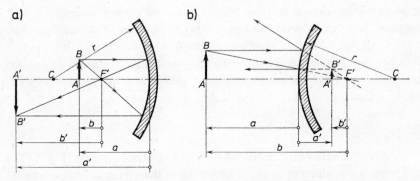

Fig. 2.79. Geometrical construction of the image produced by a concave mirror (a) and convex mirror (b).

In general, concave and convex mirrors act, respectively, as positive and negative lenses. Their reflecting surface can be *spherical* or *aspherical*. Aspherical mirrors can have *ellipsoidal*, *paraboloidal*, or *hyperboloidal* surfaces. Single aspherical mirrors are mainly used in some special illuminating systems, for dark-field microscopy, for instance, or epi-illumination [2.50]. Catoptric image-forming systems (objectives), on the other hand, make use of spherical mirrors. Plane mirrors are, however, optical elements universally employed in microscopical systems. The plane mirror is, in fact, a unique optical element which is free from aberrations, while all other mirrors, like lenses, suffer from spherical aberration, coma, astigmatism, curvature of field, and distortion, but are free from chromatic aberration. This lack of chromatism is a very important feature because entirely achromatic objectives may be constructed from mirrors. Moreover, in some specific situations curved mirrors are also free from spherical aberration. An ellipsoidal mirror, for instance, is free from spherical aberration for foci of the ellipsoid. The same is true of an hyperboloidal mirror. When one of the foci is at infinity, the ellipsoid becomes a paraboloid which is free from spherical aberration for a point at the focus and its conjugate at infinity. Similarly, there is no spherical aberration for the curvature centre of the spherical mirror.

Another advantage of mirrors is the fact that they do not absorb ultraviolet and infrared light as do glass lenses. Hence catoptric optical systems would appear to be indispensable for UV- and IR-microscopy (see Vol. 2, Chapter 10).

Catoptric objectives. For microscopy these objectives have two basic advantages. The first is that they are free from chromatic aberration and from absorption of ultraviolet and infrared light. This property was the basic reason for developing *catoptric (reflecting) objectives* for UV- and IR-microscopy in the first place but also for microspectrophotometry. The second advantage is that a catoptric objective may be designed to have a much longer working distance than a *dioptric (refracting) objective* of equivalent power and numerical aperture. This latter advantage was the main reason for developing such objectives for hotstage microscopy.

There are two principal types of catoptric objectives: the Schwarzschild type (Fig. 2.80a) and the Cassegrain or Newton type (Fig. 2.80b). Both types consist of two spherical mirrors, one (M_1) being concave and the other (M_2) convex. The concave mirror is much larger in diameter and has a smaller curvature then the convex mirror. In the first (a) the rays emerging from the object O strikes first the concave mirror M_1 and then the convex mirror M_2, whereas in the second type (b) the converse is true (the order of mirrors is reversed). In both systems the smaller central mirror M_2 partly cuts-off the rays from the object and vignettes the exit pupil of the objective; hence the image O' is formed by a centrally occluded light cone. This occlusion (vignetting) is, of course, a defect. It is defined by the factor

$$w = \frac{n \sin \sigma_1}{n \sin \sigma_2} = \frac{A_1}{A_2}, \tag{2.99}$$

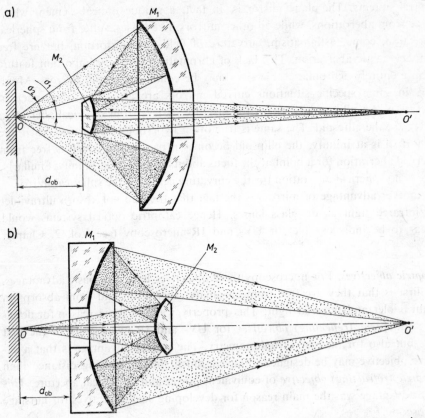

Fig. 2.80. Main types of reflecting objectives: a) Schwarzschield objective, b) Cassegrain objective.

where n is the refractive index of the medium between the object under examination and the objective, while σ_1 and σ_2 are the minimum and maximum object-side aperture angles, which determine the internal and external solid angles of the cone of rays passing through the objective; consequently $A_1 = n\sin\sigma_1$ and $A_2 = n\sin\sigma_2$ are said to be the *interal* and *external numerical apertures*. The latter is simply regarded as the numerical aperture of a catoptric objective. The *occlusion factor w* is higher in Cassegrain-type objectives, and for this reason Schwarzschild-type objectives are preferred in practice. It was stated earlier that these objectives may be free from both spherical aberration and coma if their numerical aperture A_2 is smaller than 0.50 and both mirrors, M_1 and M_2, have a common centre of curvature. For such a concentric arrangement of mirrors, the occlusion factor w cannot be smaller than 45%. It can be diminished by a non-concentric arrangement of mirrors, but in this case coma and astigmatism will appear and the field of view free from aberrations will be reduced. Schwarzschild-type objectives of magnifications $15\times$, $36\times$, $52\times$, and $74\times$ are commercially available from the Beck–Ealing company [2.51]. Their numerical apertures A_2 and working distances d_{ob} are, respectively, as follows: $A_2 = 0.28$, 0.50, 0.65, and 0.65; $d_{ob} = 24.0$ mm, 8.0 mm, 3.5 mm, and 2.0 mm.

Catadioptric obejctives. These objectives comprise one or two mirrors and a number of lenses. Such an arrangement makes it possible to achieve better correction of aberration, especially of coma and astigmatism for higher numerical apertures A_2, as well as lower occlusion factors w. In catadioptric objectives the factor w is roughly equal to 30% and can sometimes be reduced to as little as 10%. Moreover, the working distance of these objectives may be much greater than that of truly catoptric or dioptric objectives.

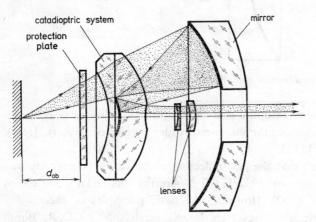

Fig. 2.81. Nomarski catadioptric objective.

Figure 2.81 shows a relatively simple catadioptric objective designed by Nomarski and manufactured by C. Reichert Wien as one of the basic components of a hotstage outfit for metallurgical microscopy. The principal parameters of this objective are as follows: magnifying power $M_{ob} = 40\times$, numerical aperture $A_2 = 0.52$, working distance $d_{ob} = 8$ mm.

Another, more complicated catadioptric objective is shown in Fig. 2.82. It is commercially available from VEB Carl Zeiss Jena. Its basic parameters are: $M_{ob} = 40\times$, $A_2 = 0.50$, $d_{ob} = 18.8$ mm, and $f' = 6.25$ mm. The laser micro-

spectral analyser LMA 10 is equipped with this objective. Apart from this, a
series of planapochromatic catadioptric objectives with standard working distances and magnifications from $16\times$ to $125\times$ is also manufactured by VEB
Carl Zeiss Jena. These objectives are apochromatically corrected within a very
large spectral range (from 250 to 1000 nm) and are eminently suitable for UV-
and VIS-microspectrophotometry. Their lens components are made of material
transparent in the spectral range mentioned above (quartz glass, fluorite).

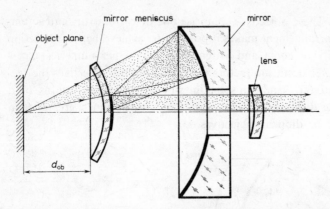

Fig. 2.82. Planapochromatic catadioptric objective (C. Zeiss Jena).

A large number of other catoptric and catadioptric objectives has been described in the relevant literature; the interested reader is referred to Refs. [1.31],
[2.8], [2.28], and [2.51]–[2.61].

It is worth noting here that the first reflecting objective for microscopy was
designed by D. D. Maksutov in 1932 [2.62]. Following this, a large variety of
mirror systems appeared [2.52]. However, the basic principles of present-day
catoptric and catadioptric objectives were largely established by C. R. Burch
(1947), D. S. Grey (1949), K. P. Norris (1951), H. Riesenberg (1959–1965),[27]
A. P. Grammatin, and V. A. Panov [2.55–2.58].

2.4.5. *Universal microscopes*

Twenty five years ago a new concept in microscope design was introduced by
E. Leitz Wetzlar. Its novelty depends on the use of basic rectilinear shapes for
the stand and component modules from which a conceptionally uniform series

---

[27] See books [2.28] and [2.54] for references.

of different instruments for more and less advanced microscopy may be easily assembled by the manufacturer or by the user himself. At one end of this series there is usually a large universal microscope, whereas at the other end there are student or even teaching microscopes.

What criteria should be considered essential to the concept of the universal microscope? The answer cannot be unanimous, but roughly speaking the following requirements must be taken into consideration:

(1) It must be capable of operating both in transmitted and reflected light with little or no changeover necessary and should have built-in dia- and epi-illuminators[28] forming a functionally integrated system of illumination.

(2) It should have unlimited versatility. In addition to standard equipment for bright-field microscopy in dia- and epi-illumination, accessories for dark-field method, phase contrast, differential interference contrast, and for qualitative observation in polarized light should be integral parts of the instrument. Switching from one microscopical method to another must be easy, fast, and with minimum effort.

(3) The stand must be adapted for optional equipment for special research in both transmitted and reflected light, such as accessories for microphotometry, microinterferometry, and fluorescence microscopy.

(4) Advanced photomicrographic equipment (automatic or semi-automatic) must be physically positioned in or on the microscope. Accessories for rear projection, CCTV and other facilities for image presentation must also be attachable to the microscope body.

(5) Perfect Köhler illumination for all objective magnifications without exchanging condensors is indispensable.

(6) The overal design should be ergonomically correct irrespective of the microscopical method being used. All controls must be accessible to the user in his normal working position, i.e., seated at the microscope.

At present almost all famous manufacturers of microscopes offer large research instruments which satisfy the requirements mentioned above. The Leitz Orthoplan (Fig. 2.67) above all has been synonymous with modern universal microscopy during the last two decades. This microscope was the first to offer a wide flat field (28 mm) to biologists, chemists, and other researchers. Present-day microscopes manufactured by Leitz, as part of a range conceptionally uniform with the Orthoplan, are Ortholux 2 (laboratory and research microscope), Dialux

---

[28] The terms "dia" and "epi" are frequently used as abbreviations for "transmitted-light" and "reflected-light".

(laboratory microscope), SM-Lux (teaching and routine microscope), and HM-Lux (student and teaching microscope) [2.16], [2.63]–[2.65].

The Vickers Photoplan appeared next on the market (Fig. 2.83). It is similar in configuration to the Orthoplan, but has a smaller field of view (25 mm). Today, the design concept and research possibilities incorporated in it are being transferred to the Vickers M17 series, which is very similar to the Leitz Dialux series.

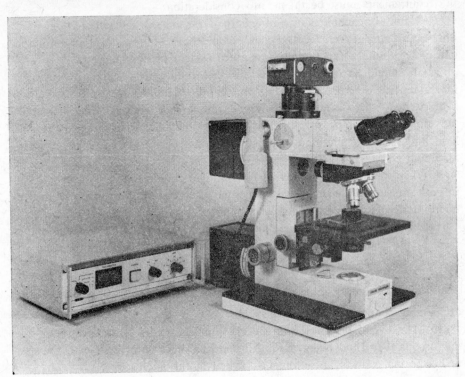

Fig. 2.83. Vickers M41 Photoplan microscope (by courtesy of Vickers).

The Nikon Apophot, Olympus Vanox, and the MBI-15 universal microscope from LOMO (Leningrad) became commercially available at almost the same time. The first of these instruments (Apophot) is no longer manufactured, but its research possibilities have been incorporated in less expensive microscopes of the Nikon Microphot V and X/Y series. The best known instruments in the Microphot range are Biophot, Labophot, and Optiphot (Fig. 2.73). They have many valuable features needed for critical observation and photomicrography; above all these microscopes are equipped with chromatic aberration-free objec-

tives (see Subsection 2.2.4). The second Japanese universal microscope, the
Olympus Vanox,[29] is also being largely replaced by less expensive models of
the BHS/BHT system, which includes a new generation of microscopes available
today from Olympus. The BHS/BHT system incorporates a new condenser for
ultra-low objective magnifications ($1\times$ to $4\times$) and ultra-low power objectives,
i.e., $1\times$ and $2\times$ magnifications, which are parfocal with standard objectives of
low-, middle- and high-power. Moreover, the BHS microscope, the most advanced model in the BH-2 series, incorporates an important feature—namely compensation for changes in supply voltage. Slight voltage variations can cause
fluctuations in illumination brightness which are generally disturbing and unpleas-
ant to the microscopist. The new Olympus supply system utilizes sophisticated
transistor technology to maintain the same level of brightness despite variations

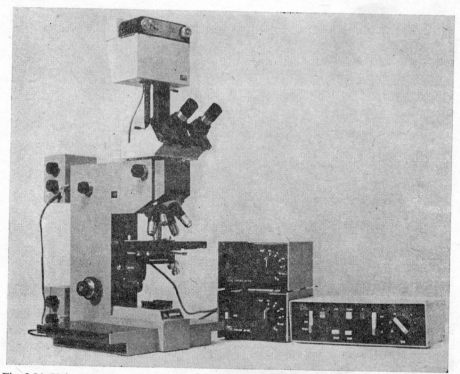

Fig. 2.84. Universal microscope Unimar (by courtesy of PZO, Warsaw).

[29] The new Vanox-S (1984) has microprocessor controls over all the main settings of the
microscope, including autofocus.

in line voltage. The desired voltage is selected by the microscopist and the selected level is maintained automatically. A unique LED voltage readout is also installed in the microscope base.

Figure 2.84 shows a universal research microscope designed in the Central Optical Laboratory, Warsaw, and developed by the Polish Optical Works for commercial production. Compared to its predecessors this microscope exhibits some features making it particularly suitable for microinterferometry (by utilizing a versatile double-refracting interference system (see Vol. 3, Chapter 16)). It is part of a range conceptionally uniform with microscopes of the Biolar and Studar series available from the Polish Optical Works.

Fig. 2.85. The Reicher UnivaR (by courtesy of C. Reichert Wien).

A conceptionally original universal microscope, the UnivaR (Fig. 2.85), was shown to the public for the first time in 1973 by C. Reichert Wien [2.66]. This large research microscope incorporates a number of technical innovations which combine the highest level of optical-mechanical engineering with advanced automation. Above all, the ray beams from the light source pass through a zoom illuminating system, which automatically optimizes the size of the light source image for dia-illumination with respect to the diameter of the objective exit pupils (Fig. 2.86). A servomotor is electronically coupled to the objective nose-

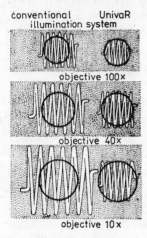

Fig. 2.86. Optimization of the size of the light source image observed in the exit pupil of microscope objectives.

piece and adjusts the zoom illuminating system to the correct position when objectives are changed.[30] The UnivaR is equipped with infinity corrected planachromatic or planapochromatic objectives and a relay system, which reimages the exit pupils of objectives in a secondary, easily accessible plane. Like the Leitz Orthoplan, it comprises widefield oculars, whose field number is 28 mm (these are only oculars of 6.3 × magnification, while those of higher magnification have a smaller field number). The relay system means that in any basic change of technique, e.g., from bright field to phase contrast, the objective can remain the same because phase rings are mounted on a turret inserted in the secondary pupil image plane of the relay system. A magnification changer is also added to the imaging system. This permits the change of final magnification by steps of fac-

[30] A nearly identical zoom illuminating system was formerly used in the Nikon Apophot.

tors $1\times$, $1.6\times$, and $2.5\times$ or continuously (zoom system) between $1\times$ and $2.5\times$. Moreover, the UnivaR incorporates some very useful features for photomicrography [2.66]. The design concept of this microscope has also been utilized in the construction of less expensive microscopes of the Polyvar series, which, together with the Neovar and Metavar laboratory microscopes, constitutes the most representative generation of instruments available today from C. Reichert Wien (or from Reichert–Jung). The Polyvar family consists of seven microscopes—each specifically designed for maximum performance in their own individual applications, yet all built round the same basic design. The Polyvar range is as follows: Polyvar (biological applications), Polyvar Met (metallurgical applications), Polyvar Met 66 (electronics applications), Polyvar IR (infrared microscope for investigations of semiconductors), Polyvar Pol (polarizing microscope), Polyvar PU (hot box microscopy), and Polyvar Telematic which incorporates an automated nosepiece—available in versions of the applications mentioned above.

The mechanical stability of the microscope presents a serious problem for both manufacturer and user. The majority of microscopes are designed in such a way that the condenser, stage, objectives, and tube are attached, by means of appropriate holders, to one side of the stand. All forces which affect the holders act as one-arm levers and greatly reduce the sharpness stability of the microscopical image. An effective solution to this probem has been achieved by Michel [2.67] with the Axiomat microscope available from Carl Zeiss Oberkochen (Fig. 2.87). The basic features of this instrument are incorporated in many (a dozen and so) square blocks (modules) which, depending on the desired function, can be assembled (screwed) together in various combinations. The result is a number of upright, inverted, and horizontal microscopes for transmitted and reflected light. Their outstanding feature is that the optical axis is coincident with the axis of symmetry of the modules. This design principle is not quite new because it was, of course, intrinsic to early microscopes and has today been successfully revived. The Axiomat has plano-objectives of magnification from $1.25\times$ to $100\times$, which are infinity corrected and its maximum image field is equal to 31 mm. Like the UnivaR, it contains a zoom illumination system, which optimizes the specimen illumination, and a relay system for the intermediate imaging of the exit pupils of objectives. Moreover, a zoom magnification changer, the Optovar zoom system, can be used to vary continuously the final magnification by factors from $0.8\times$ to $3.2\times$. There are two easily accessible real intermediate image planes, one before and the other after the Optovar zoom system, in which every possible measuring scale or graticule may be introduced. With the Optovar system magnification can be adjusted so that one interval of the scale corresponds exactly to one measuring unit (or a multiple of it) in the object plane. Focusing is done by moving the nosepiece and the infinity corrected objectives screwed to it. This may be done conventionally (by hand) and/or electronically by means of a remote control system coupled to the objective nosepiece. With its cubic

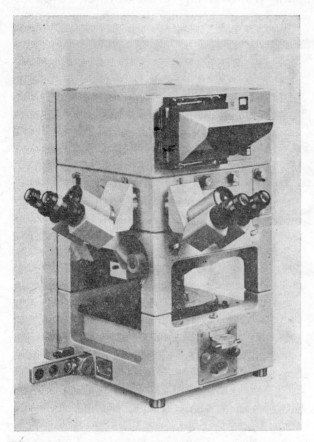

Fig. 2.87. Photo showing one of several possible arrangements of the Axiomat (by courtesy of C. Zeiss Oberkochen).

form the Axiomat is extremely suitable for use with space-saving attachments such as monochromators, TV cameras, photometers, etc. It offers bright-field and dark-field microscopy, phase contrast, differential interference contrast (Nomarski system), fluorescence, polarization, multiple-beam interference, object scanning, photometry, spectrofluorometry, TV image analysis, UV- and IR-microscopy. At the present time, the Axiomat represents the most universal system for advanced microscopy. After very careful examination McCrone [2.68] stated that its only disadvantage was its price.

Another solution to the stability problem was proposed by Nomarski and implemented by Nachet–Sopelem (Paris) in the NS800 universal microscope illustrated in Fig. 2.88. As can be seen, in appearance this instrument resembles

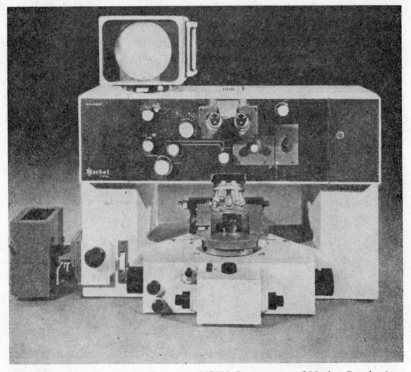

Fig. 2.88. Universal research microscope NS800 (by courtesy of Nachet-Sopelem).

a bridge. Like the Axiomat it contains a zoom illuminating system, which optimizes specimen illumination, a zoom viewing system for changing final magnification continuously by factors $0.8 \times$ to $2 \times$, and a relay system, which transfers the exit pupils of objectives to an easily accessible plane. A very important feature of this microscope is that the specimen stage has the greatest mechanical stability because it is mounted directly on a massive base. Focusing is done by moving the nosepiece, to which infinity corrected planachromatic or semiplanapochromatic objectives are attached. This focusing mechanizm does not differs from that of the UnivaR and Axiomat, the focusing process is, however, divided into three steps. Two of them are conventional (coarse and fine focusing), but the third step consists of an ultrafine adjustment of image sharpness within the depth of focus equal to 2 μm. This ultra-fine focusing is particularly suitable for the examination of transparent specimens when they are photographed step by step at slightly different depths (optical sectioning). The NS800 microscope offers bright-field (dia and epi), dark-field (dia and epi), phase contrast (dia), Nomarski differential interference contrast (dia and epi), microinterferometry (epi), fluorescence (dia and epi), polarization (dia and epi). Unfortunately, this instrument is not commercially produced (its initial price was very high) and exists only as a prototype (made in triplicate). A number of its research potentialities have been incorporated in the Nachet–Sopelem NS400 universal microscope (today also distributed by Vickers). The overall design of the latter is, however, traditional.

For the sake of completeness, we should also mention the universal NU2 research microscope manufactured by VEB Carl Zeiss Jena, as well as the Universal Microscope and Photomicroscope III available from Carl Zeiss Oberkochen. In their construction and individual features these three instruments are traditional and do not make any new contributions to the problem under discussion. In any case, the NU2 microscope and the whole Mikroval series from VEB Carl Zeiss Jena are described in detail in Ref. [2.28].

Somewhat more space should, however, be devoted to the Jenaval transmitted-light research microscope (Fig. 2.89), which belongs to the new (1980s) generation of 250-CF Jena microscopes [2.69]. This is equipped with wide-field optics (field number 25 or 32 mm) free from lateral chromatism and corrected for infinite tube length. A set of CF objectives also contains a $1 \times /0.03$ low-power wide-field planachromat that is parfocal with objectives of all other magnifications. An achromatic-aplanatic condenser unit includes a macro condenser that can be flipped into the path of rays for illuminating a 32-mm object field according to the Köhler principle. A $0.8/1/1.25 \times$ magnification changer allows the 32-mm intermediate image size to be utilized fully. A so-called contrast tube incorporates a relay loop optical system borrowed from the Peraval interference microscope; thus various contrasting techniques can be undertaken with normal bright-field objectives, whose exit pupils are reimaged in a secondary, easily accessible plane. Here there is a revolving barrel which houses the various modulators required (phase rings, diaphragms, Wollaston prisms). The modulators, of course, correlate to other elements (annular diaphragms, Wollaston compensators) on the illumination side of the optical system. The contrast tube permits

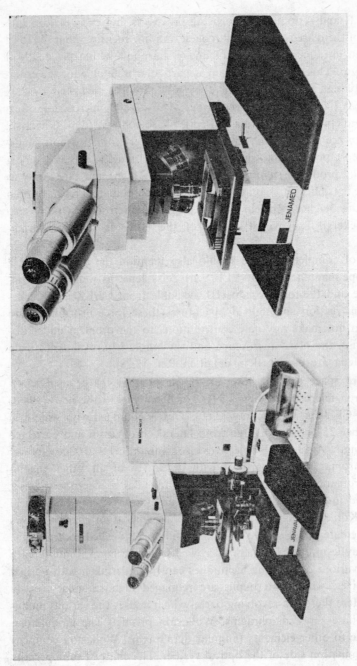

Fig. 2.89. Generation of 250-CF microscopes (VEB Carl Zeiss Jena): Jenaval and Jenamed.

observation in wide fields, up to field number 32, by positive and negative phase contrast, differential interference contrast (Normarski DIC system), central dark-field, and qualitative polarization. The stand available in a transmitted-light version provides a basis for a variety of sophisticated research techniques including reflected-light, fluorescence, microphotometry, interferometry, and image analysis techniques.

The Jenaval microscope is the most universal instrument within the generation of 250-CF microscopes available from VEB Carl Zeiss Jena. A simpler series of this generation is the Jenamed series (Fig. 2.89), which contains several instruments such as Jenamed-cytology, Jenamed-histology, Jenamed-hematology, and Jenamed-variant, all designed mainly for routine medical and biological microscopy. The Jenamed-cytology microscope, following recent trends, is equipped with a set of objectives for specimens (cell smears) covered by a spray coat instead of a cover slip, whereas the Jenamed-histology model, a standard microscope for pathologists, has cover-slip corrected objectives including the $1 \times$ low-power planachromat mentioned previously. This latter objective allows a histological section to be surveyed as a whole and details of interest to be marked by means of an object marker fitted to the nosepiece. Moreover, a special high-power dry objective $(100 \times /0.90)$ is included to the Jenamed-hematology microscope; hence troublesome immersion procedure is avoided in the study of blood smears covered by the spray coat mentioned above. The most versatile instrument within the Jenamed series is the Jenamed-variant microscope. Its modular configuration is suitable for all basic techniques of qualitative microscopy: bright-field, dark-field, phase contrast, qualitative polarization, Nomarski DIC method, and incident-light fluorescence.

A very useful unit of the Jenamed microscope is its specimen cassette system. Its principle is borrowed from the inverted microscope; the specimen slide is positioned so that the object under examination is always in the focal plane of the objective. This system means that coarse focusing control is no longer required for varying specimen slide thickness; fine focusing control, which acts on the nosepiece only, is quite sufficient for the relatively small focusing intervals resulting from possible differences in specimen thickness as well as from unavoidable tolerances of the parfocal lengths of objectives. The specimen cassette can be slid over the microscope stage in three ways: by free-hand guiding, manual movement in $x-y$ directions by control knobs attached to a mechanical stage, and motorized movement at discrete intervals or according to an electronically programmable meander course.

The advantages mentioned above show that the new generation of 250-CF

Jena microscopes represents a big step forward in instrument design for modern light microscopy.

Quite recently (1986), C. Zeiss Oberkochen presented new research microscopes—the Pyramids series—with ICS optics (Infinity Colour-Corrected System[31]) and SI (System Integration) design [2.70]. The infinity corrected objectives and SI architecture permit insertion of all illuminating and contrast-enhancing techniques (bright-field and dark-field microscopy, phase contrast, DIC, fluorescence and qualitative polarized-light microscopy) into the microscope housing without impairing the integrity of the total system. The pyramid shape of the microscopes and low centre of gravity of the stand lead to high mechanical and thermal stability. The ICS optics comprises Plan-Neofluar and plan-apochromat objectives of parfocal length of 45 mm. Their standard magnifying powers are $1.25\times$, $2.5\times$, $5\times$, $10\times$, $20\times$, $50\times$, and $100\times$. The optical system maintains a field-of-view number of 25 and a constant magnification factor $q = 1\times$, regardless of techniques used.

At the present time, the Pyramids series consists of several microscopes. Axioplan (a universal research instrument), Axiophot (a photomicroscope with a microprocessor-controlled camera system), Axiotron (a specialized instrument for semiconductor industry), and Axioskop are the most representative types. The Axioskop is a more-than-routine instrument which bridges the gap between routine and research work. However, its field-of-view number is equal to 20. In addition to the Plan-Neofluar and planapochromat objectives, C. Zeiss Oberkochen also offers achrostigmat objectives for the Axioskop family. These achromatic objectives are free from astigmatism and also offer satisfactory colour correction, but are cheaper than the Plan-Neofluars or planapochromats.

In the preceding text only microscopes designed by famous optical firms have been taken into consideration. It is, however, important to note that some sophisticated instruments have also been developed in various research laboratories. Frequently, biologists and other researchers need highly unconventional microscopes that are not intended for serial production. A versatile apparatus built by Malý and Veselý (Czechoslovak Academy of Sciences) for the study of living cells [2.71] is a good example of instruments designed by the users themselves.

The microscopes described in this section give some insight into the actual state of modern design of microscopes for general use. Some further remarks regarding the state-of-the-art and the future of light microscopy may be found in two interesting articles by Haselmann [2.72].

---

[31] Colour-corrected system = chromatic-aberration-free optics (CF).

2.4.6. Requirements of ergonomics

The light microscope has become a well-established instrument used for routine work in research, clinical examination, materials sciences, and industrial quality control by large numbers of researchers and technicians. Progress in instrument design therefore results not only from the experience and initiative of manufacturers but also from the consideration of user requirements. Among other factors, problems of ergonomics and operating convenience are of primary importance.

One of the trends in light microscopy today is wide-field optics, which offers to the microscopist a large object field and a plane image of excellent quality over the entire field of view. Optical and physiological studies have revealed that a diameter of 250 mm is the optimum size of the apparent field of view for visual microscopy [2.69]. For this reason, a $10 \times$ wide-field ocular, whose field number is 25, provides just that apparent field of view (10×25 mm $= 250$ mm). This requirement is entirely fulfilled by the new generation of 250-CF Jena microscopes, whose factory symbol "250-CF" stands for the diameter of the apparent field of view (250 mm) and optics free from both longitudinal and lateral chromatic aberrations (CF).

The maximum object field required today by histologists and some other microscopists is, however, 32 mm. This value is obtained by a 0.8 demagnification factor, which is specially introduced by means of a magnification changer as in the Jenamed microscopes (0.8/1/1.25). Thus a neat compromise is achieved between these two parameters (either 250 mm apparent field of view or 32 mm maximum object field since $25/0.8 \approx 32$ mm for $1 \times$ objective). However, it should be pointed out that for the demagnification factor $q = 0.8$, the field number (FN) becomes equal to 32. Conversely, if q is greater than 1, the field number falls below 25, e.g., $FN = 20$ for $q = 1.25$. In addition, the exit pupil of the microscope (Ramsden disc) must be conveniently situated to permit fatigue-free observation (see paragraph "Eye relief" in Subsection 2.3.3).

All these conditions must be considered part of the basic optical parameters of a modern microscope. There are also several basic requirements regarding the mechanical construction. It has always been assumed, for instance, that for convenience of focusing, the coarse and fine focusing controls should be placed in the vicinity of the microscope base. Another very important mechanical requirement is the overall height of the microscope, i.e., the distance between the base and the Ramsden disc (see Fig. 2.76). In principle, this parameter cannot be greater than 450 mm.

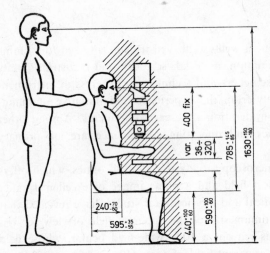

Fig. 2.90. Installation of a microscope according to requirements of ergonomics [2.73].

The requirements mentioned above are the province of the manufacturer. However, there are some details which the user must attend to himself. First of all, the microscope should be situated on a convenient table suitably adapted to the user's shape and height (Fig. 2.90). This is of the utmost importance for anyone who spends much time doing microscopical observations.

For more detailed information regarding the installation, use and conservation of common microscopes, the reader can also consult other books, e.g., [2.12], [2.48], [2.74]–[2.92].

* * *

The problems discussed in this chapter are part of general microscopy and fall within the scope of geometrical optics and of modern instrumentation. Some basic properties of microscopical imaging such as resolution, perception limit, depth of field, or contrast transfer have been omitted because they can be explained satisfactorily only by referring to wave optics or Fourier optics, which is dealt with in the following chapter.

3. Image Formation in the Microscope within the Scope of Diffraction Theory and of Fourier Optics

In the preceding chapter the microscopical image of objects was described according to laws of geometrical optics. Such a description and its interpretation can only be treated as an approximation since light is a wave motion and cannot exactly be considered solely in terms of rays. In order to estimate the distribution of light in the microscopic images and the resolving power of a microscope, the laws of physical optics, especially phenomena of diffraction and interference, must also be taken into account.

3.1. Diffraction image of a luminous point

Separate luminous points and their images play a fundamental role in the theory and practice of light microscopy. Theoretically, the point light source is a dimensionless object described by the *Dirac delta function* or *impulse function*, but in practice its size may be assumed to be smaller than the limit of the resolving power of the microscope objective. Because of diffraction and aberration, the microscopical image of the luminous point is never a punctual spot, but manifests itself as a diffraction pattern whose intensity distribution is described in terms of the *point spread function*. If an optical system is perfect or suffers from very small aberrations, the image of the luminous point is known as the *Airy pattern*.

3.1.1. Ideal Airy pattern

Let us consider an axial luminous point P (Fig. 3.1), coincident with the object plane Π of an aberration-free objective Ob. This point emanates monochromatic light with a spherical wavefront Σ_p, whose centre of curvature is at P. The objective Ob transforms Σ_p into another spherical wavefront Σ, whose centre of curvature coincides with the geometrical image P' of P. However, the image-forming wavefront Σ is perturbed by the finite aperture of the objective, which diffracts

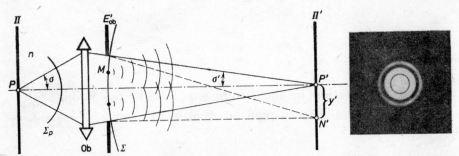

Fig. 3.1. Illustrating the formation of Airy pattern.   Fig. 3.2. Airy pattern.

the light like a baffle opening (compare Fig. 1.84b). According to the Huygens principle, each point M of the objective exit pupil E'_{ob} can be treated as a new source of spherical wavelets, which mutually interfere, and produce a Fraunhofer diffraction pattern of the luminous point P in the image plane Π' of the microscope objective Ob. A basic feature of this pattern is that the luminous energy is not only concentrated in the geometrical image P' of P, but is also spread beyond P'. The light intensity I' in any point N' of the image plane Π' depends primarily on phase differences of the wavelets, which come together to N'. All these wavelets are mutually coherent because they originate from a single primary source, and can thus interfere constructively or destructively. The result is a diffraction–interference pattern, known in this case as the Airy pattern.

If an aberration-free objective with a circular exit pupil is correctly mounted and exactly focused, then the Airy pattern has a symmetrically circular form, as shown in Fig. 3.2. This is an ideal diffraction image of a single luminous point; it consists of a central bright patch and several concentric dark and bright rings. The normalized distribution of light intensity I' across this image is expressed by

$$I' = \left[\frac{2J_1(Y)}{Y} \right]^2, \tag{3.1}$$

where $J_1(Y)$ denotes the first-order Bessel function of the first kind, and Y is a parameter known as the *Airy optical unit* and defined by

$$Y = \frac{2\pi}{\lambda} n \sin \sigma \frac{y'}{M_{ob}}, \tag{3.2}$$

where λ is the light wavelength, n the refractive index of the space between the object point P and the objective Ob (Fig. 3.1), σ the object-side aperture angle, M_{ob} the magnification of the objective, y' the distance from the centre of the Airy pattern.

A plot of Eq. (3.1) and the distribution of light amplitude, i.e., A' = $2J_1(Y)/Y$, are shown in Fig. 3.3. The central region enclosed by the first minimum is called the *Airy disc*. This contains 84% of the total luminous energy belonging to the whole Airy pattern. The secondary maxima of intensity decreases as Y increases and contain only 16% of the luminous energy of the Airy pattern.

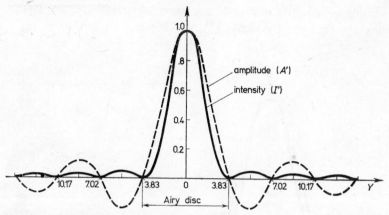

Fig. 3.3. Light distribution in the Airy pattern (the distribution such as shown here is normalized to unity at the central maximum, i.e., $A' = 1$ and $I' = 1$ for $Y = 0$).

This energy portion is, however, high as regards the formation of the image of an object consisting of multiple luminous points because the bright rings of the Airy pattern produce only a light background which diminishes the contrast of the actual image. In other words, only the Airy disc, i.e., the central circle of the Airy pattern, is a useful presentation of the luminous object point, whereas the rings which surround this circle are injurious.

The positions of the black rings of the Airy pattern correspond to $Y = 3.83$, 7.02, 10.17, ... and those of the bright rings to $Y = 5.14$, 8.46, 11.62, ..., whereas the central intensity peak occurs for $Y = 0$ (Fig. 3.3). Since the first dark ring corresponds to $Y = 3.83$, it follows from Eq. (3.2) that in the image plane Π' the radius $y_1' = r'_{\text{Airy}}$ of the Airy disc is given by

$$r'_{\text{Airy}} = \frac{1.22\lambda}{2n\sin\sigma} M_{\text{ob}}. \tag{3.3}$$

As can be seen, for a given objective magnification M_{ob} the radius r'_{Airy} of the Airy disc is the smaller the larger is the numerical aperture $n\sin\sigma$ of objective and the smaller is the wavelength λ. Assuming $\lambda = 0$ yields $r'_{\text{Airy}} = 0$, and this is in fact an approach to geometrical optics. In practice, this situation is, however, unrealisable and in any case a luminous point is viewed as a small patch of radius

$$r_{\text{Airy}} = \frac{r'_{\text{Airy}}}{M_{\text{ob}}} = \frac{1.22\lambda}{2n\sin\sigma}.$$

(3.4)

Taking for example an objective of magnifying power/numerical aperture $100 \times /$ 1.30 and putting $\lambda = 0.55$ μm, the radius $r_{\text{Airy}} = 0.26$ μm.

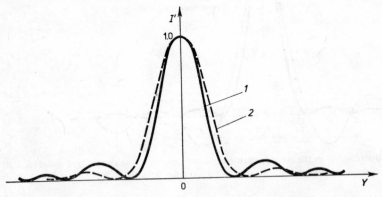

Fig. 3.4. Effect of the apodization on intensity distribution in the Airy pattern: curve *1*—unapodized objective, curve *2*—apodized objective.

In general, the higher the luminance of the central circle of the Airy pattern and the weaker its bright rings the better the microscope objective. The Airy pattern can be changed if the distribution of the amplitude and/or phase of light are modified in the objective exit pupil. In particular, the intensity of the bright rings may be lowered (Fig. 3.4) if the exit pupil transmittance is reduced gradually (e.g., exponentially) from the centre to the edge of the exit pupil. This procedure is known as *apodization* and can be used for improving the quality of an optical system [3.1, 3.2]. On the other hand, the intensity of the bright rings increases if the central part of the exit pupil is occluded (Fig. 3.5). This occurs in the catoptric or catadioptric objectives described earlier in Subsection 2.4.4.

An ideal Airy pattern appears only when an aberration-free objective is correctly focused at the luminous point. Any small defocusing causes rapid changes in this pattern (Figs. 3.6 and 3.7).

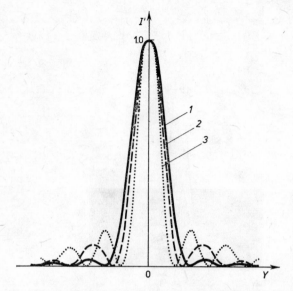

Fig. 3.5. Effect of the central occlusion on the intensity distribution in the Airy pattern: curve *1*—normal objective, curves *2* and *3*—central area of the exit pupil of objective is occluded by a circular opaque screen of radius $r_s = \frac{1}{2} r_{E'_{ob}}$ and $r_s \to r_{E'_{ob}}$ ($r_{E'_{ob}}$ is the radius of the exit pupil of objective). If the occluded area does not exceed 10% of the whole exit pupil area, the Airy pattern is not substantially altered.

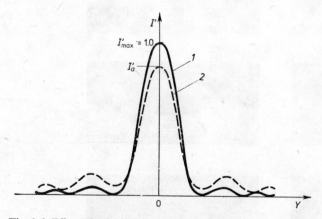

Fig. 3.6. Effect of spherical aberration or defocusing on the intensity distribution in the Airy pattern: curve *1*—microscope objective is free from spherical aberration and is correctly focused, curve *2*—objective suffers from a slight spherical aberration or its focusing is slightly defective.

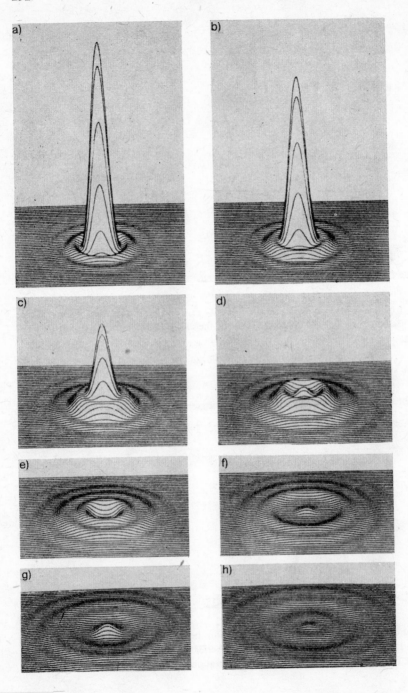

3.1.2. Aberrated Airy patterns

Hitherto we have supposed an objective free from both geometrical and chromatic aberrations. An optical system whose aberrations are reduced so that its performance depends only on the effect of light diffraction is called a *diffraction-limited system*. Hight quality modern microscope objectives belong to this category. Each geometrical aberration causes the object spherical wavefront to be trans-

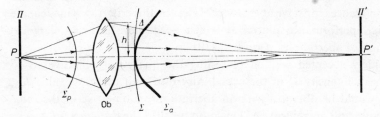

Fig. 3.8. Wavefront deformation by a microscope objective which suffers from spherical aberration.

formed by the objective into an aspherical wavefront. Figure 3.8 shows schematically a situation when spherical aberration is present. In order to evaluate the magnitude of this defect, the aberrated wavefront Σ_a is referred to a sphere Σ, which represent the ideal wavefront that would be achieved if the objective Ob were free from aberration. Both wavefronts touch each other at the axial point, whereas beyond this point they are variously seperated. The distance Δ between Σ_a and Σ is the optical path difference which is a measure of the aberration of the microscope objective, and is termed the *wavefront aberration* or *wave aberration*. If the actual wavefront Σ_a is a surface of revolution about the optical axis PP', then only spherical aberration is present and Δ at any point of this surface is merely a function of the radius h of lens zones and may generally be expressed as

$$\Delta = C_5 h^4 + C_6 h^6 + C_8 h^8 + ..., \tag{3.5}$$

where C_4, C_6, C_8, ... are the constant factors. The successive terms in Eq. (3.5) represent the primary, secondary, tertiary, ... spherical aberration.[1] For

---

[1] There seems to be no general agreement regarding this nomenclature. The terms: tertiary quinary, septenary,... aberrations are also in use.

Fig. 3.7. Tomographic (spatial) representation of the intensity distribution in the Airy patterns: a) perfectly focused, b) to h) defocused by $\Delta z = 0.5\lambda/A^2$, λ/A^2, $1.5\lambda/A^2$, $2\lambda/A^2$, $2.5\lambda/A^2$, $3\lambda/A^2$ and $3.5\lambda/A^2$, respectively (A is the numerical aperture of objective). By courtesy of Galbraith [3.3.].

objectives of moderate numerical aperture only the primary aberration is important.

Spherical aberration, like any other, causes a portion of the luminous energy of the Airy disc to be transferred to the Airy rings which spread and blur the image of the point object (Fig. 3.6). Lord Rayleigh stated that the diffraction pattern of a luminous point does not differ materially from the ideal Airy pattern if $\Delta \leqslant \lambda/4$. This is *Rayleigh's criterion*, which is generally used as a measure of permissible tolerances for wave aberration. However, this criterion is inadequate for today's high performance objectives which need more critical tolerances for the correction of aberrations, e.g., $\Delta \leqslant \lambda/10$.

Another widely accepted method of testing the quality of objectives is to compare the peak intensity I'_a in the actual Airy disc with the ideal value I'_{max} (Fig. 3.6) that would be obtained with an aberration-free objective of the same numerical aperture and magnification. This ratio is called the *definition brightness*, *Strehl intensity*, or *Strehl ratio*, and may be written as

$$Q_S = \frac{I'_a}{I'_{max}}. \tag{3.6}$$

If $Q_S \geqslant 0.8$, the objective is considered to belong to the class of high performance optical systems. The Strehl ratio is closely related to the Rayleigh criterion because spherical aberration, which produces a wavefront deformation $\Delta \approx \lambda/4$, reduces the peak intensity of the Airy disc by a factor 0.2 approximately. The *Maréchal criterion* is also worth noting [3.4, p. 108]. This states that the second root of the mean square value of actual wavefront deformations in the objective exit pupil cannot be larger than $\lambda/14$ if high performance imaging is desired. This criterion is less useful than Rayleigh's for an experimental assessment of the quality of microscope objectives; it is, however, frequently used in the theoretical calculations of optical systems.

An optical system is said to be *stigmatic* with respect to a luminous point and its image if the latter forms an ideal Airy pattern. A microscope objective

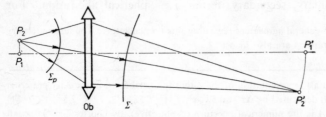

Fig. 3.9. Illustrating the formation of the Airy pattern of an off-axis luminous point.

which is *axially stigmatic* (or free from spherical aberration) with respect to points P_1 and P_1' (Fig. 3.9) may not be stigmatic with respect to an off-axial point P_2 and its image P_2'. This defect results from comatic aberration which affects the off-axial image-forming wavefront Σ so that it is not spherical, the Airy pattern of the luminous point P_2 is therefore distorted and not of revolution symmetry. It has only one plane of symmetry and its diffraction rings are no longer concentric (Fig. 3.10). If the axial astigmatism requirements are also ful-

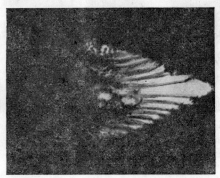

Fig. 3.10. Airy patterns deformed by comatic aberration. By courtesy of K. Patorski (Technical University of Warsaw).

filled for off-axial points P_2 and P_2' (Fig. 3.9), the microscope objective meets Abbe's sine condition (compare Eq. (2.33)) and is free from coma.

When the image P_2' of the luminous point P_2 (Fig. 3.9) is distant from the centre of the field of view, no focusing adjustment gives the standard Airy pattern, but will provide a small streak of light and another one at right angles to the former (Figs. 3.11a and c). These two streaks are surrounded by diffraction fringes and lie in two different focusing planes. Between these planes there is an intermediate plane in which the Airy disc is more or less square (Fig. 3.11b). This defect is caused by astigmatism and the streaks of light are astigmatic image lines (compare Fig. 2.27). Their diffraction fringes are feeble and those shown in Fig. 3.11 are normally not visible in the microscope; it is rather Figs. 4.2c, d and 4.3 (see Chapter 4) that illustrate true Airy patterns of off-axial luminous points observed through a microscope with objectives affected by coma and astigmatism.

So far we have confined ourselves to monochromatic light, but extending the discussion to polichromatic light presents only little difficulty. Formula (3.3) shows that the diameter $2r'_{\text{Airy}}$ of the Airy disc and that of the surrounding diffraction rings are the larger the longer is the wavelength λ; hence it appears that diffraction gives rise to some slight chromatism. It is clear that this defect cannot be confused with the chromatic aberration discussed previoulsy in Subsection 2.2.4;[2] the latter may be corrected or compensated while *diffraction chromatism* is not reducible. It is hardly perceived in the central disc of the Airy

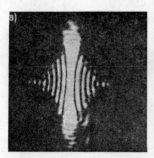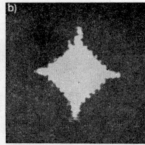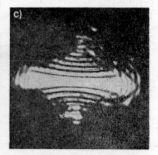

Fig. 3.11. Off-axis Airy patterns deformed by astigmatism: a) and c) diffraction appearance of tangential and sagittal line images (compare Fig. 2.27), b) image at mid-point between line images. By courtesy of K. Patorski (Technical University of Warsaw).

pattern, but may be easily observed in the first bright ring which is white at the central zone but red along the outer and blue along the inner edge. The succeeding rings are practically colourless and blurry, and in white light only two or three diffraction rings are observed.

3.2. Diffraction image of two close luminous points

Two luminous points create a more complicated situation than does a single point; to begin with, the coherence relations between light waves emerging from both points must be taken into account.

3.2.1. Coherence degree of illumination

Coherence relations in the object plane of an microscope objective can easily be examined by using the set-up shown in Fig. 3.12. An opaque screen with two pinholes P_1 and P_2 is placed in the object plane Π of the microscope objective Ob. The images P_1' and P_2' of P_1 and P_2 are observed through an ocular Oc.

---

[2] A detailed discussion of the intensity distribution in the Airy patterns formed by apochromatic, semiapochromatic, and achromatic objectives is given by Françon in Ref. [1.31].

An additional swing-out lens *BL*, the Bertrand lens, enables objective exit pupil E'_{ob} and the image plane Π' to be observed alternately. The pinholes P_1 and P_2 are illuminated by a condenser C equipped with an aperture diaphragm D_c, whose opening Q can be continuously varied. This diaphragm is situated in the first focal plane of the condenser and illuminated by a high-power lamp.

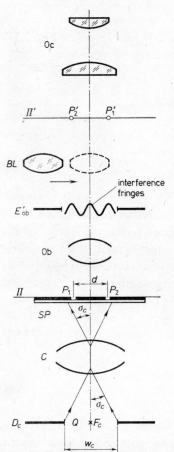

Fig. 3.12. Illustrating the discussion on the coherence degree of illumination in microscopy.

It is easily to seen that this set-up is similar to that used to observe the Young interference fringes (see Fig. 1.55). Now these fringes are observed in the objective exit pupil E'_{ob} and their visibility V (see Eq. (1.119)) may be used as a measure of the coherence degree γ_{12} of light waves emerging from the pinholes P_1 and P_2. If the aperture diaphragm opening is a uniformly luminous circle, γ_{12} is expressed by Eqs. (1.120) and (1.121), in which the denotations σ_c, f'_c, and r_c from Fig. 3.12 must be substituted for σ, z_1, and r, respectively. From Eq. (1.122) it results that the pinholes will be illuminated coherently if their distance

$$d = d_{coh} \leqslant \frac{0.16\lambda}{\sin\sigma_c},$$
(3.7)

whereas completely incoherent illumination occurs when the optical unit $Y = (2\pi/\lambda)d\sin\sigma_c$ takes the following values: 3.83, 7.02, 10.17, ... (see Fig. 1.57). Formula (1.123) shows that this kind of illumination first occurs when

$$d = d_{incoh} = \frac{0.61\lambda}{\sin\sigma_c}.$$
(3.8)

Otherwise, the illumination is treated as partially coherent. In practice, however, we assume that such illumination occurs when

$$\frac{0.16\lambda}{\sin\sigma_c} < d = d_{pcoh} < \frac{0.5\lambda}{\sin\sigma_c}.$$
(3.9)

The distance d_{coh} between two object points for which Eq. (3.7) is satisfied defines the diameter of the *coherence circle* (or coherence patch) in the object plane II. In the microscopical image, as it will be shown further on, it is impossible to resolve two object points that are coherently illuminated if their separation d is smaller than the radius r_{Airy} of the Airy disc. This radius is defined by Eq. (3.4); hence it appears that one can speak of coherent illumination of the object plane II when $d_{coh} \geqslant r_{Airy}$. By combining Eqs. (3.4) and (3.7), we obtain

$$\frac{0.16}{\sin\sigma_c} \geqslant \frac{0.61}{\sin\sigma},$$
(3.10a)

or

$$c = \frac{\sin\sigma_c}{\sin\sigma} \leqslant 0.26,$$
(3.10b)

where σ is the object-side aperture angle of the objective Ob (Fig. 3.12). If the objective and condenser are the immersion systems, $n\sin\sigma_c$ and $n\sin\sigma$ should be used instead of $\sin\sigma_c$ and $\sin\sigma$, respectively, in Eqs. (3.7) to (3.10b). Thus the latter formula may be written in more general form

$$c = \frac{A_c}{A} \leqslant 0.26,$$
(3.10c)

where A_c and A are the numerical apertures of the condenser and objective, respectively.

From the above condition it results that the specimen is coherently illuminated if the numerical aperture of the condenser is much smaller than that of the objective ($A_c \leqslant 0.26A$). As the ratio c drops to zero ($c \to 0$), the diameter d_{coh} of the coherence circle increases and becomes much larger than the radius r_{Airy} of the Airy disc. Conversely, as c increases, d_{coh} decreases and when $c = 1$, $d_{\mathrm{coh}} = 0$. In this case $A_c = A$ and Eq. (3.4) may be written as $r_{\mathrm{Airy}} = 0.61\lambda/A_c$, but Fig. 1.57 shows that two minute pinholes separated by $d = 0.61\lambda/A_c$ are illuminated incoherently ($\gamma_{12} = 0$). If, however, $c > 1$, the illumination is no longer incoherent within the area of radius r_{Airy} because γ_{12} reaches several extrema for $d > 0.61\lambda/A_c$ (see Fig. 1.57). These secondary extrema are much smaller than the principal maximum, and for $c > 1$ the illumination is qualified as partially coherent. Taking into account only the range of c from 0 to 1, we can divide it into three subranges: (1) $c = 0$–0.26, (2) $c = 0.26$–0.9, (3) $c = 0.9$–1, of which the first relates to coherent, the second to partially coherent, and the third to incoherent illumination of the specimen within areas of radius r_{Airy}.

The above discussion referred to two object points whose separation d is not larger than the radius of the Airy disc. If $d > r_{\mathrm{Airy}}$, the ratios of c for coherent, partially coherent, and incoherent illumination are, of course, different than when $d \leqslant r_{\mathrm{Airy}}$. This statement may be easily proved by using two parallel slits instead of two pinholes P_1 and P_2 (Fig. 3.12) and a slit condenser diaphragm instead of an aperture diaphragm D_c with a circular opening. This modification is an advantage for the experiment because slits give more light than pinholes or circular openings of equivalent width. If, however, a slit condenser diaphragm is used, the complex degree of coherence is no longer expressed by Eq. (1.120) but by a sincus function, i.e.,

$$\gamma_{12}(D) = \operatorname{sinc} D = \frac{\sin D}{D}, \tag{3.11}$$

where

$$D = 2\pi A_c \frac{d}{\lambda} \approx 2\pi \frac{w_c}{2f_c'} \frac{d}{\lambda}. \tag{3.12}$$

Here A_c is the numerical aperture of the condenser in the plane perpendicular to the slit diaphragm, d is the distance between the object slits (or pinholes P_1 and P_2 as shown in Fig. 3.12), w_c is the width of the slit condenser diaphragm, and f_c' is the focal length of the condenser. The plot of Eq. (3.11) is very similar to that of Eq. (1.120) as shown in Fig. 1.57. In order to obtain a correct interference pattern in the exit pupil of the objective, the slit of the condenser diaphragm must be parallel to the object slits. It is also advisable to locate all these slits symmetrically with respect to the optical axis of the condenser and objective. For a given distance d between the object slits, the most contrasty interference fringes are always obtained when the condenser slit is as narrow as possible

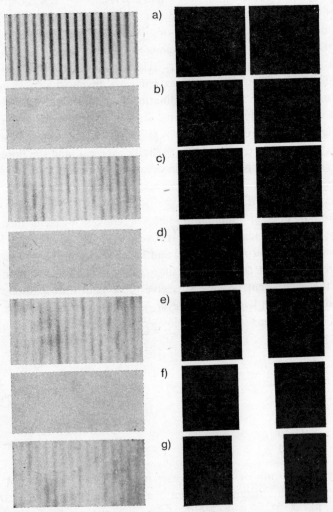

Fig. 3.13. Photomicrographs showing the complex degree of coherence in the object plane II of a microscope objective (see Fig. 3.12). The left-hand column of pictures shows a result of light interference in the exit pupil of objective for different widths of the condenser slit diaphragm (right-hand column), when a double slit is used as a test object in the object plane II.

(Fig. 3.13a). If this slit is enlarged, fringe visibility drops and for a particular slit width $w_c = w_1$ the interference fringes completely vanish in the exit pupil of the objective (Fig. 3.13b). As w_c is increased, the fringes reappear, their visibility ameliorates and reaches the second maximum value (much smaller than the first one) as shown in Fig. 3.13c. Next fringe visibility worsens and falls to zero for the second particular slit width $w_c = w_2$ (Fig. 3.13d). If w_c is made still bigger, the situation is repeated (as shown in Figs. 3.13e, f, g).

The maximum visibility of interference fringes (Fig. 3.13a) corresponds to the maximum value of Eq. (3.11), i.e., $\gamma_{12} = 1$ (or $\gamma_{12} \approx 1$); therefore the object slits are illuminated coherently. Conversely, the null-visibility of interference fringes (Figs. 3.13b, d, f) corresponds to the zero values of Eq. (3.11); therefore the object slits are illuminated incoherently. Otherwise, these slits are illuminated partially coherently.

The same series of inteference patterns as shown in Fig. 3.13 can be obtained when the distance d between the object slits is gradually increased, while the width w_c of the slit condenser diaphragm is invariable (and relatively small). This reciprocity simply results from formulae (3.11) and (3.12).

3.2.2. Two-point Airy patterns

Now we can mathematically describe the appearance of the diffraction image of two close luminous points which are represented by minute pinholes P_1 and P_2 in Fig. 3.12. The phenomenon of diffraction causes the light energy to be not only concentrated in the geometrical images P_1' and P_2' of P_1 and P_2 but also spread beyond these images as Airy patterns. Depending on their mutual coherence each pattern may interfere more or less strongly with the other, the result being a diffraction image which is no simple sum of the intensities of the component patterns. Let us assume x, y as the common Carthesian coordinates in both the object plane Π and the image plane Π', the origin of these coordinates being coincident with the optic axis of the objective Ob. This assumption simplifies the theoretical presentation of diffraction image formation and is quite valid if the Abbe sine condition is fulfilled (see Eq. (2.33)) as, of course, it is in the case of correctly designed microscope systems. It is now possible to relate the scales of measurement in the object and image planes by using the Abbe sine condition, which may now be written as $nx\sin\sigma = n'x'\sin\sigma'$ and $ny\sin\sigma = n'y'\sin\sigma'$. This condition also means that the coherence degree γ_{12}' in the image plane Π' equals that in the object plane Π, i.e., $\gamma_{12}' = \gamma_{12}$. Next, let x_1, y_1 be the coordinates of P_1 and x_2, y_2 those of P_2. Both pinholes are illuminated uniformly and their intensities are I_1 and I_2. For the sake of simplicity (but without affecting the general validity of the following considerations), it is convenient to take I_1, I_2 as equal to unity.

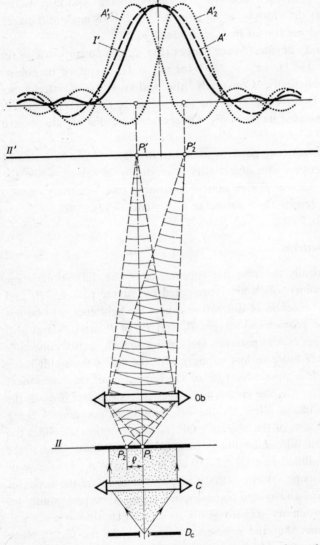

Fig. 3.14. Illustrating the formation of diffraction images of two close points, P_1 and P_2, illuminated coherently.

Under the conditions mentioned above it can be shown that the intensity distribution $I'(x, y)$ in the image plane Π' is given by

$$I'(x, y) = \left[\frac{2J_1(Y_1)}{Y_1} \right]^2 + \left[\frac{2J_1(Y_2)}{Y_2} \right]^2 + 2\gamma_{12} \frac{2J_1(Y_1)}{Y_1} \frac{2J_1(Y_2)}{Y_2}, \qquad (3.13)$$

where J_1 denotes the first-order Bessel function, Y_1 and Y_2 are the Airy optical units, i.e.,

$$Y_1 = \frac{2\pi}{\lambda} A \sqrt{(x-x_1)^2 + (y-y_1)^2}, \qquad (3.14)$$

$$Y_2 = \frac{2\pi}{\lambda} A \sqrt{(x-x_2)^2 + (y-y_2)^2}, \qquad (3.15)$$

and the degree of coherence γ_{12} is expressed by

$$\gamma_{12} = \frac{2J_1(cY_{12})}{cY_{12}}, \qquad (3.16)$$

where

$$Y_{12} = \frac{2\pi}{\lambda} A \sqrt{(x_1-x_2)^2 + (y_1-y_2)^2} = \frac{2\pi}{\lambda} Ad. \qquad (3.17)$$

Here A is the numerical aperture of the objective, λ the wavelength of the light used, d the distance between the pinholes, c the ratio of the numerical apertures of condenser and objective ($c = A_c/A$). It is worth noting that Y_1 and Y_2 are the quantities similar to Airy optical units defined by Eq. (3.2); the first of them relates to pinhole P_1 and the second to P_2.

Formula (3.13) expresses the light intensity in any point of the image plane for the most general situation when two pinholes emanate partially coherent light waves. If, however, the pinholes are coherently illuminated, $\gamma_{12} = 1$ and Eq. (3.13) takes the from

$$I'(x, y) = \left[\frac{2J_1(Y_1)}{Y_1} + \frac{2J_1(Y_2)}{Y_2} \right]^2. \qquad (3.18)$$

Note that the first and second terms in the square brackets of the above equation express the distribution of light amplitudes in the Airy patterns of the pinholes P_1 and P_2, respectively, and the resultant intensity distribution $I'(x, y)$ is the squared sum of these amplitude distributions. For illustration Fig. 3.14 shows a particular situation when two minute pinholes P_1 and P_2, separated by $d = \varrho$

and illuminated coherently, are so close together in the object plane Π of the objective Ob that the Airy disc of P_1 partly covers that of P_2.

Conversely, if the pinholes are illuminated incoherently, $\gamma_{12} = 0$ and Eq. (3.13) takes the form

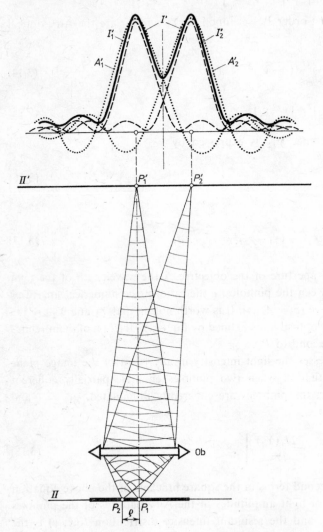

Fig. 3.15. Illustrating the formation of diffraction images of two close, mutually incoherent luminous points P_1 and P_2.

$$I'(x, y) = \left[\frac{2J_1(Y_1)}{Y_1} \right]^2 + \left[\frac{2J_1(Y_2)}{Y_2} \right]^2.$$ (3.19)

In this case the resultant intensity distribution is simply the sum of the intensity distributions inherent in the individual Airy patterns. An example illustrating this situation is shown in Fig. 3.15, where the incoherently illuminated pinholes P_1, P_2 and their geometrical images P_1', P_2' are separated by the same distances as in Fig. 3.14. Note that the resultant intensity distribution $I'(x, y)$ is different in both situations.

The latter situation (Eq. (3.19) and Fig. 3.15) is especially typical for fluorescence microscopy and for some types of dark-field microscopy, whereas the former (Eq. (3.18) and Fig. 3.14) occurs is such microscopical techniques which need coherent or nearly coherent illumination of the specimen (phase contrast, some interference methods, etc.). Conversely, common bright-field microscopes usually works with partially coherent or almost incoherent illumination.

These conclusions are valid not only for luminous points or pinholes but also for small opaque particles against a bright background; like luminous points, opaque particles produce Airy patterns.

3.3. Diffraction images of periodic objects

In 1873 Ernst Abbe (Fig. 3.16) showed that a microscopical image is formed in two stages [3.5]. First, light is diffracted by the object and then the diffracted beams are brought together again to interfere and form the object image. The Abbe theory[3] [3.6] was next refined by several scientists and especially by the Polish physicist Mieczysław Wolfke, who enriched the question of the imagery of non-self-luminous objects [3.10], examined the phenomenon of self-imaging of periodic objects in microscopy [3.11], introduced some elements of Fourier optics into microscopical imaging, and formulated the basic principles of holographic microscopy about thirty years before Denis Gabor (see Vol. 2, Chapter 11).

The Abbe principle of image formation is applicable to all objects, but can be best demonstrated by considering periodic structures, such as diffraction gratings.

---

[3] A detailed historical sketch of the Abbe theory is given by Martin in his book [2.13]. For some additional details see also Refs. [3.7]–[3.9].

Fig 3.16. Ernst Abbe (1840–1905).

3.3.1. Grating images in coherent light

Several diffraction gratings were described in Subsection 1.7.2. Some of them will now be used to interpret the Abbe theory of image formation in the microscope.

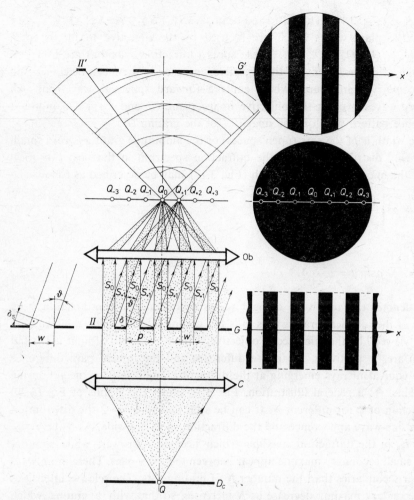

Fig. 3.17. Image formation of an amplitude grating in coherent illumination (for the sake of clarity only the direct beam S_0 and the diffracted beam S_{+1} are shown; the other diffraction beams which form spots Q_{-3}, Q_{-2}, Q_{-1}, Q_{+2} and Q_{+3} are omitted).

Amplitude gratings with very narrow slits. It was earlier shown that a parallel light beam is divided into several diffracted beams by an amplitude grating with a rectangular profile of transmittance (Figs. 1.89 and 1.90). Now, let such a grating (G) be placed in the object plane II of a microscope objective (Ob) and illuminated normally with a parallel beam of monochromatic light emerging from a small opening Q of the aperture diaphragm D_c located in the front focal plane of the condenser C (Fig. 3.17). The diffracted beams ... $S_{-2}, S_{-1}, S_0, S_{+1}, S_{+2}, ...$ can be treated as plane waves which are focused by the objective to bright spots ...$Q_{-2}, Q_{-1}, Q_0, Q_{+1}, Q_{+2}, ...$ These spots (diffraction maxima) are localized in the image-side focal plane of the objective, and each of them can be considered as a secondary light source which emits a forward spherical wavefront. All secondary waves overlap the objective image plane Π' and form a compound interference pattern which is the image G' of the grating G.[4]

If the width w of the slits (open spaces) of the amplitude grating is very small, the intensity distribution I'_{ob} of the diffraction spectrum in the posterior focal plane of the microscope objective Ob (Fig. 3.17) may be described as follows:

$$I'_{ob} = a^2 \left[\frac{\sin N\frac{1}{2}\psi}{\sin \frac{1}{2}\psi} \right]^2, \tag{3.20}$$

where

$$\psi = \frac{2\pi}{\lambda} p \sin \vartheta = \frac{2\pi}{\lambda} \delta. \tag{3.21}$$

Here a denotes the amplitude of light waves diffracted into the direction ϑ by any one of the identical grating slits, p is the grating period, N the total number of slits "viewed" by the microscope objective, λ the wavelength of light used, and ψ and δ are, respectively, the phase difference and the optical path difference between equivalent rays emerging at the diffraction angle ϑ from neighbouring grating slits. As a general illustration, Fig. 3.18 shows the graphs of Eq. (3.20) as a function of ψ for different N. It can be seen that for $N = 2$ the distribution I'_{ob} has a sine-wave appearence and the diffraction spots are wide. As N increases, so does I'_{ob} in the diffraction maxima which become narrower, while simultaneously small secondary maxima appear between the main ones. There are always two fewer secondaries than the number N of grating slits. The relative intensities of the secondary maxima decrease as N increases, so that with a grating which has a very large number of slits (and hence a small period p) the diffraction spectrum appears as a series of isolated maxima (spots) ...$Q_{-2}, Q_{-1}, Q_0, Q_{+1}, Q_{+2}, ...$ with regions of negligible intensity between them. In general, the peak intensity ($I'_{ob\,max}$) of the main maxima is proportional to N^2.

The grating image G' (Fig. 3.17) may be considered a synthesis of discrete interference processes since all pairs of diffraction spots produce interference

---

[4] Abbe called the diffraction spectrum ...$Q_{-2}, Q_{-1}, Q_0, Q_{+1}, Q_{+2}, ...$ the "primary image" and pointed out that it contains all the imaging information about the object. Consequently, the image G' was termed "secondary image". Today this nomenclature is obsolete.

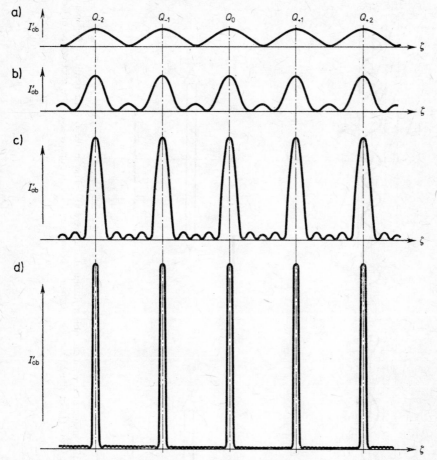

Fig. 3.18 Coherent diffraction spectra (intensity Fourier spectra) of a line grating with narrow slits when the microscope objective accepts sucessively the diffracted light from a greater and greater number (N) of grating slits: a) $N = 2$, b) $N = 3$, c) $N = 5$, d) a great number of slits. ζ indicates the direction across diffraction spots as shown in Fig. 3.17 (ζ is a function of the diffraction angle ϑ).

patterns (as in Young's experiment) which constitute components of the grating image. The theoretical study of this image will not be liable to any significant error if the image-forming wavefronts are assumed to be planes. Initially, let us consider the interference of waves emerging from the diffraction spots Q_0 and Q_{+1} (Figs. 3.17 and 3.19a). The beam S_{+1} of the first diffraction order is produced by the grating G at an angle ϑ_{+1} given by Eq. (1.162), i.e.,

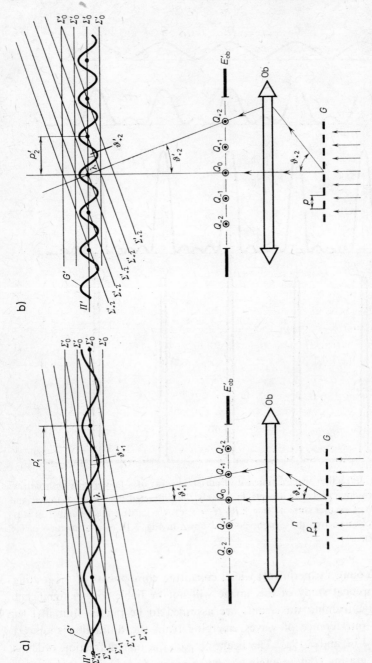

Fig. 3.19. Illustrating the formation of the image of a periodic object illuminated coherently. The image G' is a result of the interference of wavelets emerging from light spots Q_m of the diffraction spectrum occuring in the exit pupil E'_{ob} of the microscope objective Ob.

$$pn \sin \vartheta_{+1} = \lambda, \tag{3.22}$$

where n is the refractive index of the medium between the grating and the objective Ob (for dry objectives $n = 1$). After refraction this beam meets the zero diffraction order beam S_0, and both beams form wavefronts Σ'_{+1} and Σ'_0, which intersect each other under an angle ϑ'_{+1}. In the objective image plane Π' these wavefronts produce an interference pattern, whose interfringe spacing p'_1 results from the geometry of Fig. 3.19a, and is given by

$$p'_1 \sin \vartheta'_{+1} = \lambda. \tag{3.23}$$

By combining Eqs. (3.22) and (3.23) we obtain

$$p'_1 = p \frac{n \sin \vartheta_{+1}}{\sin \vartheta'_{+1}}, \tag{3.24a}$$

but from Eq. (2.33) it results that for aberration-free microscope objectives $n \sin \vartheta_{+1} / \sin \vartheta'_{+1} = M_{ob}$, and so

$$p'_1 = M_{ob} p. \tag{3.24b}$$

As can be seen, the information contained in the interference pattern resulting from the superposition of the waves of the first and zero diffraction orders is that the object is periodic and its period p can be obtained by dividing the interfringe spacing p_1 by the objective magnification M_{ob}. However, the intensity distribution in the interference pattern is rather different from that in the actual object (Fig. 3.20a). It can easily be proved that the same interference pattern, i.e., strictly speaking, pattern with the same interfringe spacing, is produced by the diffraction spots Q_0 and Q_{-1}, Q_{+1} and Q_{+2}, Q_{-1} and Q_{-3}, and so on, by any pair of two neighbouring diffraction spots (Fig. 3.17).

The situation is, however, radically different if the spots Q_0 and Q_{+2} (or Q_0 and Q_{-2}) are taken into consideration (Fig. 3.19b). In this case an interference

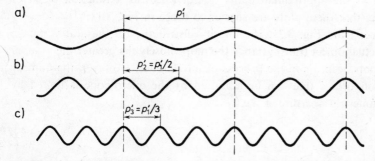

Fig. 3.20. Line grating images produced by different pairs of spots (Q_m, Fig. 3.19) of the coherent grating spectrum.

pattern also appears, but its interfringe spacing p'_2 is half that of p'_1. From the geometry of Fig. 3.19b it follows that

$$p'_2 \sin \vartheta'_{+2} = \lambda, \tag{3.25}$$

but according to Eq. (1.162),

$$pn \sin \vartheta_{+2} = 2\lambda. \tag{3.26}$$

By combining Eqs. (3.25) and (3.26) we obtain

$$p'_2 = \frac{p}{2} \frac{n \sin \vartheta_{+2}}{\sin \vartheta'_{+2}} = M_{ob} \frac{p}{2} \tag{3.27}$$

because in this case too $n \sin \vartheta_{+2}/\sin \vartheta'_{+2} = M_{ob}$. As $p = p'_1/M_{ob}$, so $p'_2 = p'_1/2$ (Fig. 3.20b). Interference patterns with the same interfringe spacing are also produced by any pair of diffraction spots Q_m and $Q_{m\pm2}$ ($m = 0, \pm1, \pm2, ...$).

Similarly, it can be shown that any pair of diffraction spots Q_m and $Q_{m\pm3}$ produces an interference pattern whose period p'_3 (Fig. 3.20c) is given by $p'_3 = M_{ob}p/3$, while the period $p'_4 = M_{ob}p/4$ is produced by the spots Q_m and $Q_{m\pm4}$, the period $p'_5 = M_{ob}p/5$ by the spots Q_m and $Q_{m\pm5}$, and so on.

If we now assume that all diffraction beams reach the image plane of the objective simultaneously, as in the case of a normal microscope, they form a highly complicated interference pattern whose intensity distribution is given by

$$I' = a'^2 \left[\frac{\sin q \frac{1}{2}\psi}{\sin \frac{1}{2}\psi} \right]^2, \tag{3.28}$$

where a' denotes the amplitude of the light waves emanated by any one of the diffraction spots (their intensity is assumed to be the same), whereas q is the total number of the image-forming diffraction spots, and ψ is defined by Eq. (3.21). Figure 3.21 shows the approximate plots of Eq. (3.28) as a function of ψ for different q. Note that these plots are similar to those in Fig. 3.18.

As we can see from Fig. 3.21, the intesity distribution in the grating image resembles the actual object transmittance the more closely the greater the number of diffraction spots occuring in the process of interference. However, the number (q) of diffraction orders that can be admitted by the microscope is limted by the objective numerical aperture A, i.e.,

$$q = \frac{p}{\lambda} 2A + 1. \tag{3.29}$$

For a given A, the number q therefore decreases as does the grating period p. Conversely, as A increases so does q (for a given p) and the image becomes more similar to the object. It is worth noting that theoretically there are $q-2$ small secondary maxima lying within each principal minimum across the grating image produced by q diffraction spots. If q is greater than 10, the secondary maxima are very small and practically unobservable. It is self-evident that any single diffraction spot does not give a periodic pattern in the image plane but merely produces a homogeneous illumination ($I' = $ const) as shown in Fig. 3.21a. In order to image any periodic structure, at least two neighbouring diffraction spots must emit their light energy to the image plane of the microscope objective.

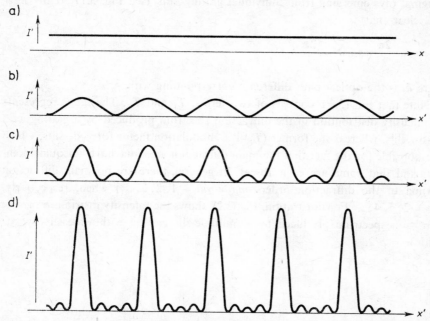

Fig. 3.21. Images of a line grating with narrow slits when the microscope objective accepts: a) one ($q = 1$), b) two, c) three, and d) five maxima of the coherent grating spectrum (x' indicates the direction perpendicular to the grating slits in the image plane Π' as shown in Fig. 3.17).

The patterns discussed above may be demonstrated by using a microscope equipped with a relay system which reimages the exit pupil of objective in an easily accessible intermediate plane. By putting on it an opaque screen one can extinguish some diffraction spots and let others continue to produce interference. This result is illustrated in Fig. V; the design of the experiment is self-explanatory.

Square amplitude grating with relatively large slits. If the width w of the grating slits is relatively large, the diffraction spectrum is somewhat different from that produced by a grating with very narrow slits. Wide slits cause interference to occur not only between the rays diffracted by adjacent slits but also between the rays emerging from the same slit. Hence, formula (3.20) takes now the form

$$I'_{\text{ob}} = a^2 \underbrace{\left[\frac{\sin \frac{1}{2}\psi_s}{\frac{1}{2}\psi_s} \right]^2}_{T_1} \underbrace{\left[\frac{\sin N \frac{1}{2}\psi}{\sin \frac{1}{2}\psi} \right]^2}_{T_2}. \tag{3.30}$$

Here ψ is the same phase difference as in Eq. (3.20), whereas ψ_s is that between marginal rays emerging from individual grating slits (see Fig. 3.17 to the left). It is clear that

$$\psi_s = \frac{2\pi}{\lambda} w \sin \vartheta = \frac{2\pi}{\lambda} \delta_s, \tag{3.31}$$

where δ_s is the optical path difference corresponding with ψ_s.

Note that Eq. (3.30) consists of two terms, T_1 and T_2. The latter represents the intensity distribution in the diffraction spectrum produced by a grating with narrow slits, whereas the former (T_1) is a modulation factor for wide slits, which is responsible for the fact that the main diffraction maxima have unequal intensities and that some are even absent. In general, there are no main diffraction maxima of the diffraction orders mp/w ($m = 1, 2, 3, ...$) when, respectively, $p/w = 2, 3, 4, ...$ For illustration, Fig. 3.22 shows the intensity distribution in the diffraction spectrum produced by a multiple-slit grating with wide slits whose width $w = p/2$.

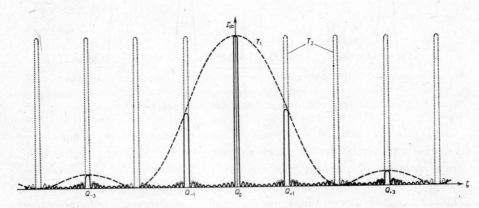

Fig. 3.22. Coherent diffraction spectrum of a square amplitude grating the period p of which is two times larger than the width w of grating slits, i.e., $w = p/2$ (compare with Fig. 3.18d, where $w \ll p$).

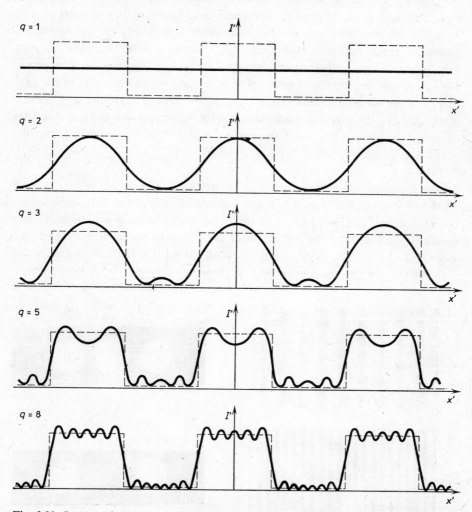

Fig. 3.23. Images of the grating the diffraction spectrum of which is shown in Fig. 3.22. The successive curves, from the top to the bottom, show the intensity distribution in the images when the microscope objective accepts successively a greater and greater number q of maxima of the coherent grating spectrum. The pecked curve shows the geometrical image; x' is the direction in the image plane Π' as marked in Fig. 3.17.

The process of image formation of a grating with wide slits and its appearance are identical with those presented above for gratings with narrow slits (compare Figs. 3.23 and 3.21). However, theoretical description of this process is now more complicated and will not be undertaken here;[5] only an approximative graphical representation is shown in Fig. 3.23. As can be seen, the intensity distribution in the grating image resembles the actual object transmittance the more closely the greater the number (q) of diffraction orders occurring in the process of image formation; this was, of course, also true of a grating with narrow slits (Fig. 3.21), the only difference being that the slit images now contain a number of secondary minima. If $q > 10$, these minima, as well as the secondary maxima within the image of opaque bars, are invisible.

Square-sine amplitude grating. It will be interesting to consider now a sinusoidal amplitude grating (see Fig. 1.93) which produces only three diffraction spots: Q_0, Q_{-1} and Q_{+1} (Fig. 3.17). The image of such a grating is completely and faithfully synthetized if the exit pupil of the objective contains just these three spots (Fig. 3.24a). The same image, i.e., with the same periodicity, occurs when the spot Q_{-1} or Q_{+1} is extinguished. If, however, the spot Q_0 is extinguished,

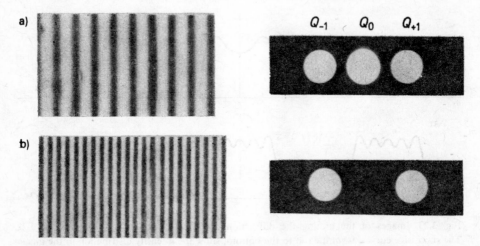

Fig. 3.24. Images and diffraction spectra of a sinusoidal amplitude grating: a) all spots, Q_{-1}, Q_0 and Q_{+1}, of the diffraction spectrum form the image, b) spot of the zero diffraction order (Q_0) is masked.

[5] For more intensive information the reader is referred to the more theoretical literature e.g., to books by Martin [2.13] and by Michel [3.12].

a fringe image with half the period appears (Fig. 3.24b). Such a grating is said to contain two spatial frequencies: the basic one $u_1 = 1/p$ and the second harmonic $u_2 = 2u_1 = 2/p$. A grating with a rectangular profile of transmittance (square amplitude grating), on the other hand, contains theoretically an infinite number of harmonic frequencies although the highest are not admitted by the microscope objective. Thus a square grating, unlike a sinusoidal one, can never be quite faithfully imaged.

Cross-line amplitude grating. It follows from the above discussion that the concept of two-stage diffraction also applies to more complicated gratings and even to non-periodic structures since any object can be treated as the summation of a series of simple square and/or sinusoidal gratings. In illustration Fig. VI shows different spatial frequencies of a cross-line amplitude grating whose diffraction spectrum is filtered by means of proper opaque screens. The pictures are self-explanatory and need no commentary.

3.3.2. Grating images in incoherent light

So far we have confined ourselves to gratings illuminated coherently, but there is no difficulty in extending the above principles of image formation to periodic objects illuminated incoherently. The latter situation will occur when the opening Q (Fig. 3.17) of an aperture diaphragm D_c is enlarged so that the numerical aperture of the condenser C and objective Ob are identical. In this case any infinitesimal element of the opening Q can be treated as a source of a light wave which is diffracted by the grating G in the manner described previously. As a result, an infinite number of diffraction spectra occur in the rear focal plane of the objective; their intensity summation produces a resultant distribution without discrete maxima and minima. However, each component diffraction spectrum can be treated as a secondary source of coherent wavelets which interfere and produce an elementary grating image in the objective image plane. By incoherent summation, an infinite number of such images produces the resultant image of the actual grating. An example is shown in Fig. 3.25. As can be seen, the intensity distribution (I') in the incoherent image of an amplitude square grating largely resembles a sine-wave without secondary maxima and minima but with a light background of some intensity I'_b which reduces the contrast of the grating image. As a rule, the contrast of incoherent images is always smaller than that of coherent images.

The above remarks also apply to the diffraction process of image formation when the grating is illuminated with polichromatic or white light. When the opening Q of the condenser diaphragm D_c (Fig. 3.17) is small, however, the diffraction spots $...Q_{-2}, Q_{-1}, Q_0, Q_{+1}, Q_{+2}, ...$ become extended and coloured. According to Eq. (1.159), the angular separation of the diffracted beams is proportional to the wavelength λ, so that the red diffraction spots lie further from the optic axis of the objective than the blue and violet spots of the equivalent diffraction orders. Diffraction spots of a given colour produce a coherent image of the grating (provided that the opening Q of the condenser diaphragm is small), but this image is incoherent with regard to that formed by diffraction spots of another colour. Hence, an incoherent summation of grating images produced by waves of different wavelengths occurs and the resultant image is incoherent. This image is, of course, free from colouration in spite of the coloured diffraction spectrum.

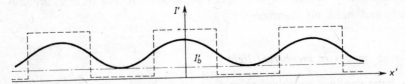

Fig. 3.25. Intensity distribution in the image of a square amplitude grating illuminated incoherently. The pecked curve shows the geometrical image; x' is the direction in the image plane Π' as marked in Fig. 3.17.

So far we have always treated the objective Ob (Fig. 3.17) as a system free from the aberrations described previously in Subsections 2.2.4 and 3.1.2. This assumption means that the phase relations between diffraction beams of successive orders are determined only by the object under examination. Otherwise, if a microscope objective suffers from aberrations, the phase relations will be disturbed and hence the intensity distribution across the object image will be distorted by comparison with that obtained with an aberration-free objective.

3.4. Diffraction imagery of extended non-periodic objects

Any extended object can be considered as consisting of infinitesimal elements acting as discrete point sources of light. Each of these sources is imaged as an Airy diffraction pattern. Depending on the degree of mutual coherence of the light waves emerging from individual points of the extended object, the resultant object image is an outcome of the coherent, partially coherent, or incoherent summation of all diffraction patterns.

Another approach to the diffraction theory of image formation of extended objects consists in viewing them as simple square and/or sinusoidal gratings which produce diffraction spectra in the exit pupil of the objective. Each of these spectra can be treated as secondary sources of wavelets which interfere and form an elementary grating image in the image plane of the objective. Depending on the degree of coherence of object illumination, the resultant object image is an outcome of the coherent, partially coherent, or incoherent summation of all elementary grating images.

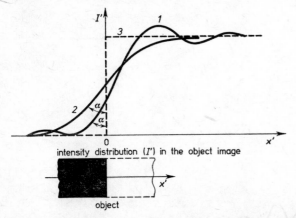

Fig. 3.26. Edge of an opaque half-plane imaged in coherent illumination (curve *1*) and incoherent illumination (curve *2*). The pecked curve (*3*) shows the geometrical image. Here and in the following figures, x and x' are the mutually conjugate coordinates in the object plane and the image plane, respectively.

In illustration Figs. 3.26–3.28 show light distribution in the images of some extended objects which tend to be of great importance in quantitative microscopy, especially in high-precision measurements of linear quantities with microscopes comprising a micrometer eyepiece or microphotometric device (see Vol. 3, Chapter 13). One such object is the half-plane (Fig. 3.26). As can be seen, its actual images (graphs *1* and *2*) significantly differ from the ideal (geometrical) image (graph *3*); the change of intensity I' in the diffraction images of the half-plane edge is not abrupt but inclined at an angle α. The smaller this angle the sharper the edge image. As a rule, for incoherent illumination the angle α is greater than for coherent illumination. The latter, however, produces diffraction fringes which surround the image of the half-plane edge (graph *1*). If the microscope is slightly defocused or suffers from aberrations, the angle α increases;

hence measuring α provides information about the performance of the objective. Suppose C' denotes the image contrast and x' the coordinate axis perpendicular to the half-plane edge in the microscope image plane, then the function $C'(x')$ or $I'(x')$ is sometimes called the *spread function*, which is a better measure of the objective performance than the angle α. It is worth noting that a straight edge and its image fulfil an important role in both the theory [3.13–3.15] and practice of imaging (see Vol. 3, Chapter 13).

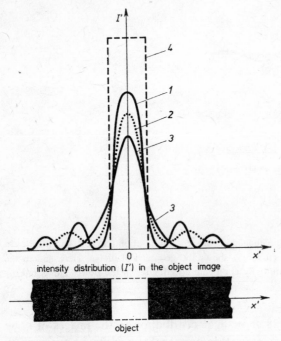

Fig. 3.27. Intensity distribution in the image of a long transparent streak on dark ground illuminated (*1*) coherently, (*2*) partially coherently and (*3*) incoherently. The pecked curve *4* shows the geometrical image.

Figure 3.27 shows the intensity distribution in the image of a long transparent streak against a dark background. It follows that the intensity I' is constant along the streak and the information required is merely the intensity changes in the direction x' perpendicular to the streak. Graphs *1*, *2*, and *3* refer to coherent, partially coherent, and incoherent illumination, respectively. In the second and third cases there are no zero secondary minima in the diffraction pattern, whereas in the first case the minima attain zero values.

A narrow transparent streak is obviously regarded as a slit. The slit-like straight streak is a standard object used for the study of the imaging properties of optical systems [3.16, 3.17].

Merely reversing Fig. 3.27 will illustrate what happens to a black streak against a white background. As the condenser aperture is increased, illumination becomes more and more incoherent, and diffraction fringes become less prominent.

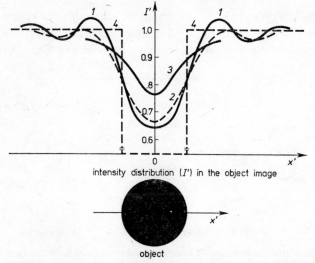

intensity distribution (*I'*) in the object image

object

Fig. 3.28. Intensity distribution in the image of a small opaque disc illuminated (*I*) coherently, (*2*) partially coherently and (*3*) incoherently. The pecked curve *4* shows the geometrical image.

The curves in Fig. 3.28 show how the intensity distribution across the image of a black circle on a bright background evolves as the coherence degree is changed by varying the aperture of the condenser. The situation will almost be reversed for a bright circle on dark background. When the circle is very small, its image is a typical Airy pattern. If the circle diameter increases, the graph showing the intensity distribution across the circle image gradually flattens for incoherent illumination.

These and related problems are discussed in more detail by Françon in his book *Progress in Microscopy* [1.31].

3.5. Fourier transform representation of microscopical imaging

Fourier transform optics was almost entirely formulated by Duffieux in 1946 [3.18]. Nowadays the Fourier analysis is largely used for the description and processing of images, and is especially useful for the representation of micro-scopical imaging. This representation is, however, on the whole mathematically complicated, therefore only some of its principles will be given here.

3.5.1. *A general approach to the description of mircoscopical imaging in coherent light using Fourier transforms*

Suppose the object O (Fig. 3.29) is so thin that its amplitude transmittance τ_a can be expressed as a two-dimensional function $f(x, y)$ of the Cartesian coordinates x, y associated with the object plane Π of the microscope objective Ob. The

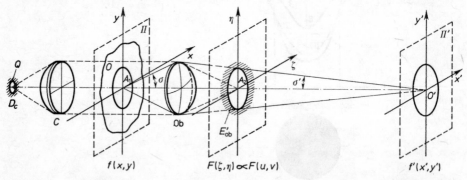

Fig. 3.29. Illustrating the Fourier transform theory of microscopical imagery in coherent light.

object O is transilluminated by a parallel beam of coherent light emerging from the small opening Q of the aperture diaphragm D_c located in the first focus of the condenser C. Next, let u, v denote the components of spatial frequencies cor-responding to x, y. If under these conditions the function $f(x, y)$ is continuous or sectionally continuous in every finite region over the plane Π and is integrable over this plane, then the following relations are valid [3.19, 3.20]:

$$F(u, v) = \iint\limits_{-\infty}^{+\infty} f(x, y) \exp[-i2\pi(xu+yv)]\,dx\,dy, \qquad (3.32a)$$

$$f(x, y) = \iint\limits_{-\infty}^{+\infty} F(u, v) \exp[i2\pi(xu+yv)]\,dx\,dy, \qquad (3.32b)$$

where $F(u, v)$ represents the *Fourier spectrum of spatial frequencies*.[6]

[6] There are also other definitions of the Fourier transforms: the factor 2π may be excluded from the exponent in Eqs. (3.32) and a factor $1/2\pi$ then appears before the integral sign in Eq.

The first of the above formulae is called the *Fourier transform* and the second is known as the *inverse Fourier transform*. Both formulae are said to be a *two-dimensional Fourier transform pair*.[7] For brevity, Eq. (3.32a) is written as $F(u, v) = T[f(x, y)]$ and Eq. (3.32b) as $f(x, y) = T^{-1}[F(u, v)]$, where $T[\ ]$ and $T^{-1}[\ ]$ indicate the operation of the direct and inverse Fourier transformations, respectively. It may be noted that $F(u, v)$ is, in general, a complex function, i.e., $F(u, v) = |F(u, v)|\exp[i\phi(u, v)]$, where the modulus $|F(u, v)|$ and argument $\phi(u, v)$ express the amplitude and phase of the Fourier spectrum, respectively.

The operations described by Eqs. (3.32) evidently occur in the microscopical imaging: the first expresses the amplitude distribution of light in the objective exit pupil E'_{ob} (Fig. 3.29), and the second, after a little modification, describes the amplitude distribution in the image plane Π'. In order to show this fact, it is sufficient to apply the diffraction formulae, which describe the distributions of light amplitude in the exit pupil and image plane of the objective Ob. The first formula has the form [1.1, 3.21]:

$$F(\zeta, \eta) = D'_1 \iint_{A_1} f(x, y)\exp\left[-i\frac{2\pi}{\lambda}\left(\frac{\zeta}{f'_{ob}}x + \frac{\eta}{f'_{ob}}y\right)\right]dx\,dy, \qquad (3.33)$$

where the integration takes place over the whole area A_1 of the field of view of the objective Ob (Fig. 3.29), D'_1 is a constant factor, λ the wavelength of light, f'_{ob} the focal length of the objective, and ζ, η are the Cartesian coordinates in the back focal plane of the objective. These exit pupil coordinates are assumed to be parallel to the object coordinates x, y. From Eq. (1.160c) it results that

$$\zeta = f'_{ob}u\lambda, \qquad \eta = f'_{ob}v\lambda, \qquad (3.34)$$

and Eq. (3.33) can be rearranged as

$$F(u, v) = D_1 \iint_{A_1} f(x, y)\exp[-i2\pi(ux + vy)]dx\,dy, \qquad (3.35a)$$

where D_1 is a constant adequately modified with respect to D'_1.

(3.32b) or this sign is preceded by a factor $1/\sqrt{2\pi}$ in both equations. These modifications are physically equivalent as long as one of them is used consistently.

[7] An one-dimensional Fourier transform pair is defined as

$$F(u) = \int_{-\infty}^{+\infty} f(x)\exp[-i2\pi ux]\,dx, \qquad (3.32c)$$

$$f(x) = \int_{-\infty}^{+\infty} F(u)\exp[i2\pi ux]\,du. \qquad (3.32d)$$

Comparing Eq. (3.32a) with (3.35a) we can state that the latter is, in fact, a Fourier transform of $f(x, y)$. Another range of integration is in this case unimportant as the integral in Eq. (3.35a) has zero values outside the area A_1 (light from area A_1 only enters the objective and function $f(x, y)$ is assumed to be zero outside A_1); hence the expression (3.35a) may also be rewritten as

$$F(u, v) = D_1 \iint_{-\infty}^{+\infty} f(x, y) \exp[-i 2\pi(ux+vy)] \mathrm{d}x \mathrm{d}y. \tag{3.35b}$$

The distribution of light amplitude, $f'(x', y')$, in the image plane Π' (Fig. 3.29) results from the superpostition of wavelets emerging from individual points (ζ, η) of the exit pupil E'_{ob}. This distribution is defined by the following diffraction formula:

$$f'(x', y') = D_2 \iint_{A_2} e(\zeta, \eta) \exp\left[-i\frac{2\pi}{\lambda}\left(\frac{x'}{g}\zeta + \frac{y'}{g}\eta\right)\right] \mathrm{d}\zeta \mathrm{d}\eta, \tag{3.36}$$

where $e(\zeta, \eta)$ is a function which describes the distribution of light amplitude in the objective exit pupil E'_{ob}, x' and y' are the Cartesian coordinates in the image plane Π' (parallel to the object coordinates x, y), D_2 is a constant factor, and g is the distance between E'_{ob} and Π'. The integration takes place over the whole area A_2 of the pupil E'_{ob}, and $e(\zeta, \eta)$ is in this case equivalent to the function $F(\zeta, \eta)$ expressed by Eq. (3.33); thus we can assume $e(\zeta, \eta) = F(\zeta, \eta)$ or $e(u, v) = F(u, v)$. Moreover, from Eqs. (2.14a) and (2.44) it results that $x' = M_{ob}x$, $y' = M_{ob}y$, and $g = -M_{ob}f'_{ob}$, where M_{ob} is the transverse magnification of the objective. By using the above relations and Eqs. (3.34), we can rewrite Eq. (3.36) in the form

$$f'(x'/M_{ob}, y'/M_{ob}) = D_2 f'^2_{ob} \lambda^2 \iint_{A_2} F(u, v) \exp[i 2\pi(ux+vy)] \mathrm{d}u \mathrm{d}v. \tag{3.37a}$$

As the function $F(u, v)$ has practically zero values outside the area A_2, the integration limits in the above expression may accordingly be written $\pm\infty$. Then Eq. (3.37a) becomes

$$f'(x'/M_{ob}, y'/M_{ob}) = D_2 f'^2_{ob} \lambda^2 \iint_{-\infty}^{+\infty} F(u, v) \exp[i 2\pi(ux+vy)] \mathrm{d}u \mathrm{d}v, \tag{3.37b}$$

and so the distribution of light amplitude in the image plane is the Fourier transform of $F(u, v)$. On the other hand, from Eq. (3.35b) it results that the inverse Fourier transform of $F(u, v)$ is

$$f(x, y) = \frac{1}{D_1} \iint_{-\infty}^{+\infty} F(u, v) \exp[i 2\pi(ux+vy)] \mathrm{d}u \mathrm{d}v. \tag{3.38}$$

Note that the last two expressions have the same integral and their combination yields

$$f'(x'/M_{ob}, y'/M_{ob}) = D_1 D_2 f_{ob}'^2 \lambda^2 f(x, y) = Df(x, y), \tag{3.39}$$

where D is resultant constant factor ($D = D_1 D_2 f_{ob}'^2 \lambda^2$). Hence it appears that the image O', represented by the function $f'(x'/M_{ob}, y'/M_{ob})$, is similar to the object O described by the function $f(x, y)$. The image is, however, enlarged by the factor M_{ob} and inverted (with respect to the object) as the objective magnification M_{ob} is formally negative, thus $x' = -x|M_{ob}|$ and $y' = -y|M_{ob}|$. The constant factor D only causes the value of the light amplitude of any image point to be changed (reduced) with respect to that of the conjugate object point. This statement is true only when the Abbe sine condition (see Eq. 2.33) is fulfilled. Finally, the process of image formation may be schematically represented as shown in Fig. 3.30: the direct Fourier transformation of the object function $f(x, y)$ gives the spectrum of spatial frequencies $F(u, v)$, and the direct Fourier transformation of $F(u, v)$ yields the image function $f'(x', y')$; the image resembles the object but is inverted, i.e., $f'(x', y') \propto f(-x, -y)$.

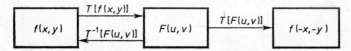

Fig. 3.30. Schematic representation of the process of microscopical imagery in the notation of Fourier transform.

It is important to note at this point that in optical systems only the direct and not the inverse Fourier transform can be obtained. The operation $T^{-1}[F(u, v)]$ marked in Fig. 3.30 can never be done optically, but may be obtained mathematically if the Fourier spectrum $F(u, v)$ is known; the only optical possiblity is the direct Fourier transform of $F(u, v)$, which represents an inverted image of the object—that is $T[F(u, v))] \rightarrow f(-x, -y)$.[8]

As is well known, the human eye and other light receptors record the intensity of light but not the light amplitude. The intensity is, however, defined by the squared amplitude of a resultant light vibration. The amplitude distribution $f'(x', y')$ expressed by Eq. (3.36) is, in general, a complex function, thus its squared modulus represents the intensity distribution $I'(x', y')$ in the image plane, i.e., $I'(x', y') = |f'(x', y')|^2$.

---

[8] These and other subtleties of Fourier transformations in optical systems are discussed in greater detail by Shulman in his excellent book *Optical Data Processing* [1.41].

The description of imaging by means of Fourier transforms is especially advantageous when dealing with some special microscope techniques (phase contrast, amplitude contrast, central dark-ground, strioscopy, etc.) in which the spectrum of spatial frequencies is modified by using spatial filters. If a function $M(\zeta, \eta)$ represents the amplitude and phase transmission factor, then the image $f'(x', y')$ is given by Eqs. (3.36), when $e(\zeta, \eta)$ is replaced by the product $M(\zeta, \eta) e(\zeta, \eta)$.

3.5.2. *Image of a luminous point as a Fourier transform of the pupil function*

In general, the image of a luminous point is described by a two-dimensional distribution called the *point spread function*, which directly results from the Fourier transform of the pupil function $e(\zeta, \eta)$. The latter expresses the changes of the amplitude and phase of light in the exit pupil of the objective, thus $e(\zeta, \eta)$ is usually complex but real if the objective is free from aberrations.

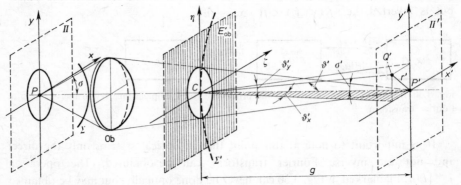

Fig. 3.31. Illustrating the Fourier transform imagery of a luminous point.

Let us take an axial luminous point P (Fig. 3.31). This point is exactly coincident with the object plane Π, and emits a spherical wavefront Σ, which is transformed into a convergent wavefront Σ', whose centre of curvature is coincident with the geometrical image P' of P.

What we want to know is the intensity distribution in the diffraction image or Airy pattern of the point P. This problem was discussed earlier in Subsection 3.1.1, and will now be considered again, using Fourier transformation. As was previously stated, the luminous energy is not only concentrated in the geometrical image of P, but is also spread beyond P' to an arbitrary point Q' separated from P' by r'. The Cartesian coordinates of Q' in the image plane Π' are x' and y'.

For the sake of simplicity, let these coordinates be expressed by the angles ϑ'_x and ϑ'_y subtended from the centre C of the exit pupil E'_{ob}. Note that P' and C are simultaneously the origins of the coordinates x', y' and ζ, η, respectively. Both coordinate systems are parallel to each other. As r' is normally much smaller than the distance between E'_{ob} and Π' ($r' \ll g$), so ϑ'_x, ϑ'_y, and ϑ' are also small and one can assume $\tan\vartheta'_x = \vartheta'_x$, $\tan\vartheta'_y = \vartheta'_y$, and $\tan\vartheta' = \vartheta'$. Consequently, $x' = g\vartheta'_x$, $y' = g\vartheta'_y$, and $r' = g\vartheta'$. Under these conditions, the distribution of light amplitude a' in the image plane Π' is then found to be [2.13]

$$a'(\vartheta'_x, \vartheta'_y) = a_0 \iint_{A_2} e(\zeta, \eta)\exp\left[-i\frac{2\pi}{\lambda}(\vartheta'_x\zeta + \vartheta'_y\eta)\right]d\zeta\,d\eta, \tag{3.40a}$$

where a_0 is a constant factor independent of the coordinates in the image plane Π' and the exit pupil E'_{ob}. Integration in the above formula takes place over the whole area A_2 of the exit pupil E'_{ob}. As the function $e(\zeta, \eta)$ has zero values outside the area A_2, the integration limits may accordingly be written $\pm \infty$. Then Eq. (3.40a) becomes

$$a'(\vartheta'_x, \vartheta'_y) = a_0 \iint_{-\infty}^{+\infty} e(\zeta, \eta)\exp\left[-i\frac{2\pi}{\lambda}(\vartheta'_x\zeta + \vartheta'_y\eta)\right]d\zeta\,d\eta, \tag{3.40b}$$

and the conclusion is that $a'(\vartheta'_x, \vartheta'_y)$ is a Fourier transform of $e(\zeta, \eta)$, i.e.,

$$a'(\vartheta'_x, \vartheta'_y) = a_0 T[e(\zeta, \eta)]. \tag{3.40c}$$

Thus the intensity distribution in the luminous point image is given by

$$i'(\vartheta'_x, \vartheta'_x) = |a'(\vartheta'_x, \vartheta'_y)|^2 = a_0^2 |T[e(\xi, \eta)]|^2. \tag{3.41}$$

If an objective is free from aberrations, and its exit pupil is circular, the diffraction image of a luminous point has the form of an ideal Airy pattern, the intensity distribution of which is given by

$$i'(\vartheta') = I'_0\left[\frac{2J_1(Y)}{Y}\right]^2, \tag{3.42a}$$

or

$$I'(\vartheta') = \frac{i'(\vartheta')}{I'_0} = \left[\frac{2J_1(Y)}{Y}\right]^2. \tag{3.42b}$$

Here I'_0 is the maximum value of intensity in the centre of the Airy disc, $J_1(Y)$ denotes the first-order Bessel function, and

$$Y = \frac{2\pi}{\lambda}r_{E'_{ob}}\vartheta', \tag{3.43}$$

where $r_{E'_{ob}}$ is the radius of the objective exit pupil E'_{ob}, and ϑ' is the angular distance in the image plane Π' "viewed" from the centre C of the pupil E'_{ob} (see Fig. 3.31).

Note that Eqs. (3.42) and (3.1) are identical and so are optical units Y. From the geometry of Fig. 3.31 it follows that $\vartheta' = r'/g$, $r_{E'_{ob}} = g\sigma'$, and $\sigma' = n \sin \sigma / M_{ob}$, where n is the refractive index of the space between the luminous point P and the objective Ob. Substituting the above quantities in Eq. (3.43) yields Eq. (3.2).

3.5.3. *Microscopical imagery described by convolution*

In what follows we shall consider an extended amplitude object to be self-luminous or illuminated incoherently. Thus the light from any object point cannot interfere with that from any other point, so that the object image results simply from overlapping point spread functions (Airy patterns) by adding their intensities.

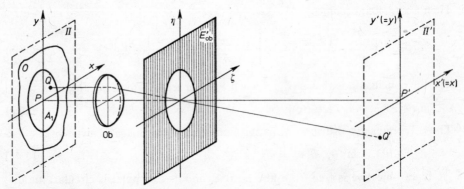

Fig. 3.32. Illustrating the Fourier transform theory of microscopical imagery in incoherent light.

For the sake of simplicity, let x' and y' be reduced coordinates in the image plane Π' of the microscope objective Ob by taking the objective magnification $M_{ob} = 1$ (Fig. 3.32); hence $x' = x$ and $y' = y$, where x and y are the Cartesian coordinates in the object plane Π. Note that the generality of the discussion below is not reduced by this assumption because it only means that the image coordinates must be rescaled—if necessary—according to the objective magnification.

When the microscope objective Ob fulfils the Abbe sine condition, then the spread function of any off-axial object point Q is identical (or nearly so) with that of the axial point P (Fig. 3.32). Consequently, the distribution of intensity in the diffraction image of Q can be described by the same spread function as that which describes the intensity distribution in the diffraction image of P, provided that the origin of the coordinates in the image plane Π' is translated from P' to Q' (P' is the geometrical image of P and Q' is that of Q). In other words, if $i'(x, y)$ denotes the intensity distribution in the Airy pattern of the axial point (P), then $i'(\tilde{x}-x, \tilde{y}-y)$ is that of an off-axial point (Q), whose coordinates are $\tilde{x}$ and $\tilde{y}$.

Now, let the function $f(x, y)$ describe the intensity distribution in the object plane Π (Fig. 3.32). An infinitesimal element, say $dxdy$, of the object O emits the light of intensity $f(x, y)dxdy$. This element is imaged as an elementary diffraction pattern which redistributes the light over the image plane Π', according to the point spread function. Thus the contribution of intensity at an arbitrary point $(\tilde{x}, \tilde{y})$ of the image plane Π' is given by $f(x, y)i'(\tilde{x}-x, \tilde{y}-y)dxdy$. But this point also receives the light from any other infinitesimal element of the object. Consequently, the total intensity I' at the point $(\tilde{x}, \tilde{y})$ is obtained by summing all the component spread functions produced by all the elements $dxdy$ of the object area A_1 "viewed" by the microscope objective Ob (Fig. 3.32), i.e.,

$$I'(x, y) = \iint\limits_{A_1} f(x, y)i'(\tilde{x}-x, \tilde{y}-y)dxdy. \tag{3.44a}$$

As the integral in the above formula has zero values outside the area A_1, the integration limits may accordingly be written $-\infty$ to $+\infty$, then

$$I'(x, y) = \int\limits_{-\infty}^{+\infty}\!\!\int f(x, y)i'(\tilde{x}-x, \tilde{y}-y)dxdy. \tag{3.44b}$$

The last formula has the form of a convolution of two functions,[9] one of which is the object function $f(x, y)$ and the other is the point spread function $i'(x, y)$. According to the convolution theorem,[10] Eq. (3.44b) can be expressed as

$$J(u, v) = F(u, v)J'(u, v), \tag{3.45}$$

[9] The convolution of two functions $g(x, y)$ and $h(x, y)$ is defined as

$$g(x, y) * h(x, y) \equiv \int\limits_{-\infty}^{+\infty}\!\!\int g(\tilde{x}-x, \tilde{y}-y)h(x, y)dxdy = \int\limits_{-\infty}^{+\infty}\!\!\int g(x, y)h(\tilde{x}-x, \tilde{y}-y)dxdy,$$

where $*$ indicates the operation of the convolution (see Ref. [1.5] or [3.4] for details).

[10] The convolution theorem states that the Fourier transform of the convolution of two functions is the product of their Fourier transforms.

where $J(u, v)$, $F(u, v)$, and $J'(u, v)$ are the Fourier transforms of $I'(x, y)$, $f(x, y)$, and $i'(x, y)$, respectively. In particular,

$$f(x, y) = \int\int_{-\infty}^{+\infty} F(u, v)\exp[i2\pi(xu+yv)]\,du\,dv, \tag{3.46}$$

$$I'(x, y) = \int\int_{-\infty}^{+\infty} J(u, v)\exp[i2\pi(xu+yv)]\,du\,dv, \tag{3.47}$$

and

$$i'(x, y) = \int\int_{-\infty}^{+\infty} J'(u, v)\exp[i2\pi(xu+yv)]\,du\,dv. \tag{3.48}$$

In the discussion presented above we considered the point spread function centred at the axial point P'. However, such a selection of this function is irrelevant. The conclusions formulated are also valid when the point spread function is centred at any paraxial point of the image plane provided that the objective satisfies the condition of isoplanatism.

3.5.4. Concept of optical transfer function

Equation (3.45) is extremely important in the theory of imaging of incoherent objects; it tells us that the Fourier transform of the intensity distribution in the image plane Π' (Fig. 3.32) is a product of the Fourier transforms of the point spread function and the intensity distribution in the object. Moreover, Eq. (3.45), when written in the form $J'(u, v) = J(u, v)/F(u, v)$, expresses the potentiality of the microscope objective for transferring spatial frequencies (strictly speaking, spatial frequency components in the x and y directions) from the object to its image. For this reason, the function $J'(u, v)$ is called the *optical transfer function*, usually abbreviated to *OTF*. Its role and properties will be discussed further on, and it is only important to note once more, to sum up this paragraph, that the Fourier transform of the pupil function $e(\zeta, \eta)$ gives the point spread function $i'(x, y)$ or, as defined earlier, $i'(\vartheta'_x, \vartheta'_y)$, and the Fourier transform of the point spread function yields the optical transfer function, i.e., $T[e(\zeta, \eta)] \to i'(x, y)$ and $T[i'(x, y)] \to OTF$.

3.5.5. Imagery of partially coherent objects

Partially coherent illumination is most frequently met with in light microscopy. An important contribution to the theory of partially coherent imagery was made above all by H. H. Hopkins in the 1950s [3.22]–[3.25]. This is the most general theory, but its mathematical description is very complicated. For this reason, only a basic outline will be given here, while for more detailed information the reader is referred to the cited papers or to Martin [2.13].

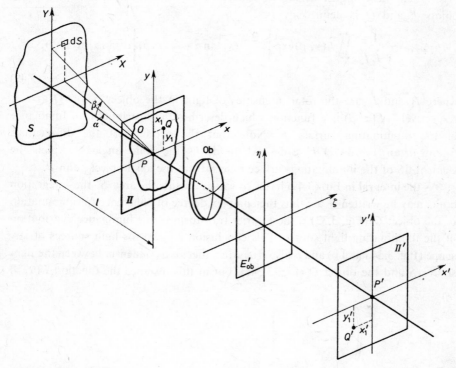

Fig. 3.33. Illustrating the Fourier transform theory of microscopical imagery in partially coherent light.

Let us consider Fig. 3.33. The object plane Π is illuminated by an extended light source whose luminous surface S is plane and perpendicular to the optic axis of the microscope objective Ob. Conformal Cartesian coordinates X, Y and x, y, as well as ζ, η and x', y', are coupled with the illuminating surface S, object plane Π, exit pupil E'_{ob}, and image plane Π', respectively. The origins of all these coordinates are coincident with the optical axis of the objective. An arbitrary infinitesimal element dS of the luminous surface S has the coordinates X_1 and Y_1, which are viewed under angles α and β from the axial point of the object plane Π. Two arbitrary object points, P and Q, are selected to define their degree of coherence. For the sake of simplicity, the point P is assumed to be coincident with the optical axis of the objective, hence the coordinates of P are $x = 0$ and $y = 0$, whereas those of Q are $x = x_1$ and $y = y_1$. If under these conditions the distance l between S and Π is much larger than the diameter of S, then the complex degree of coherence in the illumination of the object points P and Q is defined by

$$\gamma_{12} = \frac{1}{\sqrt{I_1 I_2}} \iint_S L(\alpha, \beta) \exp\left[i \frac{2\pi}{\lambda}(x_1 \tan \alpha + y_1 \tan \beta)\right] d(\tan \alpha) d(\tan \beta),$$

(3.49)

where I_1 and I_2 are the total intensities of light in the object points Q and P, respectively; $L(\alpha, \beta)$ is a function which describes the distribution of luminance of the illuminating surface S. Note that the product of the differentials $d(\tan \alpha) d(\tan \beta) = dX dY / l^2 \approx d\omega$, where $d\omega$ is the solid angle at which the element dS of the illuminating surface is seen from the axial object point P.

As the integral in Eq. (3.49) has zero values outside the area S, the integration limits may be written $\pm \infty$; thus the complex degree of coherence for illumination of the object O (Fig. 3.33) is the Fourier transform of the luminance distribution of the illuminating light source. This conclusion is valid for light sources of any shape (Fig. 3.34a and b) and is also true when there is a condenser between the light source S and the object O (Fig. 3.34c), but in this instance the function $L(\alpha, \beta)$

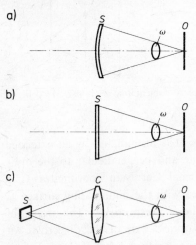

Fig. 3.34. Different ways of illuminating an object O with the same coherence degree by using (a) concave or (b) plane light diffusing surface S, and (c) condenser C between a source of diffuse light S and the object.

must refer to the aperture of the condenser; the degree of coherence is primarily conditioned by the solid angle ω at which the light source S (Fig. 3.34a and b) or the condenser aperture (Fig. 3.34c) are seen from the axial point of the object plane. It is worth noting that if the aperture of the condenser is circular and is illuminated uniformly by a primary source, Eq. (3.49) takes the form of Eq. (1.120).

Now, let $a_0(x, y)$ denote the local amplitude of a light field formed by all elements dS of the luminous surface S in the object plane Π (Fig. 3.33). This amplitude is weakened by the object O, so close behind O it is equal to $\tau_a(x, y) a_0(x, y)$; here $\tau_a(x, y)$ is the local amplitude transmittance. If the object plane is illuminated uniformly, then $|a_0(x, y)| = $ const. Let us assume this quantity is unity; then the light amplitude close behind the object O is simply equal to $\tau_a(x, y)$.

Given the conditions mentioned above and provided that the microscope objective satisfies the criterion of isoplanatism, the intensity distribution in the image plane Π' is given by

$$I'(x', y') = \int\!\!\!\int\!\!\!\int\!\!\!\int_{-\infty}^{+\infty} \gamma_{12}\, \tau_a(x_1, y_1)\, i'_a(x'-x_1, y'-y_1) \times$$

$$\times\, \tau_a^*(x_2, y_2)\, i'_a{}^*(x'-x_2, y'-y_2)\, \mathrm{d}x_1\, \mathrm{d}y_1\, \mathrm{d}x_2\, \mathrm{d}y_2, \tag{3.50a}$$

where x_1, y_1 and x_2, y_2 are, respectively, the coordinates of the object points P and Q (Fig. 3.33), γ_{12} is the degree of coherence of illumination defined by Eq. (3.49), $\mathrm{d}x_1 \mathrm{d}y_1$ and $\mathrm{d}x_2 \mathrm{d}y_2$ express the infinitesimal areas the centres of which are P and Q, respectively. The integration in the above formula is formally taken from $-\infty$ to $+\infty$, but in reality these limits can be replaced by an area identical with the field of view of the objective. In general, the functions τ_a and i'_a are complex, while τ_a^* and $i'_a{}^*$ are complex conjugate. The function i'_a or $i'_a{}^*$ is of the same importance as i' in Eqs. (3.44), but it now refers to light amplitude, while formerly it referred to intensity.

Expression (3.50a) is the most general representation of imaging. It applies not only to objects illuminated partially coherently but also to ideally coherent or completely incoherent illumination. The latter causes the function γ_{12} to be zero for any two different object points P and Q but equal to unity if Q and P coincide. In this instance, however,

$$\tau_a(x_1, y_1)\tau_a^*(x_2, y_2) = |\tau_a(x, y)|^2,$$

$$i'_a(x'-x_1, y'-y_1)\, i'_a{}^*(x'-x_2, y'-y_2) = |i'_a(x'-x, y'-y)|^2,$$

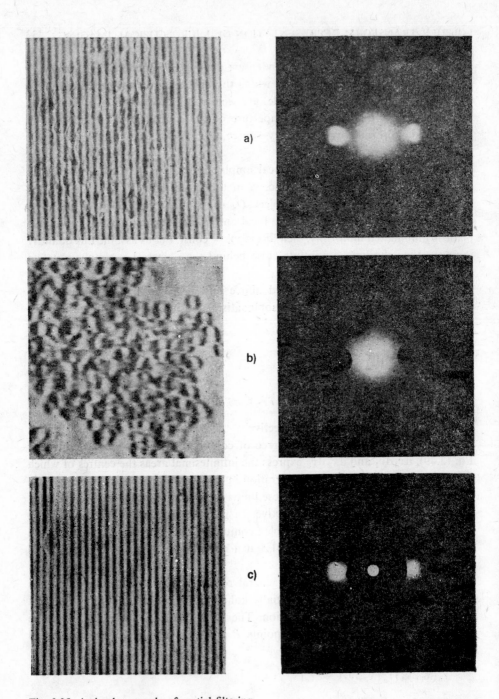

Fig. 3.35. A simple example of spatial filtering.

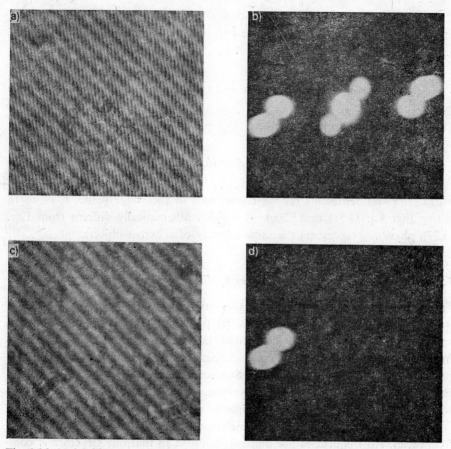

Fig. 3.36. Moiré fringes produced by two slightly crossed line gratings of spatial frequency 350 and 450 lines per millimetre.

$$dx_1 = dx_2 = dx, \qquad dy_1 = dy_2 = dy,$$

and Eq. (3.50a) takes the form

$$I'(x', y') = \iint\limits_{-\infty}^{+\infty} |\tau_a(x, y)|^2 |i'_a(x'-x, y'-y)|^2 dx\,dy, \tag{3.50b}$$

where $|\tau_a(x, y)|^2$ represents the distribution of light intensity close behind the object. Returning to Eqs. (3.44), it is evident that $|\tau_a(x, y)|^2 = f(x, y)$ and $|i'_a(x'-x, y'-y)|^2 = i'(x'-x, y'-y)$; hence Eqs. (3.50b) and (3.44b) are equivalent.

If, however, the illumination is ideally coherent, the degree of coherence $\gamma_{12} = 1$ independently of the distance between the object points P and Q (Fig. 3.33), and Eq. (3.50a) takes the form

$$I'(x', y') = \left| \iint\limits_{-\infty}^{+\infty} \tau_a(x, y) i'_a(x'-x, y'-y) \mathrm{d}x\mathrm{d}y \right|^2. \tag{3.50c}$$

This equation can also be written as

$$A'(x', y') = \iint\limits_{-\infty}^{+\infty} \tau_a(x, y) i'_a(x'-x, y'-y) \mathrm{d}x\mathrm{d}y, \tag{3.51}$$

where $A'(x', y')$ represents the light amplitude in the image plane. It is worth noting that Eq. (3.51) or (3.50c), although mathematically diffrent from Eqs. (3.37), physically represents the same imagery of coherent objects.

3.5.6. Analysis of microscopical images by filtering Fourier spectra

From what was said in the preceding subsection it is clear that Fourier spectrum is a distribution of spatial frequencies. By putting spatial filters (stops, opaque screens) in the Fourier plane (exit pupil of the objective), one can eliminate some frequencies from the object image and reinforce the visibility of others. This procedure was discussed earlier in Subsection 3.3.1, where periodic objects (amplitude gratings) and their images were analysed. Now, some further examples will be given of spatially filtered images of more complicated objects.

Frequently, there are specimens which contain low and high or middle spatial frequencies. All frequencies are mutually coupled so that one structure is completely masked by another. An illustration of such a situation is shown in Fig. 3.35. Photomicrographs a present the microscopical image and Fourier spectrum of a specimen which has been artificially assembled out of a linear grating and a layer of yeast cells. The spatial frequencies of the grating are higher than those of the cell layer. These two groups of frequencies are well enough separated in the Fourier spectrum (photo a to the right); the two side-spots primarily contain the spatial frequencies of the grating, whereas the halo which surrounds the central spot contains the spatial frequencies of the cell layer. But this layer is invisible (or nearly invisible) in the microscopical image (photo a to the left). If, however, the side-spots are masked (photo b to the right), the grating structure vanishes, and the cell layer appears clearly (photo b to the left). Conversely, if only the halo around the central spot is masked by means of an opaque ring (photo c to the right), the image of the cell layer completely disappears, while the grating structure is more clearly revealed (photo c to the left).

Another interesting result can be obtained when a specimen is composed of two linear gratings, each of which consists of alternate opaque and transparent streaks. If these gratings are set face to face with their streaks inclined at a small angle and viewed in transmitted light, a series of alternate dark and bright fringes appears (Fig. 3.26a). These are known as *moiré fringes*. They stand out more obviously (Fig. 3.36c) if the spatial frequencies of individual gratings are eliminated by masking some diffraction spots as shown in Fig. 3.36d. It is obvious that the whole Fourier spectrum, as shown in Fig. 3.36b, enables one to determine the periods of the component gratings, their inclination angle, and the period of the moiré fringes. The moiré phenomenon can easily be observed when some types of diatoms are studied.

Spatial filtering may be performed in coherent, partially coherent, and incoherent illumination [1.41, 3.26, 3.27]; the filtering mechanism is, however, simplest when coherent or quasi-coherent light is used (Figs. V, VI, 3.35, 3.36). The Fourier spectra presented in these figures are relatively simple in comparison with those produced by biological specimens and other natural objects. In any event, each Fourier spectrum provides much qualitative and quantitative information about the structure of a given specimen, its spatial frequencies, directional orientation of structural elements, shape of isolated objects, etc. The quantitative analysis of the Fourier spectra of microscopic objects is dealt with in what is known as *optical diffractometry* which is covered in more detail by Chapter 17 (Vol. 3).

3.6. Resolving power, perception limit, and useful magnification

Resolving power is the most important feature of any microscope system. This parameter depends primarily on the numerical aperture of the objective (A), but is in no way a quite unambiguous quantity because it is also affected by the type of object, the coherence degree of illumination, the correction of aberrations, and other factors. As far as the object structure is concerned, we distinguish *two-point*, *two-line*, and *grating resolving powers*.

Resolution is directly related to the useful magnification of the microscope and the perception limit. The latter quantity is frequently used as a synonym for resolving power, but this practice should be avoided since the words "resolution" and "perception" have quite different meaning.

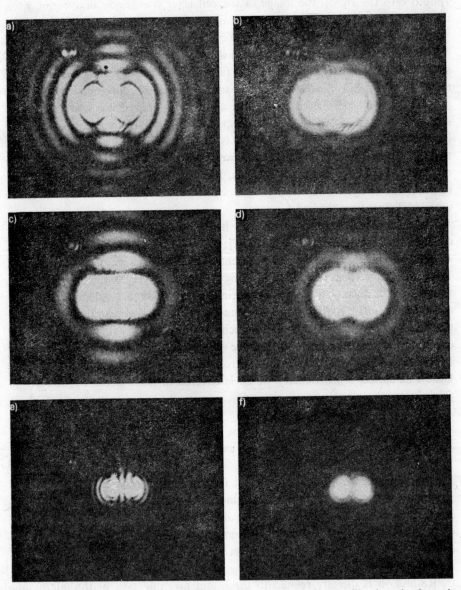

Fig. 3.37. Diffraction images of two close pinholes in an opaque screen illuminated coherently (a, c, e) and incoherently (b, d, f), and imaged with the same objective of larger and larger numerical aperture (top: minimum aperture, bottom: maximum aperture). By courtesy of K. Patorski (Technical University of Warsaw).

3.6.1. Two-point resolution

Two luminous points on a dark background can be obseved separately if their diffraction discs (Airy patterns) are not too close to one another. Let us first assume the points P_1 and P_2 are two mutually incoherent sources of light of equal intensity (Fig. 3.15). According to the Rayleigh criterion, these points are considered to be resolved when the centre of one of the Airy patterns coincides with the first dark ring of the other, and vice versa, i.e., when the intensity plots I'_1 and I'_2 are as shown in Fig. 3.15. It is clear that in this situation the distance ϱ between the points P_1 and P_2 must be equal to the radius of the Airy disc related to the object plane II, i.e., $\varrho = r_{Airy}$. From Eq. (3.4) it results that

$$\varrho = \frac{1.22\lambda}{2n\sin\sigma} = \frac{0.61\lambda}{A}, \qquad (3.52)$$

where λ is the wavelength of light and A the numerical aperture of the microscope objective ($A = n\sin\sigma$). As A increases and/or λ decreases, the distance ϱ shrinks, hence the resolving power of the microscope improves and the closer details of the specimen are resolvable.

Given the conditions mentioned above, it can be seen that the resultant intensity I' (Fig. 3.15) at the mid-point between the centres of the Airy discs is equal to 74% of the maximum value at the Airy disc centres (it is assumed that the microscope objective is free from aberrations and the two luminous points are perfectly coincident with the object plane). In order to achieve the same intensity loss of 26% for the mutually coherent luminous points P_1 and P_2 (Fig. 3.14), the distance ϱ between them must be somewhat greater, and namely

$$\varrho = \frac{1.63\lambda}{2n\sin\sigma} = \frac{0.82\lambda}{A}. \qquad (3.53)$$

As can be seen, the resolving power for coherent is worse than for incoherent light. Besides, two luminous sources or pinholes on a dark background are not as discrete in coherent as in incoherent illumination (compare Fig. 3.37); the images of two mutually coherent points are always surrounded by more intensive diffraction rings.

3.6.2. Two-line resolution

Let us now see what happens when a double slit acts as two incoherent light sources against a dark background. It can be found that by applying the same criterion as for two points, the minimum distance

$$\varrho = \frac{\lambda}{2n\sin\sigma} = \frac{0.5\lambda}{A} \tag{3.54}$$

for resolved images in incoherent illumination, and

$$\varrho = \frac{1.5\lambda}{2n\sin\sigma} = \frac{0.75\lambda}{A} \tag{3.55}$$

in coherent illumination. This shows that the two-line resolving power of the microscope is better than the two-point resolving power. This fact is well known in practice.

3.6.3. Sparrow's criterion

This criterion states that two luminous points can be treated as already separated if the second derivative d^2I'/dx'^2 is zero at the mid-point between the centres of the Airy discs (Figs. 3.14 and 3.15). According to this assumption, the two-point resolution is given by [3.28, 3.29]:

$$\varrho = \frac{0.47\lambda}{A} \quad \text{for incoherent light,} \tag{3.56}$$

and

$$\varrho = \frac{0.73\lambda}{A} \quad \text{for coherent light,} \tag{3.57}$$

whereas the two-line resolution is expressed by

$$\varrho = \frac{0.41\lambda}{A} \quad \text{for incoherent light,} \tag{3.58}$$

and

$$\varrho = \frac{0.66\lambda}{A} \quad \text{for coherent light.} \tag{3.59}$$

The above values of ϱ are, of course, somewhat smaller han those given by Eqs. (3.52)–(3.55).

3.6.4. Effect of the condenser aperture on resolution

Hopkins and Barham [3.24] showed how two-point resolution varied in relation to the condenser numerical aperture A_c. In order to make the comparison, the Rayleigh criterion was used, namely that resolution is reached when the drop in intensity at the mid-point between the centres of the Airy discs amounts to 26%. Given this assumption, the following formula was found:

$$\varrho = K\frac{\lambda}{2n\sin\sigma} = K\frac{\lambda}{2A},\qquad(3.60)$$

where K is a variable coefficient which depends on the ratio $c = A_c/A$. This coefficient is plotted in Fig. 3.38. If the condenser is almost fully diaphragmed, $A_c = 0$ and the illumination is completely coherent. In this case $K = 1.62$ and Eq. (3.60) takes the form of Eq. (3.53). When the numerical apertures of both condenser and objective are the same $(A_c = A)$, the illumination is incoherent

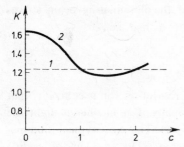

Fig. 3.38. Effect of the condenser aperture on the two-point resolution [3.24].

and $K = 1.22$. In this case Eq. (3.60) takes the form of Eq. (3.52). If c increases from zero to unity, the resolution limit ϱ decreases from $0.82\lambda/A$ to $0.61\lambda/A$, and drops to a minimum value of about $0.57\lambda/A$ when $c \approx 1.5$, but increases again thereafter. In practice, owing to stray light, the ratio c higher than unity is, however, not applicable. Experience shows that for bright-field microscopy $c \approx 0.65$ is the best compromise between image contrast and resolution (compare Fig. IV).

The theoretical conclusions reached by Hopkins and Barham on the basis of coherence theory were experimentally supported by Arnulf [3.30], though he found K-factors somewhat smaller than those resulting from the theoretical curve shown in Fig. 3.38. Thereafter Tsujichi [3.31] analysed the influence of condenser quality on microscope performance and stated that only relatively great aberrations of the condenser evidently degrade microscopical images.

Next, many other researchers [3.32–3.50] investigated how two-point resolution and/or two-line resolution vary in relation to the coherence degree γ_{12}. Beran and Parrent employed Sparrow's criterion and showed that ϱ increases (roughly linearly) when γ_{12} does [3.32].

DIFFRACTION THEORY AND FOURIER SPECTRA

3.6.5. Grating resolution

For the resolution of a grating structure, the objective must at least accept two successive diffraction beams, and especially one of the zero order and one of the first orders. By substituting $q = 2$ and $p = \varrho$ in Eq. (3.29) we obtain

$$\varrho = \frac{\lambda}{2n\sin\sigma} = \frac{\lambda}{2A}. \tag{3.61}$$

This value can only be obtained when a parallel illuminating beam strikes the grating obliquely at an angle of incidence θ (see Fig. 1.91). In this case the diffraction spots Q_0 and Q_{+1} (or Q_{-1}) occur at opposite sides of the edge of the objective exit pupil (Fig. 3.17). Consequently, if the illuminating beam strikes the grating normally ($\theta = 0$), the resolution is twice as bad, i.e.,

$$\varrho = \frac{\lambda}{n\sin\sigma} = \frac{\lambda}{A}. \tag{3.62}$$

Formulae (3.61) and (3.62) express the grating resolution for a coherent light beam which strikes a grating obliquely and normally. For incoherent light the grating resolution is found to be given by Eq. (3.61). The most typical illumination in bright-field microscopy is partially coherent. The problem of grating resolution for this type of illumination is, however, more complicated than for either coherent or incoherent illumination and has been discussed by many authors [3.51–3.59]. Here, it is not necessary to go into details and it will be enough to give only the following formula:

$$\varrho = \frac{\lambda}{A+A_c}, \tag{3.63}$$

which is generally used for the resolving power definition in partially coherent light when the numerical aperture A_c of the condenser does not exceed that (A) of the objective ($A_c \leqslant A$). If $A_c = 0$, the illumination is coherent and $\varrho = \lambda/A$. Consequently, if $A_c = A$, the illumination is incoherent and $\varrho = \lambda/2A$.

3.6.6. Effect of the eye on resolution

It is well known that the human eye does not always cooperate perfectly with the microscope [3.60, 3.61] because we spend most of our time observing objects in average illumination which corresponds to an eye pupil of approximately 3 mm. Under these conditions two objects can be observed as resolved when they subtend an angle of 1.5 minutes at the eye.[11] If the objects are at the distance of distinct vision (250 mm), the *transverse resolving power* ϱ_e of the eye is equal to 0.1 mm (ϱ_e for two line objects is somewhat smaller than 0.1 mm).

When the illumination intensity is modified, eye pupil size varies and so does the resolution capability of the eye. As has already been stated (see Subsection 2.3.2), the diameter $2r_{E'}$ of the exit pupil of a common microscope is less than 3 mm, therefore the size of the eye pupil is artificially reduced to $2r_{E'}$ irrespective of the iris aperture, and this reduction modifies the ability of the eye to produce resolved images. The problem of resolution versus the diameter of the microscope exit pupil was examined by Arnulf using the *Foucault test*, which consists of several alternately bright and dark parallel streaks of identical length and width with a transmittance distribution described by a square function $\tau(x)$ such as shown in Fig. 3.39.

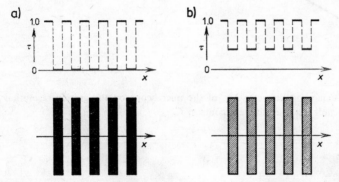

Fig. 3.39. Foucault test and distribution of its light transmittance τ: a) object contrast $C = 1$, b) $C < 1$.

The results obtained by Arnulf are shown in Fig. 3.40. As can be seen, the best resolving power (in this case the grating resolution) occurs when the diameter $2r_{E'}$ of the exit pupil of the microscope (and, consequently, that of the eye iris) is equal to 0.7 mm. For $2r_{E'} > 0.7$ mm, resolution deteriorates due to the aberrations of the eye,[12] while for $2r_{E'} < 0.7$ mm the loss of resolution is caused by the *entoptic phenomenon*. This injurious effect is generated by debris in foreign

[11] Commonly 1 minute of arc is given as the angular resolving power of the eye (or visual acuity), but a value of 1.5 minutes is closer.

[12] Statements can be found in literature to the effect that aberrated images formed at the eye retina are corrected (compensated) by the brain. The maximum mental compensation of the eye aberrations corresponds to an eye pupil of approximately 3 mm.

matter of the eyeball and is perceived as filamentary striae which mask the actual image of the object under examination (compare Figs. 3.41a and b). To overcome this defect, Lau designed a sophisticated system consisting of two microscopes axially arranged one over the other [3.62, 3.63]. The first microscope produces an intermediate image on a rotatable ground glass at which the second microscope is focused to observe the final image. The role of the ground glass is to spread the exit pupil of this two-storied microscopical system which has only been developed as a prototype and has never been available commercially.

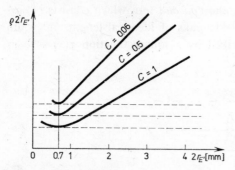

Fig. 3.40. Effect of the exit pupil diameter $2r_{E'}$ of the microscope on the grating resolution ϱ when the grating (Foucault test) has different contrast C.

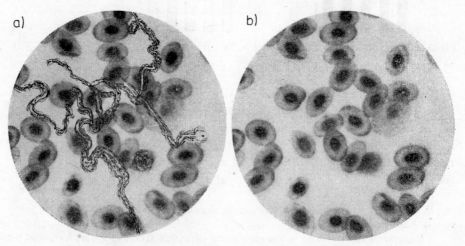

Fig. 3.41. Microscopical image (b) perturbed by the entoptic phenomenon (a). The entoptic striae (a) cannot be photographically recorded and here they are sketched by hand.

3.6.7. Perception limit

Unlike resolution, the perception limit is related to single small objects and directly associated with image contrast. The latter quantity is defined as

$$C' = \frac{I_b' - I_o'}{I_b' + I_o'},$$ (3.64a)

or

$$C' = \frac{I_b' - I_o'}{I_b'},$$ (3.64b)

where I_b' is the light intensity (or brightness) of the background and I_o' is that of the object image. Here the background is regarded as an empty region of the field of view surrounding the image. Note that Eqs. (3.64) are similar to Eqs. (2.4), which describe object contrast. The contrast definition expressed by Eq. (2.4b) and Eq. (3.64b) will be considered below.

First, let us take a small dark disc of diameter d and contrast $C = 1$; the disc is illuminated either incoherently or coherently (Fig. 3.28). Owing to diffraction, as d varies so does the contrast C' of the image disc, and when $d \to 0$ so also $C' \to 0$. A minimum value (d_{min}) of d, for which the contrast C' becomes so small that the disc cannot be perceived, is called the *perception limit*. The *lowest* or *threshold contrast* C_{min} (or C_{min}'), which can be perceived, depends on the object shape. For a dark disc on a bright ground, $C_{min} = 0.04$, and for most other objects this value drops to 0.02. Consequently the perception limit, d_{min}, is different for objects of different shapes, and also depends on the coherence degree of illumination. In general, d_{min} can be expressed as

$$d_{min} = \varkappa \frac{\lambda}{2n \sin \sigma} = \varkappa \frac{\lambda}{2A},$$ (3.65)

where $\varkappa$ is a coefficient which depends on the object shape and the coherence degree of illumination. For a dark disc on a bright ground, $\varkappa = 0.26$ in incoherent illumination and $\varkappa = 0.16$ in coherent illumination, whereas for a dark line on a bright ground, $\varkappa = 0.02$ in incoherent illumination and $\varkappa = 0.01$ in coherent illumination [1.31].

The data given above show that the perception limit, unlike the resolution limit, is better in coherent illumination. By substituting $\varkappa = 0.26$ (or $\varkappa = 0.16$) in Eq. (3.65) we can see that the diameter of a dark disc which it is still possible to perceive is much smaller than the radius of an ideal Airy disc (compare Eq. (3.4)). Moreover, the perceived width of a dark line is approximately one twentieth of that of a dark disc. Finally, the perception limit is absolutely smaller than the resolution limit.

The above remarks refer to dark objects against a bright background. The perception of a small white disc or line on a dark background, on the other hand, involves somewhat different problems. As a small dark object dwindles in size, the undesirable share of diffracted light increases until the object image completely vanishes. Conversely, as a small bright disc on a dark background dwindles in size, the ultimate image takes the form of a typical Airy pattern. If the size of the disc shrinks still further, the spread function showing the intensity distribution in the Airy pattern gradually vanishes, but the size of the Airy disc remain constant. Hence, it appears that the perception or visibility of a small bright object on a dark background merely depends on the luminous flux emerging from the object. The smaller a pinhole in an opaque screen, the more intensive the illumination needed to perceive the pinhole image. The same applies to a slit, but very small slits need less light than pinholes to render them visible.

3.6.8. Useful magnification

To see an object structure, the bright and opaque streaks of a grating for instance, it is not enough to use an objective of proper numerical aperture; the intermediate image of the object structure must also be presented at a sufficiently large angle which should be somewhat larger than the eye's angular resolving power. As was stated earlier, this quantity or visual acuity (α) is normally equal to 1.5 minutes of arc. Is this value sufficient to allow the naked eye to see two images P_1' and P_2' of two object points P_1 and P_2 (Fig. 3.15), whose lateral separation is equal to the objective resolving power ϱ? To answer this question, let us look at the images P_1' and P_2' from the distance of distinct vision (250 mm) using an objective of magnifying power $M_{ob} = 5\times$ and of numerical aperture $A = 0.12$. From Eq. (3.52) it follows that in this case the images P_1' and P_2' are separated by $\varrho' = \varrho M_{ob}$ $= 0.61\lambda M_{ob}/A = 0.014$ mm, whereas the linear resolving power ϱ_e of the eye, corresponding to the acuity of 1.5', is equal to about 0.1 mm at the distance of 250 mm. As can be seen, ϱ' is much smaller than ϱ_e and must be multiplied about ten times in order to allow the images P_1' and P_2', and also the object points P_1 and P_2, to be seen as separated discs. Next, let us consider a high-power objective of magnifying power $M_{ob} = 100\times$ and of numerical aperture $A = 1.30$. Here the separation $\varrho' = 0.026$ mm and is also too small (by about five times) to allow the images P_1' and P_2' to be seen as separated discs. Thus the general conclusion is that the separation ϱ' must be magnified, by means of an ocular, so many times ($M_{oc\,min}$) until the images P_1' and P_2' appear to the eye at least at a separation of ϱ_e. Hence, $M_{oc\,min}\varrho' = \varrho_e$, or

$$M_{oc\,min}M_{ob}\varrho = M_{min}\varrho = \varrho_e, \tag{3.66}$$

where $M_{min} = M_{ob}M_{oc\,min}$ denotes the minimum total magnification. Taking $\varrho = 0.61\lambda/A$ and $\lambda = 0.00055$ mm, the above equation may be written as

$$M_{min} = 3000\varrho_e A \quad [\varrho_e \text{ in mm}]. \tag{3.67}$$

For comfortable microscopical observation one takes ϱ_e somewhat larger than 0.1 mm, viz., $\varrho_e = 0.15$ mm (corresponding to $\alpha = 2'$). By substituting this value in Eq. (3.67), the desired minimum magnification will be found to be

$$M_{min} = 450A. \tag{3.68}$$

These considerations apply to objects of moderate contrast. For lower magnifications than $450A$, the image of a specimen under examination is more brillant, but the resolving power of the microscope objective is not made best use of and the observer's eye can no longer resolve a fine structure whose period p is $\varrho \approx \lambda/2A$ (in incoherent illumination). Conversely, for higher magnifications than $500A$, image contrast drops and the resolution of details no longer improves. However, experience has shown that maximum magnification (M_{max}) for objects of high contrast may be increased up to $1000A$. Observation at greater magnification than $1000A$ does not reveal new details, but only increases the size of those already observed with a drop in sharpness. Furthermore, the exit pupil of the microscope decreases with increasing ocular magnification, hence minute inhomogeneities in the observer's eyeball give rise to obtrusive entoptic disturbances (see Fig. 3.41a), and dust particles on the optical system become trublesome.

Magnification (M_u) higher than $500A$ but smaller than $1000A$ is said to be *useful magnification*, i.e.,

$$500A \leqslant M_u \leqslant 1000A. \tag{3.69}$$

Values above $1000A$ are called *empty magnification*, that is, magnification which enlarges the image size, but does not give more information. In practice, this would appear to mean an upper limit of about $1400\times$ for the final magnification in light microscopy (with high-power objectives $100\times/1.40$).

Nevertheless, in some special techniques of visual microscopy magnifications beyond the upper limit of useful magnification may be employed [2.27]. In particular, this includes techniques that spread the microscope exit pupil, and thus do away with entoptic disturbances. Lau, for instance, cites many examples of the

advantages of visual magnifications up to $6000\times$ offered by his double microscope system [3.62, 3.63]. The Dynascopic system too (see Subsection 2.3.8)
allows one to take advantage of the value of visual magnification beyond the
upper limit of useful magnification.

The above considerations apply to visual observation for which the resolution
is limited by the visual elements of the retina. If the final image is recorded on
a photographic emulsion, the useful magnification is conditioned by the resolving
power ϱ_p of the photosensitive material. The latter quantity is mainly determined
by the silver halide grains which are of different sizes depending on the speed
of photographic films or plates. In fine-grain films the grain size is less than
1 μm and equal to about 2 μm in films of moderate speed. In the exposed and
processed film the silver halide grains build up the photographic picture whose
resolution is, however, worse than the size of the unexposed grains because
during exposure light scatters from the individual grains and produces a blackened
halo called the *circle of diffusion*, the diameter of which determines the *photographic
resolving power* ϱ_p. In this case Eq. (3.67) takes the form

$$M_{\min} = 3000\varrho_p A, \tag{3.70}$$

where $M_{\min}$ is the minimum magnification of the image on the photographic
film or plate.

The resolving power (or graininess) ϱ_p of the negative emulsion after development varies from about 5 μm (slow-speed emulsion) to 20 μm (fast-speed emulsion). In these circumstances the minimum desirable magnification $M_{\min}$ on the
negative is $15A$ (slow-speed emulsion) and $60A$ (fast-speed emulsion). These
values, as can be seen, are much smaller than for visual microscopy. However,
experience has shown that minimum photographic magnification is more fa-
vourable if the images corresponding to two object points just resolvable by the
microscope objective are separated on the emulsion by about ten times the
emulsion resolving power; hence it appears that the ten times rule requires $M_{\min}$
$\approx 600A$ for fast-speed emulsions and $M_{\min} \approx 150A$ for slow-speed emulsions.
In general, the same value as for visual observation, i.e., $M_{\min} = 500A$ is roughly
accepted for photomicrography as well. Consequently, the maximum desirable
magnification on the negative is also of the same order as for visual observation,
i.e., $1000A$. However, this limit may even be increased beacuse photomicrography
obviates several disturbances occuring in visual observation, and thus permits
much larger microscopical magnification than $1000A$, say up to $3000\times$, using
not only high-power objectives and oculars but also intermediate magnification
changers or large extensions of a bellow camera. It is useful, of course, to take

greatly enlarged photomicrographs if the microscopical images are to be observed from a great distance. In general, the useful magnification M_u for photomicrography is defined as

$$\frac{250+d}{250}\,500A \leqslant M_u \leqslant \frac{250+d}{250}\,1000A \quad [d \text{ in mm}], \tag{3.71}$$

where $250+d$ is the distance (in mm) from which a photo is observed. If photos are destined for observation at a distance of distinct vision for instance, then $d = 0$ and M_u is the same as for direct visual microscopy. Conversely, when photomicrographs are destined for remote observation from a great distance, say up to 2000 mm, then $250+d = 2000$ mm and the range of useful magnification is $4000A \leqslant M_u \leqslant 8000A$. Of course, such a great magnification range is normally obtained by producing enlarged prints from negatives of smaller magnification. When subsequent enlargements are required, fine grain emulsions and fine grain developers are necessary; otherwise, the grain of the emulsion will be visible.

Finally, it is important to note that optimal magnification in visual microscopy is highly dependent on the performance of the eye-microscope system, the visual contrast threshold, the spatial frequency structure of the object under study, the coherence degree of illumination, etc. For more intensive information the reader is referred to special papers, e.g., [3.64] and [3.65].

3.7. Three-dimensional imagery in microscopy

The discussion of microscopical imagery and resolution presented so far has assumed that the object field was a flat plane perpendicular to the optic axis of the microscope objective and that the object image was similarly contained in a single plane. In reality, however, both the object space and the image space have three dimensions. The problem of three-dimensionality will briefly be considered in the following pages.

3.7.1. Depth of field and depth of image

If one object plane Π (Fig. 3.42) is sharply imaged, details distant from this plane by $-\Delta z_1$ and $+\Delta z_2$ may still be observed or recorded with satisfactory sharpness in the image plane Π'. The range $2\Delta z = |-\Delta z_1| + \Delta z_2$ is called the *depth of field* when measured in the object space, whereas conjugate distances $-\Delta z_1'$ and $+\Delta z_2'$ from the image plane Π' determine the *depth of image* $2\Delta z' = |-\Delta z_1'| +$

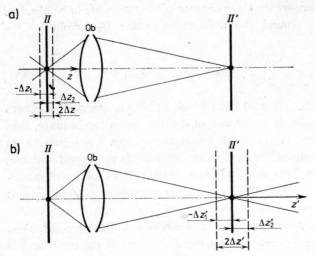

Fig. 3.42. Depth of field (a) and depth of image (b) of a microscope objective.

$+\Delta z_2'$. In general, $|-\Delta z_1| \neq \Delta z_2$ and $|-\Delta z_1'| \neq \Delta z_2'$, but in microscopy the difference between $|-\Delta z_1|$ and Δz_2 as well as that between $|-\Delta z_1'|$ and $\Delta z_2'$ is very small, and one can take $|-\Delta z_1| = \Delta z_2 = \Delta z$ and $|-\Delta z_1'| = \Delta z_2' = \Delta z'$.

From Eqs. (2.17a) and (2.17d) it results that

$$2\Delta z' = M_l 2\Delta z = M_{ob}^2 2\Delta z \frac{n'}{n}, \tag{3.72a}$$

where M_l and M_{ob} are, respectively, the longitudinal and transverse magnifications of the objective, n is the refractive index of the object space, and n' is that of the image space. As the image space in microscopy is usually the air, the refractive index $n' = 1$, and Eq. (3.72a) takes the forms:

$$2\Delta z' = M_{ob}^2 \frac{2\Delta z}{n} \tag{3.72b}$$

for immersion objectives, and

$$2\Delta z' = M_{ob}^2 2\Delta z \tag{3.72c}$$

for dry objectives.

Depth of field is sometimes used instead of *depth of focus*. This is bad practice; the term "depth of focus" signifies the latitude of focusing defined as an axial distance along which the microscope objective or object can be moved without noticeable degradation of *image sharpness* or *definition*.[13] In this case the depth of focus may be quantitatively expressed in divisions of the graduated fine adjustment knob.

In Subsection 3.1.1 we considered the distribution of light across the Airy pattern restricting ourselves, however, to the image plane perpendicular to the optic axis of the microscope objective. Now, for what follows, it is also necessary to know the axial distribution of intensity $I'(z')$ in the diffraction image of a lu-

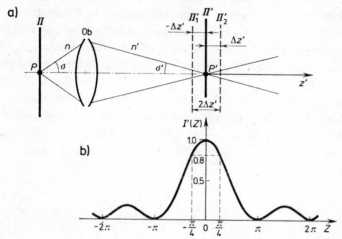

Fig. 3.43. Axial distribution of intensity in the diffraction image of a luminous point.

minous point P situated in the object plane Π of an aberration-free microscope objective Ob (Fig. 3.43a). We find that the distribution $I'(z')$ normalized to unity for $z' = 0$ may be expressed as

$$I'(Z) = \left(\frac{\sin Z}{Z}\right)^2,$$

(3.73)

where

$$Z = \frac{\pi n'}{2\lambda}(\sin^2\sigma')z' = \frac{\pi}{2\lambda n'}A'^2 z'.$$

(3.74)

Here n' is the refractive index of the image space ($n' = 1$ as is usual in microscopy),

---

[13] In general, the terms "sharpness" and "definition" are used synonymously. However, definition has sometimes been used mistakenly instead of resolution. The latter term means the ability to separate two similar objects, whereas definition is the ability to show clearly the form of an object.

λ the wavelength of light, σ' the image-side aperture angle, A' the image-side numerical aperture of the objective ($A' = n'\sin\sigma' = \sin\sigma'$), z' the Cartesian coordinate measured along the optic axis of objective starting from the geometrical image P' of the object point P. The plot of Eq. (3.73) is shown in Fig. 3.43b. As can be seen, a central maximum of predominant intensity and a series of secondary maxima occur. The secondary maxima of gradually decreasing intensity are separated by minima of zero intensity which appear for $Z = \pm\pi$, $\pm 2\pi$, $\pm 3\pi$, ... By comparing Eqs. (3.73) and (3.1), as well as Figs. 3.43b and 3.3, we find that the axial distribution of intensity in the Airy pattern is different from the lateral distribution. Hence, it appears that the aberration-free diffraction image of a luminous point is spherically asymmetric; thus the *longitudinal (axial) resolving power* of the microscope differs from the transverse resolving power which was discussed earlier.

If we consider the situation shown in Fig. 3.43, we see that the depth of image is determined by planes Π_1' and Π_2', distant symmetrically from the image plane Π' by $-\Delta z'$ and $+\Delta z'$, for which the intensity $I'(Z)$ falls to 80% of its maximum value. Such a decrease of intensity occurs for $Z = \pm\pi/4$. Substituting this value in Eq. (3.74) yields

$$\Delta z' = \frac{0.5\lambda}{n'\sin^2\sigma'} = \frac{0.5 n'\lambda}{A'^2}, \tag{3.75}$$

and we obtain

$$2\Delta z' = \frac{\lambda}{n'\sin^2\sigma'} = \frac{n'\lambda}{A'^2} \tag{3.76a}$$

for the depth of image, and, according to Eqs. (2.53) and (3.72),

$$2\Delta z = \frac{\lambda}{n\sin^2\sigma} = \frac{n\lambda}{A^2} \tag{3.76b}$$

for the depth of field. Here n is the refractive index of the object space and σ is the object-side aperture angle, whereas A denotes the numerical aperture of the objective ($A = n\sin\sigma$). As the angle σ' is normally quite small, the distance $2\Delta z'$ is relatively large by comparison with $2\Delta z$. This is especially advantageous for photomicrography because the camera position is not critical after focusing has been adjusted.

Almost identical formulae for depth of image and depth of field result from the Rayleigh quarter-wave criterion (see Subsection 2.3.6). They have the form

$$2\Delta z' = \frac{\lambda}{4n'\sin^2(\frac{1}{2}\sigma')}, \tag{3.77a}$$

and

$$2\Delta z = \frac{\lambda}{4n\sin^2\left(\frac{1}{2}\sigma\right)}.$$ (3.77b)

If σ and σ' are small, one can take $\sin\sigma = \sigma$, $\sin\sigma' = \sigma'$, $\sin(\sigma/2) = \sigma/2$, and $\sin(\sigma'/2) = \sigma'/2$. Under these assumptions, there is no difference between Eqs. (3.76) and (3.77). It is also worth noting that Eq. (3.77b) is similar to Eq. (2.72), which was derived for dry objectives; here Eq. (3.77b) is also valid for immersion objectives.

Some other criteria are also available for depth of field. One of them is based on the optical transfer function [3.66]. The first zero value od this function for a defocused optical system defines the depth of field.

Formulae (3.76) and (3.77) are true for aberration-free objectives the resolving power of which depends primarily on diffraction. Such optical systems are said to be *diffraction-limited*; their point spread function results simply from the intensity distribution in an ideal Airy pattern. If, however, the point spread function is deformed (enlarged) by some residual aberrations, the depth of field results not only from diffraction phenomena, but is also, or even primarily, generated by geometrical paths of image-forming rays. In this case we talk about the *geometrical depth of field*.

Besides, Eqs. (3.76) and (3.77) do not include some factors which result from properties of visual observation. Therefore, other formulae derived mainly by Berek [3.67, 3.68] are used to express the *visual depth of field*. The most popular of them has the form

$$2\Delta z = n\frac{0.5\lambda}{A^2} + n\frac{340}{AM_m} \quad [\mu m],$$ (3.78)

where M_m is the total visual magnification of the microscope defined by Eq. (2.43). A graphical representation of Eq. (3.78) for a refractive index n of the object space equal to 1.45 is shown in Fig. 3.44, which yields an approximate visual depth of field $2\Delta z$ if the numerical aperture A of the objective and the magnification M_m are known.

The values of $2\Delta z$ resulting from Fig. 3.44 must be multipled by a factor $n/1.45$ when the refractive index n of the object space is not equal to 1.45. The diagram is after all only approximate because Eq. (3.78) does not include a term for eye accommodation. If accommodation is undertaken, the last equation must be rewritten as

$$2\Delta z = C_1\frac{4\lambda n}{A^2} + C_2\frac{n}{AM_m} + n\frac{a_v^2}{M_m^2}\left(\frac{1}{a_1} - \frac{1}{a_2}\right) \quad [mm],$$ (3.79)

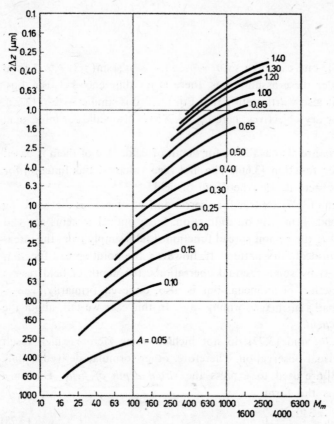

Fig. 3.44. Diagram for a rough evaluation of the depth of field $2\Delta z$ (M_m denotes the total magni-fication of microscope, A the numerical aperture of objective).

where C_1 and C_2 are the constant factors, a_v is the distance of distinct vision ($a_v = -250$ mm), a_1 the distance of the *near point*, and a_2 the distance of the far point of the eye. The difference $1/a_1 - 1/a_2$ is the *accommodation range*. This quantity depends on the individual properties of the observer's eye. The constant factor C_1 is dimensionless and usually takes a value of 0.125, whereas $C_2 = 0.34$ mm (after Berek). The first term of Eq. (3.79) results from the diffraction theory of microscopical imaging, and the second refers to the properties of the eye. The third term results from the accommodation; it must be added for the visual depth of field when observation is performed without a micrometer eyepiece.

The depth of field is different for various spatial frequencies. This problem was examined by Hopkins [3.69], who stated that the depth of field varies with spatial frequencies (u) as follows:

$$2\Delta z = \frac{0.4}{u\sin\sigma},\qquad(3.80)$$

where u is expressed in cycles per millimeters, and σ is the object-side aperture angle of the microscope objective.

It is well known, moreover, that the depth of field can be increased if the condenser aperture is diaphragmed. This practice, however, results in a reduction of the resolving power of the microscope and is only admissible when resolution is not of paramount importance. In general, maximum resolution and great depth of field are simply not possible at the same time.

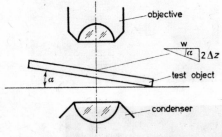

Fig. 3.45. Set-up for an experimental evaluation of the depth of field by using a slant test object.

Depth of field may be experimentally determined by using an arrangement as shown in Fig. 3.45. A test object, e.g., a line grating, is placed on the microscope stage in a sloping position at an angle α to the object plane. The observer sees a sharp image only along a strip of width w. Knowing the angle α and measuring the width w of the sharply imaged object strip, we obtain the depth of field as

$$2\Delta z = nw\tan\alpha,\qquad(3.81)$$

where n is the refractive index of the medium in which the test object is embedded. If the embedding medium is the air, $n = 1$. This procedure is especially suitable for photomicrography (Fig. 3.46).

3.7.2. Stereopsis and depth perception

Stereopsis is the capacity to perceive depth of field and to recognize depth difference in binocular vision. The smallest perceivable depth difference is called *depth perception*. This quantity primarily depends upon the *stereo angular resolving*

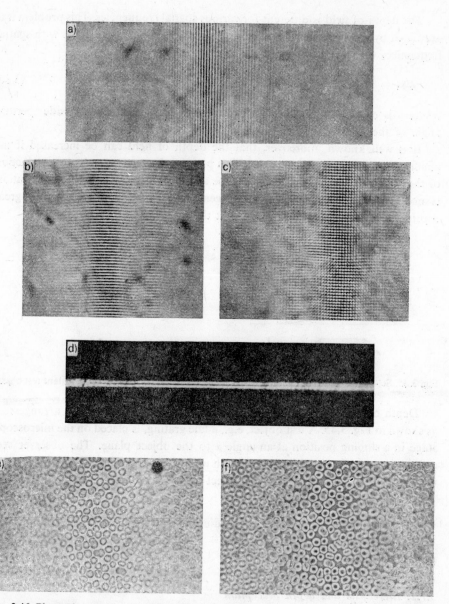

Fig. 3.46. Photomicrographs illustrating the experimental evaluation of the photomicrographic depth of field by using different test objects slanted at an angle α to the optic axis of microscope objective (see Fig. 3.45): a) and b) line grating, c) cross-line grating, d) double slit, e) and f) smear of human blood in positive and negative phase contrast.

power[14] α_s and the smaller it is the better. Depending on the brightness in the object space and some other factors [3.70, 3.71], α_s = 5 to 60 angular seconds. For daylight vision, $\alpha_s \approx 10''$ is generally accepted. For binocular visual microscopy, depth perception δz is expressed by the following equation:

$$\delta z = \frac{250\alpha_s}{M_m \tan \gamma} \quad [\text{mm}], \tag{3.82}$$

where M_m is the total magnification of the microscope, and γ is the convergence angle of the binocular vision. For normal binocular microscopy, $\gamma = 0$; thus $\delta z = \infty$ and the observer sees the whole depth of field in the same plane without any stereopsis. The situation is quite different if a stereoscopic microscope is used; a basic feature of stereoscopic microscopes is their convergence angle γ equal to about 14°. If, e.g., $M_m = 100$ and α_s is assumed to be $10''$ (or 0.00005 rd), then the depth perception $\delta z = 0.5$ μm. It will be interesting to compare this value with the depth perception of the naked eye. In the latter case the distance eye-object is assumed to be 250 mm and the interpupillary distance $y = 65$ mm. From these two distances it follows that the angle of convergence γ of the eyes is equal to 15° and $\tan 15° = 0.268$. By substituting this value and $M_m = 1$ into Eq. (3.82), we obtain $\delta z = 0.046$ mm (for $\alpha_s = 10''$). This gives much worse depth perception than when a stereoscopic microscope is used.

In his book [3.70] Schober discusses the theoretical principles of stereopsis in detail. See also, for instance, Ref. [3.72] for details on optical features of stereomicroscopes, and compare Chapter 13 (Vol. 3), in which a stereoscopic method of depth measurements under the microscope is described.

3.7.3. Shallow and deep microscopy

From Equations (3.76)–(3.80) it follows that depth of field rapidly decreases when the numerical aperture of the objective increases. Is this an advantage or not? If the object under study is thiner than $2\Delta z$, there is no problem; if, however, the object is thicker than $2\Delta z$, a conflict occurs between the requirements of satisfactory resolution and depth of field. To obtain maximum resolution, an objective of high numerical aperture must be used, so that the depth of field becomes extremely small, and only a very thin layer of a transparent object can be observed. This is not generally desirable, though in some instances it may be helpful for optical sectioning. This special technique allows the object to be studied visually or photographically by successsive focusing on a series of optical

---

[14] This parameter must not be confused with the angular resolving power of the eye or visual acuity (α) which was defined in Subsection 3.6.8.

sections. To obtain extremely effective sectioning, a method with greatly reduced depth of field, i.e., *shallow microscopy*, must be used. Nomarski's DIC system belongs to this kind of microscopy (see Vol. 2, Chapter 7).

Many biologists and other microscopists are, however, also interested in *deep microscopy*, and here a problem arises if high magnification as well as a satisfactory resolution of fine details are required. The resolving power and depth of field of a microscope both depend on the numerical aperture of the objective but in such a way that high resolution and large depth of field (deep microscopy) are incompatible; the higher the numerical aperture of an imaging system the better the resolving power ϱ but also more shallow the depth of field $2\Delta z$. Moreover, there is a kind of *uncertainty relation* between ϱ and $2\Delta z$. By combining, e.g., Eqs. (3.61) and (3.76b) we obtain

$$\frac{2\Delta z}{\varrho^2} = \frac{4n}{\lambda}, \tag{3.83}$$

and the relation of uncertainty mentioned above can be expressed as

$$\frac{2\Delta z}{\varrho^2} \leqslant \frac{4n}{\lambda}, \tag{3.84a}$$

or more generally as

$$\frac{2\Delta z}{\varrho^2} \leqslant U\frac{1}{\lambda}, \tag{3.84b}$$

where U is a constant which depends on the combination of specific equations that express the resolving power and depth of field. For instance, Eqs. (3.61) and (3.76b) give $U = 4n$, while Eqs. (3.62) and (3.76b) give $U = 1$.

For these and other reasons, effective deep microscopy with high resolution is impossible by any conventional method. Some unconventional procedures have, however, been described in the literature and the most interesting will be surveyed here.

Figures 3.47 and 3.48 show the principle of deep microscopy (or *three-dimensional microscopy*) proposed by the authors of the paper [3.73]. The method uses multiple-image storing and decoding. A fine amplitude grating with black-and-white bars is placed in front of a photographic plate. The plate is successively exposed, while the microscope is appropriately re-focused by Δz and the coding grating is rotated (Fig. 3.47). After processing, the coded photographic plate is replaced in its recording position. The specimen is removed, and the coded multifocus photograph illuminated with the microscope in its original recording arrangement, which now acts as a decoding system (Fig. 3.48). An additional lens

L_1 is placed behind the photograph. In the back focal plane of this lens a coloured diffraction spectrum appears, consisting of a series of light spots ... Q_{-1}, Q_0, Q_{+1} ... A second additional lens L_2, identical to L_1, is placed with its front focus in the diffraction spectrum. If only the central diffraction spot, Q_0, is

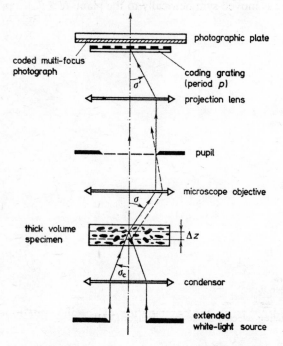

Fig. 3.47. Microscopical system for multi-focus image recording [3.73].

isolated, the grating modulation disappears in the decoded image formed by the lens L_2. The same unmodulated final image is obtained by using one of the higher order diffraction spots. For each orientation of the coding grating (Fig. 3.47), the same diffraction spectrum occurs with spots distributed according to the grating orientation. Each individual diffraction spectrum ... Q_{-1}, Q_0, Q_{+1} ... contains information from different layers Δz of the specimen under examination. The reader is referred to the original paper [3.73] for a more detailed discussion of this point.

The method mentioned above is a modification of the pure photographic superposition of successively focused images. The superposition means that parts of the image out of focus are improved; on the other hand, those parts that would be sharp in a normal photograph become worse. When suitably integrated along the optic axis, the superposed image, though somewhat better, suffers from low contrast of high spatial frequencies. To overcome this defect, Häusler combined incoherent superposition of successively focused images with coherent filtering of the integrated image in a coherent image processor [3.74]. An object with depth $2\Delta z$ is moved symmetrically to the plane Π of sharp

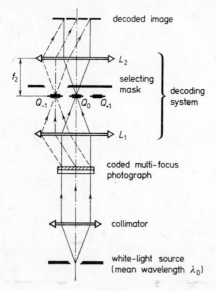

Fig. 3.48. Decoding system for extracting selected single-focus images [3.73].

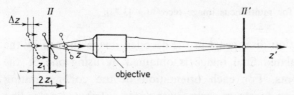

Fig. 3.49. Illustrating the principle of incoherent superposition by moving a microscopic object along the optic axis of microscope objective [3.74].

imaging by a distance $2z_1$, which is large in comparison to $2\Delta z$ (Fig. 3.49). The intensity in the image plane Π', conjugated to Π, is integrated by a photographic plate. The integrated photograph is then filtered in a conventional coherent image processor. This photographic process is, however, complicated and inferior to electronic filtering coupled with a television camera. The result is an undegraded (sharp) image with increased depth of field (see Fig. VII).

Another system for deep microscopy was developed by Gregory and Courtney-Pratt [3.75], who suggested it might be possible to increase the depth of field of a microscope by using a chromatic image-forming system which focuses different spectral colours at different distance from the system instead of a typical objective. We might therefore be able to bring into sharp images at the same time two (or more) planes of the object space by using two (or more) spectral colours.

Some further possibilities of increasing depth of field are also offered by optical scanning microscopy (see Vol. 2, Chapter 12).

3.8. Image contrast and optical transfer function

Although optical transfer function (OTF) has less practical application in microscopy than in other fields of instrumental optics, its theoretical significance is nevertheless very important in microscopical imaging; thus everyone engaged in more advanced microscopy should be familiar with the general concept of optical information transfer and some basic definitions, relations and quantities connected directly or indirectly with OTF.

In Subsection 3.5.4 it was shown that the optical transfer function was the Fourier transform of the point spread function, but it was also found (see Subsection 3.5.2) that the amplitude point spread function was the Fourier transform of the pupil function of the microscope objective. Hence, the OTF must closely be related to the pupil function itself. Before deriving this relation, it will be useful, however, to give some further definitions and the necessary nomenclature directly connected with the optical transfer function. In many books on geometrical and physical optics this nomenclature is not uniform, but that proposed here complies with British Standard BS 4779:1971 *Recommendations for Measurement of the Optical Transfer Function of Optical Devices*.

3.8.1. Extended nomenclature and definitions relating to OTF

Returning to the optical transfer function (see Subsection 3.5.4), it is important to note that this function is, in general, a complex quantity, i.e.,

$$J'(u, v) = |J'(u, v)|\exp[-i\chi(u, v)], \tag{3.85}$$

where $|J'(u, v)|$ is the modulus and $\chi(u, v)$ is the argument of $J'(u, v)$.

To explain Eq. (3.85), let us consider a sine-wave amplitude grating of spatial frequency u (or period $p = 1/u$). The transmittance of this grating (Fig. 3.50a)

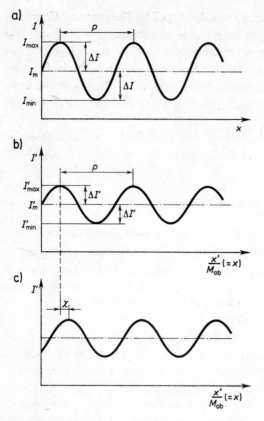

Fig. 3.50. Illustrating the concept of optical transfer function (x and x' are the mutually conjugate coordinates in the object and image planes).

varies sinusoidally in the direction perpendicular to the length of grating lines. One of the basic features of such an object is the *modulation* defined as

$$C = \frac{I_{max} - I_{min}}{I_{max} + I_{min}}, \qquad (3.86a)$$

where I_{max} and I_{min} denote the maximum and minimum values of light intensity transmitted by the uniformly illuminated grating. In other words, the modulation is the grating contrast, defined earlier by Eq. (2.4a). Note that an equivalent definition relates modulation to the ratio of the amplitude (ΔI) of variation in the sinusoidal intensity distribution to the mean intensity level (I_m), i.e.,

$$C = \frac{\Delta I}{I_m}. \qquad (3.86b)$$

The image formed by a microscope objective of a sinusoidally varying object intensity distribution is also a sinusoidally varying intensity distribution (Fig. 3.50b). Its modulation (C') is defined as for object modulation, i.e.,

$$C' = \frac{I'_{max} - I'_{min}}{I'_{max} + I'_{min}}, \qquad (3.87a)$$

or

$$C' = \frac{\Delta I'}{I'_m}, \qquad (3.87b)$$

where I'_{max} and I'_{min} are the maximum and minimum values of image intensity, and $\Delta I'$ denotes the amplitude of the sinusoidal intensity distribution across the grating image. The ratio of image modulation (C') to object modulation (C) at a particular spatial frequency is called the *modulation transfer factor*. The variation of this quantity with spatial frequency is said to be the *modulation transfer function* (*MTF*), which is in fact equivalent to the modulus of the optical transfer function, i.e.,

$$|J'(u)| = \frac{C'(u)}{C(u)} = \frac{I'(u)}{I(u)} \frac{I_m(u)}{I'_m(u)}. \qquad (3.88)$$

This function is usually normalized to unity at zero spatial frequency, i.e.,

$$|J'_N(u)| = \frac{|J'(u)|}{|J'(u = 0)|}. \qquad (3.89)$$

The grating image formed by an actual optical system may be laterally displaced (Fig. 3.50c) from a correct position (Fig. 3.50b) theoretically prescribed for an ideal (aberration-free) system. This displacement (χ) or *phase transfer value* (*PTV*) is in fact the argument of the optical transfer function defined by Eq. (3.85). It is measured in units of the image spatial frequency and quoted in radians, where 2π rad is equivalent to one period of the sinusoidal intensity distribution in the grating image (the unit of spatial frequency is cycles (c) per millimetre). The variation of the phase transfer value with spatial frequency is called the *phase transfer function* (*PTF*). This function is zero at zero spatial frequency.

The optical transfer function is in fact a complex function whose modulus is the modulation transfer function and argument the phase transfer function. If a microscope objective has a circular exit pupil, as is generally the case, and is free from aberrations, then its optical transfer function is represented by curve in Fig. 3.51. As can be seen, $J'_N(u)$ drops nearly monotonically as u increases, and there is no image contrast at all for spatial frequencies above a certain limit (u_r) which roughly correspond to

$$u_r = \frac{1}{\varrho} = \frac{2A}{\lambda}, \tag{3.90}$$

where ϱ is the grating resolving power of the objective, λ the wavelength, and A the numerical aperture of the objective. For $u \geqslant u_r$, there is no image modulation or, in other words, spatial frequencies greater than u_r are not transferred by the objective to its image plane. This situation is illustrated in Fig. 3.52, where the left-hand and right-hand sine-wave curves represent, respectively, amplitude gratings of different frequencies and their images. Although object

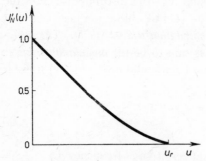

Fig. 3.51. Optical transfer function for an objective free from aberrations or focus error.

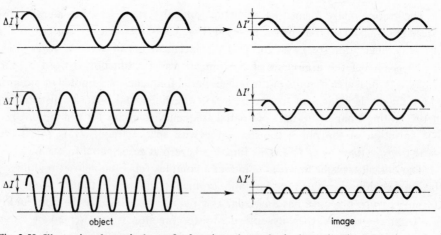

Fig. 3.52. Illustrating the optical transfer function whose plot is shown in Fig. 3.51.

modulation ΔI is the same for all frequencies, image modulation $\Delta I'$ decreases as spatial frequencies increase. This rule is general for optical imaging in incoherent light.

3.8.2. Optical transfer function as autocorrelation of pupil function

A practical expression for the optical transfer function can be given in terms of the autocorrelation of the pupil function $e(\zeta, \eta)$.[15] This operation is done by shearing the exit pupil E'_{ob} of the microscope objective (Fig. 3.53), and may be expressed as the autocorrelation integral

$$\iint\limits_{-\infty}^{+\infty} e(s_\zeta+\zeta, s_\eta+\eta)e^*(\zeta, \eta)\,\mathrm{d}\zeta\,\mathrm{d}\eta,$$

where ζ, η, are the Cartesian coordinates of the exit pupil E'_{ob}, and s_ζ, s_η are the components of the pupil shear $(\mathbf{s})$ projected at the coordinates ζ, η. Without

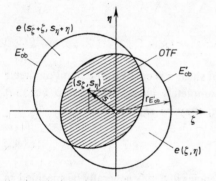

Fig. 3.53. Illustrating the concept of the optical transfer function derived from the autocorrelation of the pupil function.

restricting the general validity, we can assume that the objective magnification M_{ob} is equal to unity. Then taking into account Eqs. (2.54) and (3.34), we can express the shear components as

$$s_\zeta = \lambda f'_{ob} u, \qquad s_\eta = \lambda f'_{ob} v, \tag{3.91}$$

where λ is the wavelength of light, u, v are the spatial frequencies in the directions of coordinates ζ, η, and f'_{ob} is the focal length of objective.

Under these conditions, the above-mentioned expression for the OTF is given by

[15] See Subsection 3.5.2 for definition.

$$J'_N(u, v) = \frac{1}{N} \iint\limits_{-\infty}^{+\infty} e(\lambda f'_{ob} u + \zeta, \ \lambda f'_{ob} v + \eta) e^*(\zeta, \eta) d\zeta d\eta, \qquad (3.92)$$

where

$$N = \iint\limits_{-\infty}^{+\infty} |e(\zeta, \eta)|^2 d\zeta d\eta \qquad (3.93)$$

is a normalizing factor to make $J'_N(0, 0) = 1$. The range of integration in the above equations is only formally infinite; the integral in Eq. (3.92) is only non-zero over the area common to the exit pupil E'_{ob} and its twin shifted by $\mathbf{s}$, while the integral (3.93) is only non-zero over the circle of radius $r_{E'_{ob}}$. With reference to Fig. 3.53, the OTF values are simply equal to the values of crosshatched areas relative to that of the circle area of radius $r_{E'_{ob}}$. Figure 3.53 and Eq. (3.92) show immediately that the OTF must diminish for the increasing spatial frequencies u, v and vanish identically for

$$u = u_r = \frac{2}{\lambda f'_{ob}} r_{E'_{ob}} \quad \text{and} \quad v = v_r = \frac{2}{\lambda f'_{ob}} r_{E'_{ob}}. \qquad (3.94)$$

This extreme situation occurs when the pupil shear s is equal to the pupil diameter $2r_{E'_{ob}}$; in this case the twin exit pupil is completely clear. Taking into account Eq. (2.54), the *cut-off frequencies* u_r and v_r may be expressed as

$$u_r = v_r = \frac{2A}{\lambda}, \qquad (3.95)$$

and for a circular exit pupil the spatial frequencies u, v may generally be specified as normalized quantities

$$\bar{u} = \frac{s}{2r_{E'_{ob}}} \frac{2A}{\lambda} = \bar{s} \frac{A}{\lambda}, \qquad \bar{v} = \frac{s}{2r_{E'_{ob}}} \frac{2A}{\lambda} = \bar{s} \frac{A}{\lambda}, \qquad (3.96)$$

where $\bar{s} = s/r_{E'_{ob}}$ represents the normalized exit pupil shear expressed as a fraction of the exit pupil radius.

The above discussion is illustrated in Fig. 3.54, where only the one-dimensional spatial frequency $\bar{u}$ is taken into account, and the exit pupil shear is parallel to the coordinate ζ along which different values of $\bar{u}$ are considered. It is clear that hatched areas in Fig. 3.54 do not diminish linearly as $\bar{s}$ increases, hence the plot $J'_N(\bar{u})$ is a downward bent line as shown in Figs. 3.51 and 3.54. For comparison, Fig. 3.55 shows the OTF for an aberration-free objective the exit pupil of which

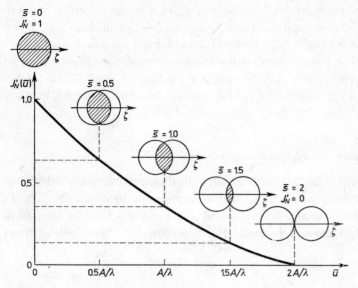

Fig. 3.54. Graphical representation of the optical transfer function of a perfect objective with circular exit pupil (A denotes the numerical aperture of objective).

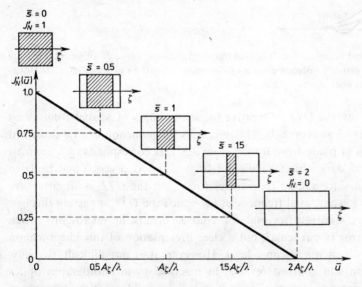

Fig. 3.55. As in Fig. 3.54, but the objective has a square exit pupil (A_ζ denotes the numerical aperture of objective in the direction of pupil shear).

is square or rectangular. In this case the plot $J'_N(u)$ is linear since the common area of the sheared exit pupil is a linear function of the pupil shear s.

The expression for the OTF in terms of the exit pupil autocorrelation constitutes a very useful and easy procedure for analysing the transfer properties of an optical system when its exit pupil is variously modified to obtain some special techniques of light microscopy, e.g., phase contrast, amplitude contrast, modulation contrast, etc., (see Vol. 2).

3.8.3. Contrast reversal and spurious resolution

Until now it has been tacitly assumed that the optical system (microscope objective) is perfect. Aberrations, focus errors, light absorption, or light scattering, however, cause the OTF to deviate (curves *2* and *3* in Fig. 3.56) from its ideal form (Figs. 3.51 and 3.54, curve *1* in Fig. 3.56). Sometimes, if aberrations or focus

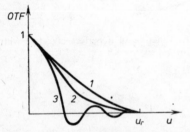

Fig. 3.56. Ideal (*1*) and actual (*2* and *3*) optical transfer functions: curve *1*—objective free from aberrations or focus error, *2*—objective suffers from small aberrations or focus error, *3*—objective suffers from great aberrations or large focus error.

errors are significant, the OTF is negative for some ranges of spatial frequencies as indicated in Fig. 3.56 (curve *3*). This injurious phenomenon can be described by a sudden jump in phase from 0 to π (Fig. 3.57). It corresponds to a *contrast reversal* (images of dark object details are bright, and vice versa) and leads to *spurious resolution* since spatial frequencies for which the OTF is negative are resolved, whereas lower spatial frequencies for which the OTF is passing through zero are invisible. Contrast reversal, of course, occurs in microscopy when there is a focus error in particular, but a clear presentation of this phenomenon tends to be difficult at microscopic level. However, it is not difficult to show spurious resolution and contrast reversal in macroscopy or photography optics by using a resolving-power test with continuously variable spatial frequencies. Such a test, known as the *Simens star*, is presented in Fig. 3.58a, whereas Fig. 3.58b

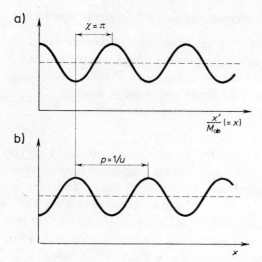

Fig. 3.57. Contrast reversal of a spatial frequency $u = 1/p$.

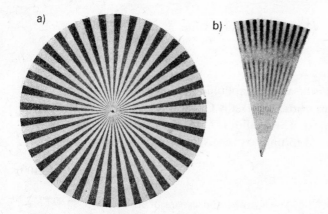

Fig. 3.58. Simens star (a) and its fragment (b) imaged by an objective which suffers from great aberrations or large focus error. The image b shows reversed contrast of some spatial frequencies and spurious resolution.

shows a fragment of this test imaged by an objective the OTF of which is represented by curve 3 in Fig. 3.56.

The loss of OTF or MTF with increasing spatial frequencies is one of the most effective measures of the quality of optical systems to be used for imaging extended objects in incoherent light.

3.8.4. Optical transfer function for coherent and partially coherent illumination

The considerations presented above refer to completely incoherent imaging, whereas for coherent and partially coherent imaging the situation is more complicated. First, for coherent light the *OTF* does not refer to the intensity modulation but to modulation of light amplitude in the object and image planes.

Using the convolution theorem gives the following equation from (3.51):

$$J_a(u, v) = F_a(u, v) J'_a(u, v), \qquad (3.97)$$

where $J_a(u, v)$, $F_a(u, v)$, and $J'_a(u, v)$ are the Fourier transforms of $A'(x, y)$, $\tau_a(x, y)$, and $i'_a(x, y)$, respectively. Here $\tau_a(x, y)$ is the amplitude transmittance of the object, $A'(x, y)$ is the light amplitude in the object image, and $i'_a(x, y)$ represents the *amplitude point spread function*. As can be seen, Eq. (3.97) is similar to Eq. (3.45) The only difference is that Eq. (3.45) has referred to light intensities, while Eq. (3.97) refers to light amplitudes. Now, $J'_a(u, v)$ is said to be the optical transfer function for coherent light. From Eq. (3.97) it follows that

$$J'_a(u, v) = \frac{J_a(u, v)}{F_a(u, v)}, \qquad (3.98)$$

and on the analogy of Eq. (3.88) one can also write

$$|J'_a(u)| = \frac{C'_a(u)}{C_a(u)} \qquad (3.99)$$

if the object has the form of a line amplitude grating with a sine-wave profile of amplitude transmittance whose contrast is $C_a(u)$; consequently $C'_a(u)$ represents the amplitude image contrast. The modulus $|J'_a(u)|$ is the *MTF* for coherent light. Usually it is also normalized to unity at zero spatial frequency, i.e.,

$$|J'_{aN}(u)| = \frac{|J'_a(u)|}{|J'_a(u=0)|}. \qquad (3.100)$$

If an objective is free from aberrations and errors of focus, then its *OTF* for coherent light is represented by curve *e* in Fig. 3.59. As can be seen, $J'_{aN}(u) = 1$ for $u = 0$ to $u = u_{rc}$, and suddenly drops to zero, while there is no image contrast at all for spatial frequencies above u_{rc}. It may be found that $u_{rc} = u_r/2$, hence in coherent light the limit spatial frequency u_{rc} transferred by the objective from the object plane to the image plane is half that transferred in incoherent light (compare curve *a* with *e* in Fig. 3.59).

Now, let us assume that the incoherent illumination of the grating under consideration (Fig. 3.50) becomes partially coherent and next increasingly coherent. In that case the *OTF* for incoherent illumination (curve *a* in Fig. 3.59) increasingly resembles (curves b, c, d) that for coherent illumination (curve e).[16] Consequently, the cut-off spatial frequency moves from u_r to u_{rc}. What is the theoretical interpretation of this phenomenon? The answer is not simple, but an approximate explanation (which applies in particular to amplitude objects of low contrast and small dark objects on a predominant bright background) is as follows.

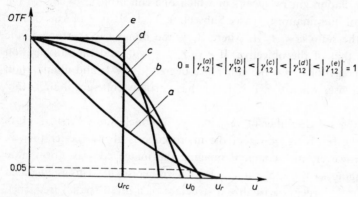

Fig. 3.59. Optical transfer functions of an objective free from aberrations and correctly focused: curve a—in incoherent light, b, c and d—in partially coherent light of greater and greater degree of coherence γ_{12}, e—in coherent light.

To begin with, in partially coherent illumination the OTF is not only a simple autocorrelation of the exit pupil of the objective, but depends also on the distribution of light intensity in the entrance pupil of the condenser. For incoherent illumiation, the autocorrelation formula (3.92) can briefly be written as

$$J'_N(\bar{u}) = e(\bar{u}) \odot e^*(-\bar{u}), \tag{3.101}$$

where $\odot$ denotes the autocorrelation operation. Partially coherent illumination modifies the OTF as follows:

$$J'_{Npc}(\bar{u}) = [a(\bar{u})e(\bar{u})] \odot e^*(-\bar{u}), \tag{3.102}$$

where $a(\bar{u})$ denotes the distribution (spectrum) of the light intensity (expressed as a function of the normalized spatial frequency $\bar{u}$) in the entrance pupil of the condenser. Equation (3.102) shows that the OTF for partially coherent illumination is given by autocorrelating the exit pupil of the objective and multiplying it by the function of light distribution in the entrance pupil of the condenser.

[16] A simple method for coherent and incoherent OTF visualization has been presented by the authors of the paper [3.76].

Hence, it appears that the *OTF* can be changed and even optimized by modifying the condenser aperture. This modification can be achieved in different ways by using diaphragms of different sizes and various shapes, for example. It has been proved among other things that for some special techniques an annular diaphragm is more advantageous than a circular one (see Vol. 2, Chapter 5 and Section 6.1).

However, the simplest way of modifying the condenser aperture depends on the use of an iris diaphragm by means of which one can arbitrarily change the coherence degree of illumination γ_{12} (see Subsection 3.2.1). It is well known that γ_{12} depends on the ratio $c = A_c/A$, where A_c is the numerical aperture of the condenser and A that of the objective. If $c = 1$, $\gamma_{12} = 0$, and the illumination is incoherent, whereas for $c = 0$ the coherence degree $\gamma_{12} = 1$, and illumination is coherent. Otherwise, when $0 < c < 1$, one has partially coherent illumination $(1 > \gamma_{12} > 0)$.

Figure 3.59 suggests a number of very important conclusions. First of all, in incoherent light the resolving power of the microscope is always better than in coherent light. However, in incoherent imaging the image contrast for spatial frequencies belonging to the range $\bar{u} = 1$ to 2 is rather small. This is qualified as a defect because in order to visualize a structure of a given spatial frequency the modulation transfer factor (C'/C) must be significantly greater than zero. Usually, a minimum value of C'/C as small as 0.05 is still perceived. Thus in incoherent light the maximum spatial frequency (u_0) which can still be observed is somewhat smaller than the cut-off value u_r. Conversely, in coherent light all structures contained within the range of spatial frequencies from zero to $\bar{u} = 1$ are imaged with the same modulation transfer factors; hence the image contrast is generally higher than in incoherent light, but the resolution limit is worse (cut-off value u_{rc} for coherent light is smaller than that, u_r, for incoherent light). Moreover, the coherent illumination produces many diffraction artifacts which greatly disturb the actual image.[17] Therefore, a compromise between resolution and contrast is offered by partially coherent illumination of the proper degree of coherence γ_{12} (compare curves *b* and *c* with *a* and *e* in Fig. 3.59). In bright-field microscopy the coherence degree of illumination is, as it has just been stated, easily controlled by means of the iris diaphragm of the condenser. Experience shows that the compromise between contrast and resolution is optimal when the ratio (c) of the numerical apertures of condenser and objective is equal to 2/3 or

$$2r_c' = \tfrac{2}{3} 2r_{E_{ob}'}, \tag{3.103}$$

[17] This problem will be discussed in more detail in Vol. 2, Chapter 11.

where $2r_{E'_{ob}}$ is the diameter of the exit pupil E'_{ob} of the microscope objective (Fig. 3.60), and $2r'_c$ is that of the image of the condenser aperture (opening of the condenser iris diaphragm) observed in the objective exit pupil.

There is an extensive literature [3.77–3.90] on the theory and properties of the *OTF* for various optical systems. In microscopy this function is primarily

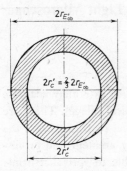

Fig. 3.60. Illustrating Eq. (3.103).

used in theoretical calculations concerning the design of microscope objectives. Sometimes it is also used to evaluate microscope systems quality by measuring the *MTF* (see Section 4.6), but here it is less effective because of the methodological and technical difficulties involved in ascertaining the same conditions of measurement under which microscope objectives really work.

*
*   *

The problems raised in this chapter refer primarily to bright-field microscopy. Naturally, however, many of the theoretical and practical problems discussed here are also valid for other, more specialized microscopical methods and techniques dealt with in the next two volumes. The chapter as it stands is intended to cover a selection of topics basic for advanced light microscopy.

4. Assessment of the Optical Performance of the Light Microscope

Modern light microscopes and their optical components need extremely accurate and sensitive methods for testing the quality of microscopical images. There are many tools and instruments for the assessment of the optical performance of the light microscope; some of them can easily be employed by amateur and professional users of microscopes, while others are only available to specialist firms. In principle, every professional microscopist should be able to evaluate qualitatively the performance of the microscope he uses and to recognize the occurrence of different types of aberrations. What is more, he should be able to look for individual aberrations, and to estimate their magnitude.

This chapter presents both well-known methods for the quality control of microscopical images and some unknown interference techniques (recently developed by the author of this book) for performance evaluation of microscope objectives.

4.1. Resolution as a quality measure of microscope optical systems

There are many simple test objects that can be used by any microscopist to study the optical performance of light microscopes; these include natural specimens such as diatoms and also artificial test objects such as amplitude gratings [3.53, 4.1–4.3].

Formerly standard test specimes were, in principle, diatoms mounted in air or Canadian balsam.[1] Today, however, these specimens are no longer much used because the results obtained in tests of the overall performance of microscope optical systems based on the observation of resolved structures of diatoms is quite inadequate. Nevertheless, preparations of diatoms can help one to assess

---

[1] Preparations of diatoms are commercially available, e.g., from Albert Elger, Eutin in Holstein, Friedrichstrasse 26, West Germany.

the quality of adjustment of the illumination system of transmitted-light microscopes. Table 4.1 contains the best-known diatoms and dimension (d) of their structural elements (striae, ovoid pearls) which should be clearly resolved in white light by a microscope objective of a given numerical aperture A when the condenser numerical aperture A_c is very small (axial illumination) or equal to A (oblique

TABLE 4.1

Diatoms for resolution assessment of light microscopes

Name of diatoms	Dimensions of periodic structures of diatoms d [μm]	Numerical aperture A of objective necessary for resolving the diatom structure	
		axial illumination	oblique illumination
Pinnularia apulenta	1.67–2.00	0.33–0.28	0.17–0.14
Navicula lyra	0.83–1.00	0.67–0.55	0.34–0.28
Stauroneis phoenicenteron	0.71	0.80	0.40
Anomoeoneis serians	0.48	1.15	0.58
Pleurosigma angulatum	0.45–0.56	1.23–1.00	0.60–0.50
Surirella gemma	0.42	1.30	0.65
Nitzschia sigmoida	0.38	1.40	0.70
Nitschia obtusa	0.33–0.36	1.67–1.53	0.84–0.77
Frustulia rhomboides	0.28	—	1.00
Amphipleura pellucida	0.24	—	1.15

illumination). As can be seen, the numerical apertures A listed in Table 4.1 are somewhat greater than those resulting from Eqs. (3.61) and (3.62) for $\varrho = d$. This is no mistake but an intentional deviation especially adopted to secure a somewhat better resolution of a given diatom structure than would results from the theoretical limit of the resolving power of the microscope.

One great disadvantage of preparations of diatoms is that a single diatom cell occupies only a small region of the field of view of the microscope; hence it is impossible to evaluate at once the imaging quality across the entire field of view. For this reason, a more convenient test preparation exists in the form of an amplitude line or cross-line grating, such as shown in Figs. III and IV. The uniform grating structure covers the whole field of view of the microscope and enables one to estimate not only resolution but also field curvature and distortion.

The resolving power of a microscope is, of course, strictly defined by the numerical aperture of the objective and by the mode of illumination. There is no need, therefore, to use the test preparations mentioned above for testing only resolution. Besides, it is risky to judge the overall performance of the optical system of a microscope by its resolving power. Sometimes the nominal value for limit resolution is observed well enough, though the microscope objective suffers from a considerable spherical aberration which means that the objective is of generally poor quality. After all, it is difficult to distinguish between correct and spurious resolution at a microscopical level (see Subsection 3.8.3).

On the other hand, the effective numerical aperture of the objective is sometimes smaller than the nominal numerical aperture or its experimental value measured by means of an apertometer when the illumination system is adjusted according to the Köhler principle (see Subsection 2.3.4). The reduction of the useful numerical aperture of the objective is caused by a badly designed optical system of illumination. The most frequent defect is that when adjusted according to the Köhler principle of illumination, the field diaphragm of the microscope illuminator vignettes the aperture of the objective.

A reasonable estimate of the effective numerical aperture of the objective is possible if a condenser with a graduated iris diaphragm is available. This diaphragm is adjusted so that the image of its opening just covers the exit pupil of the objective (observed with the naked eye or better still with the auxiliary telescope normally available as part of the standard phase contrast equipment). If the condenser has not been calibrated by the manufacturer, this operation can be carried out by the user. The calibration procedure is the same as above but in reverse, using objectives whose numerical aperture is known and free from vignetting. Otherwise, one can measure the diameter of the iris diaphragm and calculate the numerical aperture of the condenser from Eq. (2.56). In this case the focal length of the condenser must, however, be known.

4.2. Evaluation of microscope objective quality by using the diffraction star test

The Airy patterns of luminous points are commonly used by microscope manufacturers for testing the *mounting quality* of objectives [2.13, 4.4, 4.5]. The so-called *star tests* are employed for this purpose. A silver film evaporated in vacuum onto a microscope slide may exhibit minute pinholes which act as point light sources when they are illuminated through the condenser by a high-power lamp. Spherical aberration, coma, and astigmatism are all very easily detectable with a star-test object. The "mounting" spherical aberration is due to the improper axial separation of objective lenses. To inspect this aberration, the silvered slide is placed on the object stage and the objective under examination is focused on one of the pinholes. The out-of-focus images obtained by alternately fine focusing up and down are then observed. If a low- or middle-power objective is free from spherical aberration, the out-of-focus Airy patterns are nearly identical (Fig. 4.1a). When, however, a high-power objective is free from this aberration, the out-of-focus Airy patterns are no longer identical and the correct assessment of the objective quality is a more specialized procedure [2.7, 4.6, 4.7].

If a blurred disc with a bright centre is seen on focusing upwards and a bright ring on focusing downwards, the objective is spherically undercorrected (Fig. 4.1b); the reverse shows that the objective is spherically overcorrected (Fig. 4.1c).

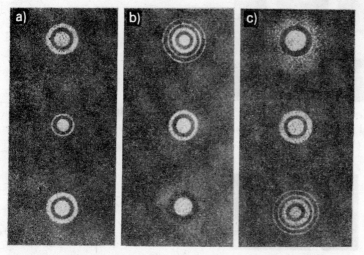

Fig. 4.1. Assessment of spherical aberration of microscope objectives by examination of Airy patterns: a) objective free from spherical aberration, b) spherical aberration undercorrected c) spherical aberration overcorrected (top — objective underfocused, middle — objective focused, bottom — objective overfocused).

As far as coma and astigmatism are concerned in the testing of the mounting quality of objectives, what one primarily looks for are so-called *axial comatic aberration* and *axial astigmatism* .These two mounting aberrations are closely associated and result from the slightly excentric or oblique mounting of the component lenses of the objective. As a result, comatic aberration (Fig. 4.2) and astigmatism (Fig. 4.3) occur not only in off-axial zones but also in the centre of the field of view. Axial coma and axial astigmatism are, of course, more troublesome than the normal off-axial variants of these two aberrations. When coma and astigmatism occur in the axial zone, the Airy pattern is no longer symmetrical in the centre of the field of view. The asymmetry shows up better if the objective is defocused. It is obvious that both the axial and off-axial variants of these aberrations are inadmissible.

If the star test is illuminated with white light, chromatic aberration can also be examined. Longitudinal chromatism reveals itself as a colour variation of the axial Airy disc when the objective under examination is alternately defocused up and down; whereas in the case of transverse chromatism the off-axial Airy

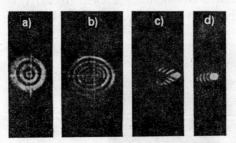

Fig. 4.2. Typical Airy patterns produced by a microscope objective which suffers from comatic aberrations: a) axial pattern, b) slightly defocused axial pattern, c) and d) focused off-axis patterns.

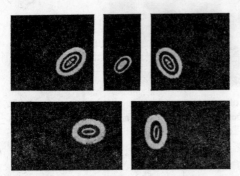

Fig. 4.3. Typical Airy patterns produced by a microscope objective which suffers from astigmatism.

patterns are asymmetrically coloured. If the objective is apochromatically corrected, the out-of-focus axial Airy discs, as well as the optimally focused disc, should be practically white. Colours resulting from chromatic aberration should not be mistaken for the very subtle irreducible colouration generated by the diffraction phenomenon. However, it should be noted in passing that the star test is not the best preparation for the quantitative study of chromatic aberration for which some other more convenient test objects exist, e.g., the stage micrometer (see Section 4.5).

Star tests are also suitable for the study of strain birefringence in objective lenses. This defect causes the Airy pattern to be geometrically deformed (Fig. 4.4). To check for the absence of strain birefringence, it is, however, better to use a conoscopic arrangement (see Section 4.8) rather than a star-test slide.

Fig. 4.4. Airy patterns deformed by strain birefringence of objective lenses.

If the objective under examination is corrected for a cover slip, the silvered surface of the star-test slide must naturally be covered with a proper cover glass. Similarly, an immersion objective requires an appropriate immersion liquid between the star-test slide and the objective front lens.

To sum up, making an exact assessment of the performance of microscope objectives is a highly specialized procedure and testing their quality be using a star test only is obviously inadequate. However, for testing the mounting quality of objectives by an experienced technician, the star test is a very useful tool indeed.

4.3. Interference methods of examining aberrations of microscope objectives

Interference methods for testing the quality of microscope objectives are widely used by manufacturers and researchers. One of these methods, developed by the author of this book [4.8–4.10], will be presented here.

The method uses a double-refracting interferometer with a slit secondary light source. The slit is placed in the object plane of the objective under examination and treated as a Dirac delta function (δ). An interference image of the Fourier transform of this function can be observed in the exit pupil of the objective to be examined. More recently a birefringent fibre was used in place of the planar slit, thus achieving a more versatile instrument [4.11, 4.12].

4.3.1. *A double-refracting interferometer for the study of aberrations of objectives*

Imagine a diaphragm with a very narrow slit placed in the object plane of an aberration-free microscope objective. The slit is illuminated with a light beam parallel to the objective axis. If, as is postulated, the slit is very narrow, the amplitude distribution of light in the object plane can be expressed by the Dirac delta function $\delta(x)$.[2] The Fourier transform of this function, $T[\delta(x)]$, arises

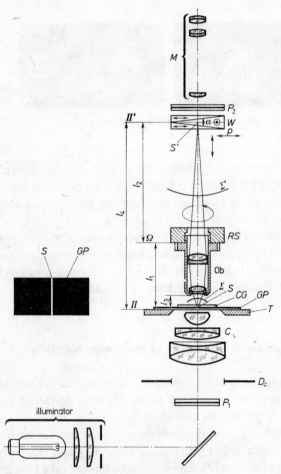

Fig. 4.5. Optical system of a double-refracting interferometer for examination of the quality of microscope objectives [4.8–4.10].

[2] See Ref. [1.5], Chapter 3 for a definition.

in the back focal plane of the objective. If u denotes the spatial frequency in the direction x at right angles to the slit, then $T[\delta(x)] = F(u) = 1$. This means that the wavefront is plane in the Fourier plane and has the same amplitude across the whole exit pupil of the objective (we assume that the objective does not absorb the light). When the objective suffer from aberrations, a deformed Fourier transform of the δ-function occurs.

The Fourier transform mentioned above can be observed by means of a wavefront shear interferometer such as shown in Fig. 4.5. A basic element of this instrument is the symmetric Wollaston prism W (see Subsection 1.6.5) placed between two polarizers (P_1 and P_2) in the image plane Π' of the objective under study Ob. The polarizers are crossed and their directions of light vibration form an angle of 45° with the principal sections of the prism W. The slit S is parallel to the wedge edges of this prism and occupies the object plane Π preceded by a condenser C. A divergent wavefront Σ leaves the slit S. When passing through the objective Ob, the wavefront Σ is subjected to phase retardations corresponding to optical path differences occurring in the optical system of the objective. The image-forming wavefront Σ' is split by the prism W into two wavefronts, Σ_1 and Σ_2 (not shown in Fig. 4.5), polarized at right angles. When passing through the polarizer P_2, both wavefronts interfere with each other. An interference pattern is observed in the Fourier plane of the objective Ob when a low-power auxiliary microscope M is focused onto this plane. If the prism W is adjusted

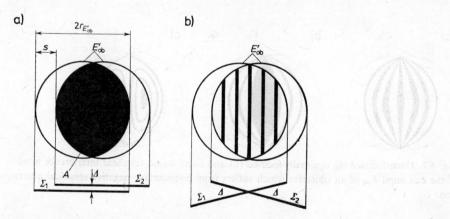

Fig. 4.6. Exactly focused (a) and slightly defocused (b) shear interference images of the exit pupil E'_{ob} of an aberration-free microscope objective (the term "focused" means that the objective under examination is focused onto the slit S of the interferometer shown in Fig. 4.5; consequently, the term "defocused" specifies that the objective is retained in either overfocused or underfocused position with respect to the interferometer slit S).

so that its centre is brought into coincidence with the slit image S' produced by the aberration-free objective, than a uniform field interference pattern is obtained over the area where the sheared wavefronts Σ_1 and Σ_2 overlap (Fig. 4.6a). Otherwise parallel interference fringes occur (Fig. 4.6b). The interfringe spacing decreases as the distance between the slit image S' and the birefringent prism centre increases.

Normally, the prism W is set at a constant distance l_2 behind the objective end surface Ω. This distance directly depends on the mechanical tube length of the objective Ob. Under these conditions, uniform field interference (Fig. 4.6a) is turned into fringe interference (Fig. 4.6b), and vice versa, by defocusing the interferometer, i.e., by changing the distance l_3 between the slit S and the objective Ob (Fig. 4.5). Distances l_1 and l_4 are changed at the same time.

When the prism W is moved transversely (p) to the objective axis, the optical path difference Δ (*bias retardation*) between the split wavefronts Σ_1 and Σ_2 is changed (Fig. 4.6); as a result the colour (if white light is used) or brightness (if monochromatic light is used) of the uniform field interference is altered or the interference fringes are moved.

The wavefronts Σ_1 and Σ_2, in fact, represent a duplicated Fourier transform of the delta function. The lateral duplication (shear) s is the larger the greater is the apex angle α of the birefringent prism W (Fig. 4.5). When the objective Ob suffers from aberrations, the plane wavefronts Σ_1 and Σ_2 (Fig. 4.6) will

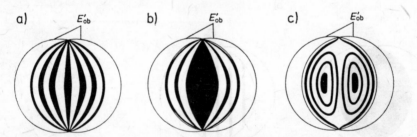

Fig. 4.7. Underfocused (a), optimally focused (b) and overfocused (c) shear interference images of the exit pupil E'_{ob} of an objective which suffers from considerable negative spherical aberration.

be deformed, and neither uniform field interference nor regular fringes will be obtained.

Figure 4.7 shows what happens when the objective under examination suffers from severe negative spherical aberration such as occurs for uncorrected positive lenses (compare Fig. 2.23a). As can be seen, no uniform interference arises for the focused position of the objective, whereas slightly defocused interference patterns contain a number of arched or oval fringes. The situation is almost identical when the objective suffers from a considerable overcorrected (positive) spherical aberration (Fig. 4.8). In this case, however, the appearance of the interference fringes is more complicated than in Fig. 4.7.

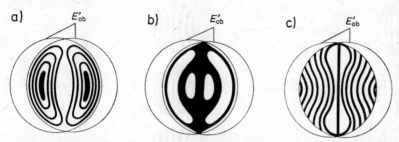

Fig. 4.8. Underfocused (a), optimally focused (b) and overfocused (c) shear interference images of the exit pupil E'_{ob} of an objective which suffers from considerable positive spherical aberration.

The interference patterns shown in Figs. 4.7 and 4.8 occur for extremely poor objectives which are quite inadequate for microscopical work. In the author's experience, however, very high quality microscope objectives have been known to display interference patterns such as shown in Fig. 4.6; their spherical aberration is not perceived if the wavefront shear s is not larger than 25% of the exit pupil diameter $2r_{E'_{ob}}$. However, such objectives are rare, and far more typical are those shown in Figs. 4.9b–i, which give a schematic of the most frequently occurring defects in the correction of spherical aberration.

When the microscope objective under examination does not suffer from comatic aberration and/or astigmatism, the interference patterns shown in Figs. 4.6–4.9 retain exactly the same appearance when the objective is rotated around its optic axis. Otherwise, a change in the configuration of the interference fringes may be observed. Figure 4.10 shows typical interference patterns with hyperbolic and elliptic fringes which indicate that the objective suffers from considerable axial coma. Rotating the objective through 360° causes the patterns shown in Fig. 4.10 to appear alternately at every 90°. For the detection of coma and other asymmetric aberrations, therefore, the interferometer (Fig. 4.5) is provided with a rotatable socket RS.

Frequently coma is coupled with astigmatism. The latter aberration manifests itself primarily by a rotation of the interference fringes when the interferometer

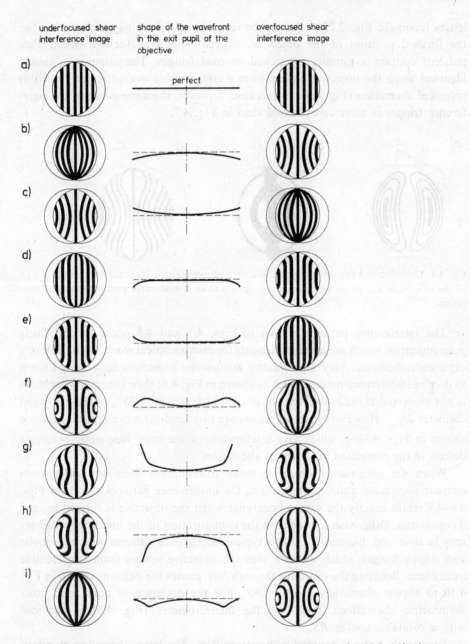

Fig. 4.9. Shear interference images showing the most frequently occuring defects in the correction of spherical aberration of microscope objectives.

a) E'_{ob} b) E'_{ob}

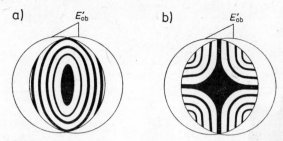

Fig. 4.10. Shear interference images (optimally focused) of the exit pupil E'_{ob} of an objective which suffers from considerable axial coma: from a) to b) one passes by rotating the objective through an angle of 90°.

is increasingly defocused. This interferometer also enables one to detect many other defects such as lack of optical homogeneity in lenses, strain birefringence, errors in the shape of lens surfaces, etc. This instrument is shown in Fig. 4.11. It comprises four interchangeable birefringent prisms with different apex angles $\alpha = 5°$, $10°$, $15°$, and $20°$, and four slit slides with slits of various width w defined by

$$w = \frac{\lambda}{8 M_{ob}(n_e - n_o)\tan \alpha} ,\tag{4.1}$$

where λ is the wavelength of used light (for white light $\lambda \approx 0.55$ μm), M_{ob} the magnifying power of the objective under examination, and $n_e - n_o$ the birefringence of the material from which the birefringent prism is made (quartz crystal). The slit S (Fig. 4.5) must be exactly parallel to the apex edge of the prism W and positioned at the optic axis of the interferometer. For this purpose, the object stage T is rotatable around the optic axis and translatable transversely to this axis in two perpendicular directions. The interferometer does not need a high-power light source, and a low-voltage lamp (6V/15W) is quite sufficient.

In order to illustrate the performance of the interferometer, a number of interferograms taken with this instrument are presented in Figs. 4.12–4.20. They show one perfect microscope objective (Fig. 4.12) and several other objectives which suffer from various defects.

4.3.2. Interferometric demonstration of the effect of cover slip thickness deviations on the spherical aberration of objectives

Returning one more to Fig. 4.5, it is important to note that the slit S is engraved in an opaque thin film of silver deposited by vacuum evaporation onto a glass

Fig. 4.11. Double-refracting interferometer whose optical system is shown in Fig. 4.5.

plate *GP*. If an objective under examination is corrected for a cover glass, the slit *S* must be covered with a cover slip *CG* of proper thickness. Otherwise, spherical aberration will occur.

In Subsection 2.3.6 we discussed the role of the cover slip and its importance for the quality of microscopical imaging. Now we can compare our previous theoretical considerations with an experiment in which the slit *S* (Fig. 4.5) was covered with cover slips of different thickness and dry high-power objectives of good quality, corrected for the standard cover glass thickness $t = 0.17$ mm, were interferometrically examined. Exemplary interferograms, all optimally focused, for two objectives $40 \times /0.65$ are shown in Figs 4.21 and 4.22. As can be seen, spherical aberration rapidly increases as the cover slip thickness t becomes

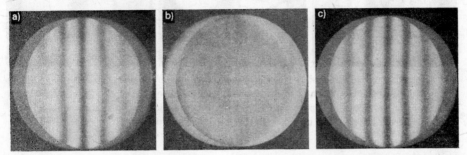

Fig. 4.12. Underfocused (a), focused (b) and overfocused (c) shear interferograms of the exit pupil of a perfect microscope objective $10 \times /0.30$.

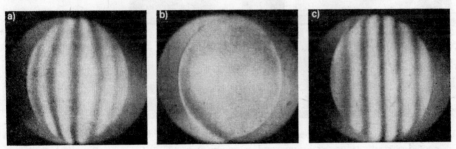

Fig. 4.13. Underfocused (a), focused (b) and overfocused (c) shear interferograms of the exit pupil of a microscope objective $40 \times /0.75$ which suffers from small (rather tolerable) spherical aberration.

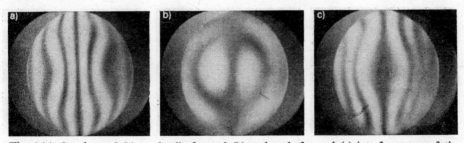

Fig. 4.14. Overfocused (a), optimally focused (b) and underfocused (c) interferograms of the exit pupil of a microscope objective $40 \times /0.65$ which suffers from significant (rather intolerable) positive spherical aberration (compare with Fig. 4.8).

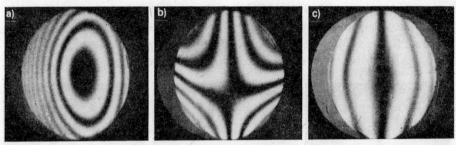

Fig. 4.15. Optimally focused shear interferograms of a low-power objective consisting of two lenses (cemented doublet): a) and b) the doublet is mounted somewhat obliquely in its mount, hence axial coma arises (compare with Fig. 4.10), c) the same doublet is mounted correctly, but suffers from considerable undercorrected spherical aberration.

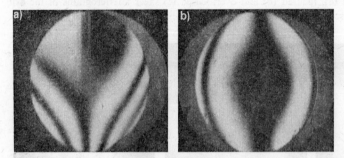

Fig. 4.16. Optimally focused shear interferograms of a cemented doublet: a) the doublet is not cemented coaxially, thus axial coma and astigmatism occur, b) the same doublet is cemented nearly coaxially.

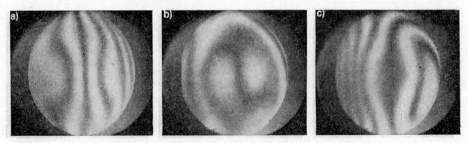

Fig. 4.17. Overfocused (a), optimally focused (b) and underfocused (c) shear interferograms of a microscope objective $40 \times /0.65$ which suffers from significant positive spherical aberration and error of coaxial mounting.

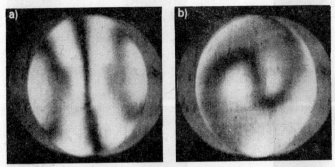

Fig. 4.18. Underfocused (a) and optimally focused (b) shear interferograms of a microscope objective $10 \times /0.25$ which suffers from an error of coaxial mounting and incorporates a lens with defectively fashioned surface.

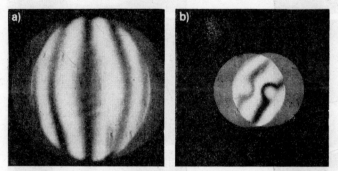

Fig. 4.19. The objective whose interferograms are shown in Fig. 4.18 consisted of two doublets. In order to localize the defective lens surface, the objective has been disassembled and separate interferograms (a and b) of its back and front doublets have been recorded. Interferogram b shows that the front doublet included the defectively fashioned (truncated) surface.

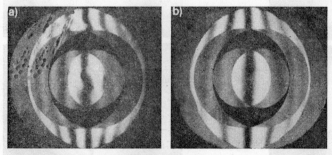

Fig. 4.20. Slightly defocused interferograms of two cemented lenses with a phase ring between them: a) the layer of cement incorporates inhomogeneities of refractive index, b) the same lenses are cemented with homogeneous cement.

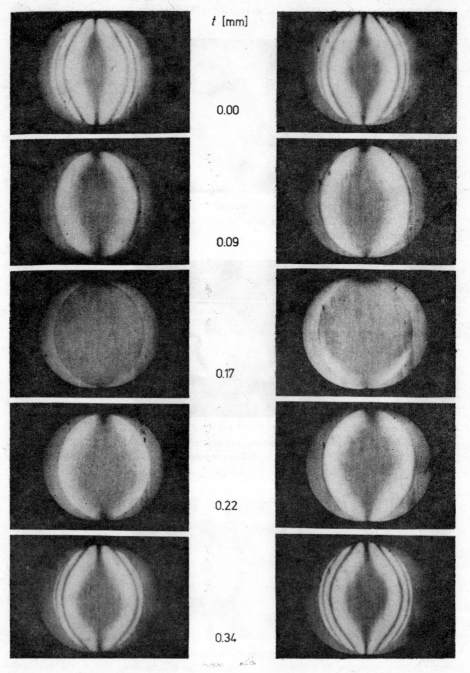

t [mm]

0.00

0.09

0.17

0.22

0.34

Fig. 4.21

Fig. 4.22

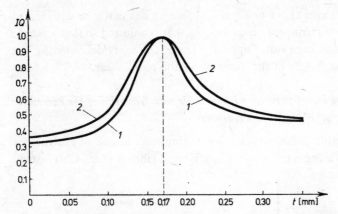

Fig. 4.23. Plots illustrating the effect of cover slip thickness deviations on the correction of spherical aberration of the microscope objectives whose shear interferograms are shown in Figs. 4.21 and 4.22. Curve *1* refers to Fig. 4.21 and curve *2* to Fig. 4.22.

smaller or greater than 0.17 mm. It has been proved that the dry objectives of numerical aperture 0.65 suffer from a spherical aberration of $\lambda/4$ when the cover slip thickness deviates by 0.01 mm from 0.17 mm. This result corresponds to the theoretical data presented in Subsection 2.3.6 (see Table 2.8).

A quantitative interferometric quality of the microscope objective can be expressed by a factor $IQ = A_u/A_0$, where A_0 is the whole area of overlapping regions of the sheared exit pupil E'_{ob}, while A_u is the area within A_0 where light intensity or colour is uniform (or nearly uniform) for the optimally focused shear interference image of E'_{ob}. If only spherical aberration is involved, as shown in Figs. 4.21 and 4.22, the interferometric quality factor may also be expressed as $IQ = w_u/w_0$, where w_0 is the maximum width of A_0 in the direction of the exit pupil shearing and w_u is a fragment of w_0 within which interference is uniform (or nearly uniform). Figure 4.23 shows the variation of $IQ = w_u/w_0$ as a function of the cover slip thickness t for microscope objectives whose shear interferograms are shown in Figs. 4.21 and 4.22. As can be seen, the maximum value of IQ, equal to unity, is for $t = 0.17$ mm, and the interferometric quality factor IQ rapidly drops when t increases or decreases. Performance of the objective is, of course, maximal if $IQ = 1$.

Fig. 4.21. Optimally focused shear interferograms of the exit pupil of a high quality planachromatic objective 40× /0.65/0.17 used with cover slips of various thickness t.

Fig. 4.22. Optimally focused shear interferograms of the exit pupil of a high quality achromatic objective 40× /0.65/0.17 used with cover slips of various thickness t.

The width w_u of the area A_u of uniform interference can easily be determined in white light when the birefringent prism W (Fig. 4.5) is adjusted so that a purple colour of the first interference order appears. The width w_u is delimited by the maximum extent of the purple in the direction of the pupil shear.

4.3.3. A lateral shearing interferometer with variable wavefront tilting for extremely accurate studies of objective aberrations

An exceptionally versatile interferometer for testing microscope objectives has been devised [4.12] by using a transparent birefringent fibre B (Fig. 4.24) instead

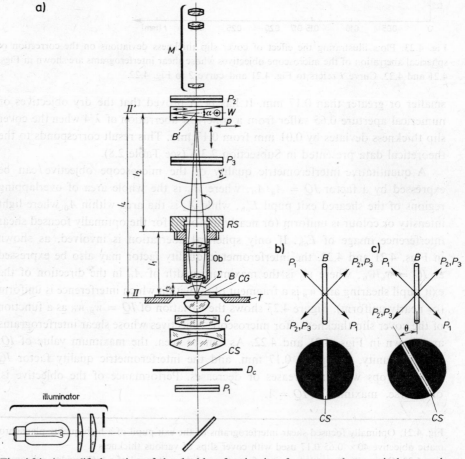

Fig. 4.24. A modified version of the double-refracting interferometer whose optical system is shown in Fig. 4.5.

of a silvered glass plate GP with a narrow slit S (Fig. 4.5) and a slit diaphragm instead of the circular diaphragm D_c. But in this case an additional polarizer P_3 should be inserted behind the objective Ob. This polarizer is crossed with P_1, and the fibre B forms an angle of 45° with the directions of light vibration $(P_1 P_1$ and $P_3 P_3)$ of both polarizers. The fibre B is parallel to the wedge edges of the birefringent prism W, whereas the condenser slit CS can be rotated around the optic axis of condenser. If CS is parallel to B (Fig. 4.24b), the interferometric system functions similarly to that shown in Fig. 4.5. If, however, the slit CS forms an angle θ with the fibre axis, the situation is quite different; uniform field interference does not occur at all and even with a perfectly focused aberration-free objective there is always fringe interference. For a particular value of θ, the interference fringes become parallel to the direction of the shear of the exit pupil E'_{ob} (Fig. 4.25b); this means that the interfering wavefront are tilted about the shear axis.

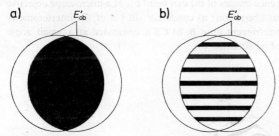

Fig. 4.25. Exactly focused shear interference images of the exit pupil E'_{ob} of an aberration-free microscope objective examined by means of the double-refracting interferometer whose optical system is shown in Fig. 4.24: a) condenser slit CS is parallel with respect to the birefringent fibre B, b) CS is orientated at such an angle θ to B to obtain interference fringes parallel with the direction of wavefront shear.

The use of tilt about the shear axis makes the interpretation of lateral shearing interferograms simpler and increases the sensitivity of the interferometer. These features are illustrated in Figs. 4.26c and 4.27b, which represent optimally focused interference images of the exit pupil of microscope objectives whose interferograms have already been shown in Figs. 4.12b and 4.14b, respectively. A very small marginal spherical aberration, which is not perceived in Fig. 4.12b (or in Fig. 4.26a), becomes distinctly visible in Figs. 4.26b and c as a small bend of interference fringes near the edges of the exit pupil E'_{ob}.

The interference quality factor (relating to spherical aberration) can now be defined as $IQ = l_u/w_0$, where w_0 is the maximum width of the interference area

A_0 along the shear axis, and l_u is the length of a straight portion of the central interference fringe.

The use of wavefront tilt about the shear axis causes coma to manifest itself as more and less arched interference fringes when the optimally focused objective is rotated about its optic axis. Astigmatism is detected by changes in fringe spacing and/or tilt when the objective is rotated through 90°.

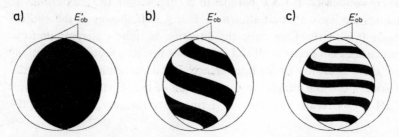

Fig. 4.26. Exactly focused shear interference images of the exit pupil E'_{ob} of a microscope objective which suffers from very small spherical aberration: a) condenser slit CS of the interferometer shown in Fig. 4.24 is parallel to the birefringent fibre B, b) CS is orientated at a small angle θ to B, c) θ is greater than in the preceding situation (b).

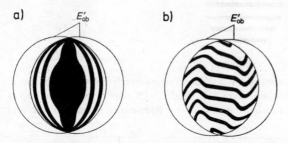

Fig. 4.27. Optimally focused interference images of the exit pupil E'_{ob} of a microscope objective which suffers from significant overcorrected spherical aberration: a) condenser slit CS of the interferometer shown in Fig. 4.24. is parallel to the birefringent fibre B, b) CS is orientated at such an angle θ with respect to B to obtain interference fringes parallel with the direction of wavefront shear in the central region of interference pattern.

4.3.4. Other interferometers for testing microscope optical systems

The interferometers described above (Figs. 4.5 and 4.24) are suitable for the examination of microscope objectives corrected at a finite tube length, while infinity corrected objectives can be tested with these instruments when used together with their tube lens (*TL*, see Fig. 2.59), which focuses the slit *S* (Fig. 4.5) or birefringent fibre *B* (Fig. 4.24) inside the birefringent prism *W*.

A polarization interferometer for testing microscope objectives was also designed by Nomarski. However, the Nomarski interferometer functions in reflected light and incorporates a mirror instead of a slit (*S*, Fig. 4.5) in the object plane (*Π*) of the objective to be tested. Light from an epi-illuminator passes twice, back and forth, through a special birefringent prism (double-refracting "diasporameter") and the objective under examination. The birefringent prism of this instrument is able to produce continuously variable shearing of the exit pupil of the objective; it is described in Ref. [4.13].

It is worth noting that it was Twyman who first suggested using an interference method for testing microscope objectives [4.14]. His instrument was based on the Michelson interferometer (see Fig. 1.54). The separation in space of the two interfering waves makes this instrument very sensitive to vibration and, for every test, the two arms of the interference system must be readjusted to equalize their optical paths. These drawbacks make the Twyman interferometer useless for testing microscope objectives today. Next, a series of common-path interferometers were developed in which both the test and reference waves follow almost the same path. These instruments are reviewed in Refs. [4.15] and [4.16]. The double-refracting systems described above belong to this category; they are insensitive to vibrations and other external disturbances. Recently (1980) an interferometer for testing microscope objectives of high quality has also been developed at VEB Carl Zeiss Jena [4.17]. The interpretation of shearing interferograms is, however, not as simple as that of Twyman interferograms because if the objective being tested suffers from aberrations, interference occurs between two imperfect wavefronts, instead of between one aberrated wavefront (from the objective) and one perfect, spherical or plane, wavefront (from a reference mirror). The use of tilt about the shear axis, first introduced by Bates in a version of Mach–Zehnder shearing interferometer (compare Fig. 7.2 in Volume 2), greatly reduces the problem [4.15]. The manipulation with the Bates interferometer is greatly complicated indeed, in comparison with the author's system shown in Fig. 4.24.

4.4. Measurement of distortion and field curvature

Distortion results when lateral magnification varies across the image field (see Subsection 2.2.4). This aberration is mainly produced by microscope oculars. It can easily be measured by projecting a stage micrometer on a screen placed at a sufficiently large distance from the microscope ocular. Distortion ΔM is determined from Eq. (2.36), in which y' is now the distance between individual graduations of the micrometer image at the central image region and $\tilde{y}'$ that at the marginal zone of the image field. The authors of papers [4.36] and [4.37] have stated that typical values of ΔM are not greater than 3% for standard achromatic, planachromatic, and planapochromatic microscope systems commercially available from famous manufacturers (VEB Carl Zeiss Jena, E. Leitz Wetzlar, Carl Zeiss Oberkochen, C. Reichert Wien, PZO Warsaw, and others).

In contrast to distortion, field curvature is primarily produced by the microscope objective. Usually this aberration is not only corrected but also compensated by using oculars with a field curvature opposite to that of the objective. Therefore, the ability of a microscope to produce a flat image field must be regarded as the joint result of objectives and oculars.

Field curvature may be evaluated in different ways. The simplest consists in using a stage micrometer which is imaged sharply in the centre of the field of view so that the diameter d_s of the central region of maximum sharpness may be assessed by counting the sharply imaged micrometer graduations. The image field flatness IF is expressed as

$$IF = \frac{d_s}{2r_m}, \tag{4.2}$$

where $2r_m$ is the diameter of the object field of the microscope. The most frequent values of IF defined by Eq. (4.2) for laboratory and research microscopes available from different manufacturers are presented in Table 4.2.

The flatness of the image field given by Eq. (4.2) is the better the smaller is the field number (FN) of an ocular. Thus as far as field curvature is concerned, more useful information on the general performance of a microscope is supplied simply by the minimum diameter $2r_{\min}$ of the object field imaged sharply on a projection screen placed behind the microscope ocular at a distance equal to or greater than 2 m. The most frequent values of $2r_{\min}$ for standard microscope systems are given in Table 4.3 [4.36, 4.37].

In visual microscopy a certain amount of field curvature can be tolerated, and when using middle- and high-power objectives it is necessary, after examining the specimen at the centre of the field of view, to refocus the microscope slightly in order to examine the specimen at the edge of the field of view. This refocusing is frequently considered to be a measure of field curvature and is expressed in graduations of the fine focusing knob. However, this method of evaluating field curvature is less accurate than those mentioned previously. In photomicro-

TABLE 4.2

Approximate values of image flatness (*IF*) of typical microscope systems, expressed by Eq. (4.2) [4.36, 4.37]

Numerical aperture of objective (*A*)	*IF* in %	
	Planachromatic and planapochromatic systems	Achromatic and apochromatic systems
$A \leqslant 0.15$	70	60
$0.15 < A \leqslant 0.40$	70	55
$0.40 < A \leqslant 0.65$	75	50
$0.65 < A$	100	35

TABLE 4.3

Minimum diameter $2r_{min}$ for the object field imaged sharply on a far distant projection screen by typical planachromatic and planapochromatic microscope systems [4.36, 4.37]

Nominal magnifying powers of ocular (M_{oc})	$2r_{min}$ in mm for objective magnifying powers (M_{ob})				
	$4 \times$	$10 \times$	$16 \times - 25 \times$	$40 \times$	$100 \times$
$10 \times$	2.4	0.9	0.6	0.30	0.11
$15 \times$	2.0	0.7	0.4	0.25	0.09
$20 \times$	1.6	0.5	0.3	0.20	0.08

graphy field curvature is particularly troublesome as a single focusing position of the microscope is used and the marginal zone of the image field is blurred. In this case plano-correction optical systems are required.

4.5. Evaluation of chromatic aberration correction

The presence of longitudinal chromatic aberration is easily detected if the image of a stage micrometer is observed through a lunette placed coaxially behind the microscope ocular. The lunette is focused at infinity and its eyepiece is fitted with a focal plate on which a cross-line or micrometer scale is engraved. The stage micrometer is successively illuminated with monochromatic light of wavelength $\lambda_{C'} = 644$ nm, $\lambda_e = 546$ nm, and $\lambda_{F'} = 480$ nm. If the optical system of the microscope under examination suffers from longitudinal chromatic aberration, the wavelengths mentioned above produce sharp images of the stage micrometer for different fine focusing adjustments of the microscope. For the quantitative evaluation of longitudinal chromatic aberration, it is sufficient to measure three values of defocusing (D) for a small fragment of the micrometer imaged sharply on the focal plate of the lunette. These values are

$$D_1 = a'_{C'} - a'_{F'}, \tag{4.3a}$$
$$D_2 = a'_e - a'_{C'}, \tag{4.3b}$$
$$D_3 = a'_e - a'_{F'}, \tag{4.3c}$$

where $a'_{C'}$, a'_e, and $a'_{F'}$ are the focusing values for $\lambda_{C'}$, λ_e, and $\lambda_{F'}$, respectively; they are taken from the graduated fine focusing knob or from another measuring instrument attached to the focusing mechanism of the microscope under examination. The sharp images of the stage micrometer are observed in the axial region of the field of view since off-axis zones are disturbed by lateral chromatic aberration. Table 4.4 shows the most frequent values of D_1 and D_2 for standard visual microscopy [4.36, 4.37]. The values of D_3 are omitted in this table as $D_3 \approx D_2$.

TABLE 4.4

The most frequently occurring values of longitudinal chromatic aberration expressed by defocusing D_1 and D_2 (see Eqs. 4.3a and b) for typical microscope systems

Numerical aperture of objective (A)	Defocusing in μm			
	D_1	D_2	D_1	D_2
	Achromatic and planachromatic systems		Apochromatic and planapochromatic systems	
$A \leqslant 0.15$	45	65	13	30
$0.15 < A \leqslant 0.40$	3	10	2	5
$0.40 < A \leqslant 0.65$	1.5	5	1.5	2
$0.65 < A$	1	2.5	1	1

Lateral chromatic aberration is present if off-axis images of stage micrometer graduations appear at different distances $y'_{C'}$, y'_e, and y'_F from the centre of the field of view when the stage micrometer is illuminated with monochromatic light of wavelength $\lambda_{C'}$, then λ_e and $\lambda_{F'}$. In this case lateral chromatism or rather chromatic difference of magnification (CDM) can be expressed as the relative difference of the distances mentioned above, i.e., $(y'_{C'} - y'_{F'})/y'_{C'}$, $(y'_{C'} - y'_e)/y'_{C'}$, and $(y'_e - y'_{F'})/y'_e$. Typical values of $CDM = (y'_{C'} - y'_{F'})/y'_{C'}$ for most standard visual microscopes are given in Table 4.5 [4.36, 4.37].

TABLE 4.5

Chromatic difference of magnification (CDM) usually occurring in typical microscope systems

Magnifying power of objectives (M_{ob})	CDM in %		
	Achromatic systems	Planachromatic systems	Apo- and planapo-chromatic systems
$10\times$	0.4	0.27	0.2
$20\times$	0.4	0.27	0.2
$40\times$	0.4	0.27	0.2
$100\times$	0.6	0.27	0.2

4.6. Measurement of the optical transfer function of microscope objectives

A defect of interferometric methods for assessing the performance of microscope objectives (see Section 4.3) is the fact that the interference image of the exit pupil of an objective under examination does not reveal degradation of the contrast of specimen images. This degradation is generally caused not only by aberrations but also by light reflections, light scattering, and other phenomena, and may be recorded by the measurement of the optical transfer function (OTF). There are two basic methods for measuring the OTF: by scanning and by autocorrelation.

4.6.1. Scanning method

The optical transfer function of an optical system is given by the Fourier transform of its point spread function. The general validity is not affected if the point spread function is replaced by a slit spread function. The latter is the intensity distribution in the image of a line source illuminated incoherently and considered along different azimuths. The OTF is a complex function of the spatial frequency u, expressed by Eq. (3.85), in which the second component (v) of spatial frequency is now omitted; $|J'(u)|$ represents the ratio of the contrast in the image to the contrast in the object, the object being a sinusoidal grating of frequency u; $\chi(u)$ represents the lateral shift of the image with respect to an ideal image produced by an aberration-free optical system.

In practice, the object slit is not a line but a streak of finite width. Its image is then the convolution of the slit function with the point spread function of the optical system. Representing the Fourier transforms of the object and the image functions by $F(u)$ and $J(u)$ respectively, we have, according to Eq. (3.45), the relation

$$J(u) = F(u)J'(u). \tag{4.4}$$

The Fourier transform $F(u)$ being known, the *OTF* or $J'(u)$ can be determined if the $J(u)$ can be measured. To determine $J(u)$, the intensity distribution in the image plane may be measured by means of a narrow slit and its Fourier transform calculated numerically. A more effective method is to scan the image of the slit object with a sinusoidal or non-sinusoidal grating. This is the principle of the scanning method of *OTF* measurement.

Various types of scanning instruments for the measurement of the *OTF* (or *MTF*) with sinusoidal and non-sinusoidal gratings (transmission functions) have been developed [4.19–4.27]. Among them the best known is the Eros,

Fig. 4.28. Double-refracting interferometer similar to that shown in Figs. 4.5 and 4.11 but additionally fitted with a photometer for measuring optical transfer function using autocorrelation method.

commercially available from SIRA (England). This instrument has been used by Sojecki [4.20] and his co-workers [4.21] for measuring the *MTF* of microscope objectives.

4.6.2. Autocorrelation method

This method is based on Eq. (3.92). The integral in this equation can be calculated if the imaging properties of an optical system are known. The integration can also be carried out experimentally by means of a shearing interferometer [4.27]. Various types of shearing interferometers have been developed to measure the *OTF* by this method [1.37, 4.28–4.30]. Popielas has successfully used the double-refracting interferometer (Fig. 4.28) described in Subsection 4.3.1 [4.29]. It can be shown that the autocorrelation method is equivalent to the scanning method, but the results obtained from both methods may be slightly different. The reproducibility of the measurements of the *OTF* or *MTF* by using different methods or even different instruments within the same method is not good. In microscopy these measurements tend to be undertaken rather by designers of microscope objectives, and may be especially useful for testing the performance of new prototypes.

4.7. Measurement of stray light inside the optical system of a microscope

It is important for microscope designers to know the magnitude of stray light inside the optical system of a microscope. This undesirable light originates from reflection and scattering of rays travelling from a light source to the image plane. An important contribution to the theory and measurement of stray light has been made by Russian researchers in particular [2.33, 2.34, 4.31–4.34]. Instruments for measuring stray light in microscope systems are of considerable sophistication. For more detailed information the reader is referred to the papers cited above.

4.8. Evaluation of strain birefringence of microscope objectives

Strain birefringence in microscope objectives is frequently considerable. This defect may be tolerable in ordinary microscopy, but is inadmissible in systems which use polarized light (polarizing microscopy, polarization interference microscopes, etc.).

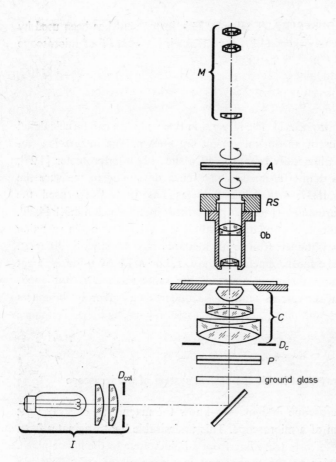

Fig. 4.29. Set-up for testing the strain birefringence of microscope objectives.

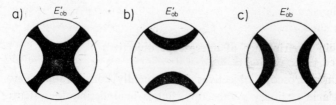

Fig. 4.30. Dark cross and hyperbola-like fringes in the exit pupil of a strain-free microscope objective.

It is easy to test for freedom from strain birefringence by using a set-up such as shown in Fig. 4.29. When the illuminator I, condenser C, and objective Ob are aligned, the polars P and A crossed, and diaphragms D_{col} and D_c properly open, a dark cross should appear in the exit pupil E'_{ob} of the objective (Fig. 4.30a). Normally the cross is observable with middle- and high-power objectives. It can be visible to the naked eye, but observation is made easier by using an auxiliary low-power microscope M focused at the exit pupil of the objective Ob. The device is adjusted according to the Köhler principle of illumination with no object slide lying between the objective Ob and the condenser C. The cross (Fig. 4.30a) is symmetrical and very dark between crossed polars if the objective being tested and, of course, the condenser C as well are free from strain birefringence. When the analyser A or polarizer P is rotated, the cross should open symmetrically into two hyperbola-like fringes such as shown in Figs. 4.30b and c. These fringes should remain dark until they disappear beyond the edge of the exit pupil E'_{ob}. The dark cross and the hyperbola-like fringes remain unchanged when the objective Ob (free from strain birefringence) is rotated round its optic axis (Fig. 4.31). For this purpose the test instrument (Fig. 4.29) is provided with a rotatable socket RS.

Fig. 4.31. Dark cross in the exit pupil of a strain-free microscope objective $40 \times /0.65$.

The situation differs when the objective suffer from radial, lateral, or point strain. The dark cross appears only for particular azimuthal positions of the objective (Fig. 4.32a) and for any other positions its arms will open (Fig. 4.32b and c) as in the case of the strain-free objective when the analyser or polarizer were rotated. The pictures shown in Fig. 4.32 illustrate quite small strains which do not cause any appreciable deformations of the Airy patterns when the objective is used without crossed polars (see Fig. 4.4). If the magnitude of strain birefringence is considerable, however, the dark cross does not appear at all or is extremely deformed (Figs. 4.33 and 4.34).

With low-power objectives free from strains the arms of the cross are so large that they fill the entire exit pupil of the objective. When the objective is rotated

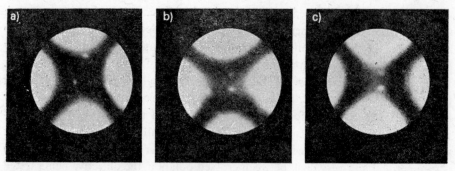

Fig. 4.32. Dark cross (a) and its slightly open arms (b, c) of a microscope objective $40 \times /0.65$ with small strain birefringence.

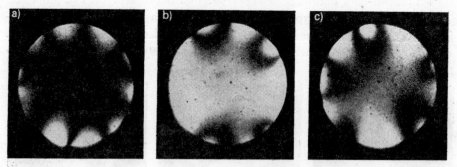

Fig. 4.33. Example of a microscope objective $20 \times /0.40$ which suffers from great point strains.

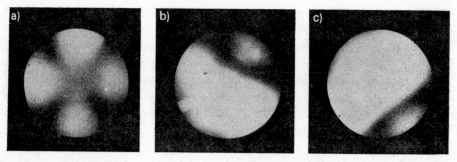

Fig. 4.34. Another example of a microscope objective $10 \times /0.25$ which suffers from great strain birefringence.

around its optic axis, the darkness of the exit pupil remains unchanged. Otherwise, if a low-power objective suffers from strain birefringence, the darkness of the exit pupil between crossed polars is incomplete and changes when the objective is turned.

Strain birefringence of microscope condensers may also be tested in the same way. In this case the condenser C (Fig. 4.29) should be fixed in a rotatable socket and a strain-free objective Ob must be used.

The possible influence of the condenser on the test for freedom from strain birefringence of the objective can be eliminated if a polarizing foil is placed between these two lens systems instead of the polarizer P lying before the condenser [4.35].

Finally, it is worth noting that all tests for freedom from strain birefringence may be performed with the interferometric instrument described previously in Subsection 4.3.1. It is only necessary to remove the birefringent prism W (Fig. 4.5) and glass plate GS with slit. Some less critical examinations of strains in objectives and condensers can also be performed with typical polarizing microscopes.

* * *

The information provided in this chapter is intended to cover a selection of indispensable tests for the evaluation of the optical performance of microscopes. It has been the author's experience that the most useful tests are those performed with the double-refracting interferometer illustrated in Figs. 4.5 and 4.11. If this instrument shows that the interference quality of an objective being tested is satisfactory, then its general performance is as a rule also satisfactory.

Some further and more sophisticated methods and instruments for testing the performance of microscope optical systems and their components may additionally be found in the book by Ivanova and Kirillovskii [4.38].

Epilogue to Volume 1

Light microscopy has recently made rapid progress in the development of new designs that improve both optical and mechanical quality of microscopes for research and general use. In the past decade, some quite new microscope families have appeared on the market. Notable among them are Polyvar series from C. Reichert Wien, 250-CF (Jenaval and Jenamed) generation from VEB Carl Zeiss Jena, Pyramid series (Axioplan, Axiophot and Axioskop) from C. Zeiss Oberkochen, Vanox microscopes (S and T) supported by the BH2 series of routine and research microscopes from Olympus, Microphot family from Nikon, and Aristoplan universal research microscope with Vario-Orthomat E automatic camera from E. Leitz Wetzlar.

In the design of new research microscopes, the main trends are as follows: (1) chromatic-free (CF) optics; (2) flat image field with high field-of-view numbers, equal to 25 and even more (up to 32); (3) infinity corrected objectives, which enable different components for specialized microscopical techniques to be located in the space between the objective nosepiece and tube lens without changing the optical performance of the microscope; (4) modular design and integrated architecture, which allow the user to switch from one microscopical technique to another with ease and, in particular, without loss of field of view and high imaging quality.

The demand for precise photographic documentation is today greater than ever before and photomicrography has made great strides in the past two decades. An example is primarily the Leitz Aristoplan with Orthomat E camera system. The automatic exposure measurement of the Orthomat E continues during each exposure or can be performed and stored before exposure. The measuring spot can be moved over the field to be photographed permitting a precise exposure of any chosen image detail without moving the specimen. Exposure times between 1/100 second and 99 minutes can be selected, and film speed from 3 to 8000 ASA accommodated. Exposure correction is automatic for the five different formats that the Orthomat E accepts. All useful photomicrographic data are digitally displayed on a separate control panel. Variable magnification tube (Variophot) gives a field-of-view number of 28 with all microscopical techniques. Its magnification factor q can be varied continuously from 1 to $2.5\times$.

Another example is the Axiophot microscope from C. Zeiss Oberkochen. The fully electronic microprocessor-controlled camera system has a highly sensitive autocontrol exposure unit and is designed for the lowest possible amount of light, and all important functions (automatic exposure, multiple exposure, camera selection with various film formats, data projection onto a large-format camera) are clearly marked and displayed digitally for verification. The system can easily be adapted to the most extreme specimen image conditions (dark-field illumination, fluorescence, polarized light).

Nearly the same documentation techniques are offered by an integral photomicrographic system of the Nikon Microphot FX microscope, and also by the Olympus Vanox-S photomicroscope. The former has, e.g., 1 per cent spot or 30 per cent average exposure measurement switching, direct image projection, and movable measurement area. On the other hand, the latter is fitted with an autofocus mechanism designed to eliminate focusing errors in low magnification photomicrography (objectives of magnifying power from $1\times$ to $10\times$), and ensure needle-sharp imagery throughout the entire magnification range.

Further progress has also been made in the development of inverted research biological microscopes. Apart from the models described earlier in Section 2.4, it is worth mentioning here the IMT-2 research inverted microscope, which was only recently developed by Olympus and is now on the market. It is equipped with a variety of specialized devices, such as for phase contrast, Nomarski differential interference contrast, dark-field microscopy, observation in polarized light, and for reflected-light fluorescence microscopy. An Olympus automatic PM-10AD photographic camera or large format camera can also be attached to this microscope (see Ref. [E1.1] for more details).

However, the most fascinating achievement in light microscopy during the past decade is Video Enhanced Contrast (VEC). This new and highly useful method was initiated by R. D. Allen and I. Inoué in 1981 [E 1.2]. It uses a high-resolution video camera (Hamamatsu CCTV camera) attached to a microscope, a digital image processor and various computer assisted accessories, which eliminate the influence of the background and are able to display extremely small details which cannot be observed by using other techniques of light micriscopy. Background artefacts are removed in real time by the processor by first digitizing the video image of the field without a specimen and then subtracting this image from each incoming specimen image. Moreover, the VEC system can also reduce limitations in the electronic processing system itself. In particular, averaging video frames reduces the video noise.

Video enhancement was originally used successfully with differential interference contrast (see Section 7.3 in Volume 2) and polarized-light microscopy (see Section 15.3 in Volume 3), and enabled the cell organelles of size below the limit of resolving power of the light microscope to be observed on a monitor screen, with a final magnification of $7000 \times$ or even more. The useful magnification is in this case five times higher than in conventional microscopy (see Subsection 3.6.8). Today, this method is particularly used for dynamic studies of living cells [E1.3–E1.5], for visualization of individual colloidal gold particles in living cells [E1.6], and also in the semiconductor industry [E1.7].

The VEC system is commercially manufactured by Hamamatsu (Japan) as the Photonic Microscope System (recently, in the C-1966 model) and represents a huge advance indeed in light microscopy.

References

Chapter 1

[1.1] Born M. and Wolf E., *Principles of Optics* (6th ed.), Pergamon Press, Oxford 1980.
[1.2] Klein M. V., *Optics*, John Wiley and Sons, New York 1970.
[1.3] Slayter E. M., *Optical Methods in Biology*, Wiley-Interscience, New York 1970.
[1.4] Longhurst R. S., *Geometrical and Physical Optics* (2nd ed.), Longmans, Green and Co., London 1967.
[1.5] Lipson S. G. and Lipson H., *Optical Physics*, Cambridge University Press, Cambridge 1969.
[1.6] Ratajczyk F., *Optyka ośrodków anizotropowych* (*Optics of Anisotropic Media*), Technical University of Wrocław, Wrocław 1982 (in Polish).
[1.7] Webb R. H., *Elementary Wave Optics*, Academic Press, New York and London 1969.
[1.8] Gaskill J. D., *Linear Systems, Fourier Transforms and Optics*, John Wiley and Sons, New York 1978.
[1.9] Peřina J., *Coherence of Light*, Van Nostrand Reinhold Co., London 1972.
[1.10] Beran M. J. and Parrent G. B., Jr., *Theory of Partial Coherence*, Prentice-Hall, Englewood Cliffs, N. J., 1964.
[1.11] Beesley M. J., *Lasers and their Applications*, Taylor and Francis, London 1972.
[1.12] Clarke D. and Grainger J. F., *Polarized Light and Optical Measurement*, Pergamon Press, Oxford 1971.
[1.13] Shurcliff W. A., *Polarized Light. Production and Use*, Harvard University Press, Cambridge, Mass. 1962.
[1.14] Shurcliff W. A. and Ballard S. S., *Polarized Light*, Van Nostrand Company, London 1964.
[1.15] King R. J. and Talim S. P., Some aspects of polarizer performance, *J. Phys. E* (*Sci. Instrum.*), **4** (1971), 93–96.
[1.16] Austin R., Thin film polarizing devices, *Electro-Optical Systems Design*, February 1974, 30–35.
[1.17] Brown E. B., *Modern Optics*, Reinhold Publishing Corporation, New York 1966.
[1.18] Jerrard H. G., Optical compensators for measurement of elliptical polarization, *J. Opt. Soc. Am.*, **38** (1948), 35–54.
[1.19] Levi L., *Applied Optics: a Guide to Modern Optical System Design*, John Wiley and Sons, Inc., New York 1968.

[1.20] Turner A. F., Some current developments in multilayer optical films, *J. Phys. Radium*, **11** (1950), 444–450.

[1.21] Hecht E. and Zajac A., *Optics*, Addison–Vesley Publishing Company, Reading, Mass. 1974.

[1.22] Ghatak A. K. and Thyagarajan K., *Contemporary Optics*, Plenum Press, New York and London 1978.

[1.23] Hunsperger R. G., *Integrated Optics. Theory and Technology*, Springer-Verlag, Berlin–Heidelberg–New York 1982.

[1.24] Harrick N. J., *Internal Reflection Spectroscopy*, Wiley Interscience, New York 1967.

[1.25] Bryngdahl O., Evanescent waves in optical imaging, in: E. Wolf (ed.), *Progress in Optics*, **11** (1973), 167–221.

[1.26] Massy G. A., Microscopy and pattern generation with scanned evanescent waves, *Appl. Optics*, **29** (1984), 658–660.

[1.27] Temple P. A., Total internal reflection microscopy: a surface inspection technique, *Appl. Optics*, **20** (1981), 2656–2660.

[1.28] Sokolov A. V., *Optical Properties of Metals*, American Elsevier Publishing Company, New York 1967.

[1.29] Bamford C. R., Optical properties of flat glass, *J. Non-Crystalline Solids*, **47** (1982), 1–20.

[1.30] Bamford C. R., *Colour Generation and Control in Glass*, Elsevier, Amsterdam 1979.

[1.31] Françon M., *Progress in Microscopy*, Pergamon Press, Oxford–London–New York–Paris 1961.

[1.32] Sládková J., *Interference of Light*, ILIFFE Books Ltd., London and SNTL-Publishers of Technical Literature, Prague 1968.

[1.33] Steel W. H., *Interferometry* (2nd ed.), University Press, Cambridge 1983.

[1.34] Scott G. D. and Shirahata R., Phase-change correction in film-thickness measurements by multiple-beam FECO fringes, *J. Opt. Soc. Am.*, **61** (1971), 131–133.

[1.35] Pluta M., On double-refracting microinterferometers which suffer from a variable inter-fringe spacing across the image plane, *J. Microscopy*, **146** (1987), 41–54.

[1.36] Nomarski G., Microinterféromètre différential à ondes polarisées, *J. Phys. Radium*, **16** (1955), 9S–11S.

[1.37] Françon M. and Mallick S., *Polarization Interferometers*, Wiley Interscience, London–New York–Sydney–Toronto 1971.

[1.38] Françon M., *Holography*, Academic Press, New York–London 1974.

[1.39] Rudolph D. *et al.*, Zone plates for use in X-ray microscopy, *80. Tagung der DGaO and 1979 European Optics Conference*, Bad Harzburg 1979, p. 54 (abstract).

[1.40] Schmahl G. *et al.*, X-ray microscopy of biological specimens with a zone plate micro-scope, *ibid*, p. 54 (abstract).

[1.41] Shulman A. R., *Optical Data Processing*, John Wiley and Sons, Inc., New York–London–Sydney-Toronto 1970.

[1.42] Baez A. V., Image formation by a Fresnel zone plate, in: Clark G. L. (ed.), *The Encyclopedia of Microscopy*, Reinhold Publishing Co., New York 1961, pp. 552–561.

[1.43] Michette A. G., *Optical Systems for Soft X Rays*, Plenum Press, New York 1986.

Chapter 2

[2.1] Welford W. T., *Aberrations of the Symmetrical Optical Systems*, Academic Press, London–New York–San Francisco 1974.
[2.2] Anzai S., Chromatic-aberration-free optics for microscopy, *Am. Lab.*, **10** (1978), No. 4, 83.
[2.3] Wakimoto Z., Chromatic aberration-free optics for microscopy, *Proc. Roy. Micr. Soc.*, **14** (1979), 327–328.
[2.4] *Nikon CF-Lenses*, Catalogue 8370-01 KEC 903-30/1.
[2.5] Riesenberg H., Chromatically corrected microscope objectives from Jena, *Jena Rev.*, **25** (1980), 158–163.
[2.6] Wagnerowski T., Tolerances of the mechanical tube length, *Pomiary. Automatyka. Kontrola*, **10** (1964), 213–214 (in Polish).
[2.7] Loveland R. P., *Photomicrography*, John Wiley and Sons, Inc., New York 1970.
[2.8] Skvortsov G. E., Panov V. A., and Polyakov N. I., *Microscopes*, Izd. Mashinostroenie, Leningrad 1969 (in Russian).
[2.9] Kalisiak S., Kosiba Z., Kozłowski T. and Praxmeyer A., A method of measurement and an apparatus for microscopic objective extra-precision length control, *Abstracts of the 1983 European Optical Conference*, 30 May-4 June 1983, Rydzyna, Poland (p. 124).
[2.10] Bradbury S., Edward Miles Nelson, a pioneer of modern microscopy, *Proc. Roy. Micr. Soc.*, **17** (1982), 188–190.
[2.11] Haselmann H., Who was August Köhler? *Proc. Roy. Micr. Soc.*, **18** (1983), 170–172.
[2.12] White G. W., *Introduction to Microscopy*, Butterworths, London 1966.
[2.13] Martin L. C., *The Theory of the Microscope*, Blackie and Sons, Ltd., London 1966.
[2.14] Turygin I. A. *Applied Optics,* Izd. Mashinostroenie, Leningrad 1965 (in Russian).
[2.15] *Optical Systems for the Microscope*, Brochure No. F41-101-e WIII/71 PTo, edited by Carl Zeiss Oberkochen 1971.
[2.16] Determann H. and Lepusch F., *The Microscope and its Application*, Brochure No. 512-69c/Engl., edited by E. Leitz, Wetzlar 1977.
[2.17] Tiedeken R., Sturm H. and Hofmann C., Über ein neues Stereoprojektionsverfahren, *Feingerätetechnik*, **20** (1971), 550–553.
[2.18] Astvatsaturov A. V. *et al.*, Microprojection instrument with a holographic screen. *OMP*,[1] **50** (1983), No. 10, 27–29 (in Russian).
[2.19] Michel K., *Die Mikrophotographie* (3. Auflage), Springer, Wien–New York 1967.
[2.20] Bergner J., Gelbke E. and Mehlies W. E., *Praktische Mikrofotografie* (2. Auflage), WEB Fotokinverlag, Leipzig 1973.
[2.21] Ericksenn L., The Wild M400 Photomikroscop, *Photomethods*, **29** (1986), 21–24.
[2.22] Kryszczyński T. and Bogdan H., Plancompensation wide-field "Fotal" projection oculars of negative power, *Optyka*,[2] **11** (1976), 73–75. (in Polish).

[1] *Optiko-Mekhanicheskaya Promyslennost'* (translation into English: *Soviet Journal of the Optical Technology*, published by Optical Society of America).
[2] A bulletin on applied optics published quarterly by Central Optical Laboratory, Warsaw, Poland.

[2.23] Boegehold H. and Köhler A., Das Homal, ein System, welches das mikrophotographische Bild ebnet, *Z. wiss. Mikr.*, **39** (1922), 249–262.

[2.24] Boegehold H., *Das optische System des Mikroskops*, VEB Verlag Technik, Berlin 1958.

[2.25] Claussen H. C., Microscope objectives with plano-correction, *Appl. Optics*, **3** (1964), 993–1003 (see also *Leitz-Mitt.*, **4** (1967), 65).

[2.26] Baensch R., Leitz Vario-Orthomat, ein neues Kamerasystem für die automatische Mikrophotographie, *Leitz-Mitt.*, **7** (1979), 187–189.

[2.27] Gabler F. and Kropp K., Photomicrography and its automation, in: R. Barer and V. F. Cosslett, *Advances in Optical and Electron Microscopy*, vol. 4, Academic Press, London–New York 1971, pp. 385–413.

[2.28] Beyer H. (ed.), *Handbuch der Mikroskopie*, VEB Verlag Technik, Berlin 1973.

[2.29] Hauser F., *Das Arbeiten mit auffalenden Licht in der Mikroskopie*, Akademische Verlagsgesellschaft, Geest und Portig K.-G., Leipzig 1960.

[2.30] Freund H. (ed.), *Handbuch der Mikroskopie in der Technik*, vol. 1, pt. 2: Instrumente für Auflicht-Mikroskopie, Umschau Verlag, Frankfurt/Main 1960.

[2.31] Samuels L. E., *Optical Microscopy of Carbon Steels*, American Society for Metals, Metals Park (Ohio) 1980.

[2.32] Schwartz M. J., Solving design problems with pellicles, *Electro-Optical Systems Design*, **2** (1970), 8.

[2.33] Grammatin A. P., Stray light in reflected-light microscopes, *OMP*, **36** (1969), No. 3, 26–29 (in Russian).

[2.34] Arlevskii A. G. and Grammatin A. P., Effect of the optical system on the amount of stray light in the image plane of reflected-light microscope, *OMP*, **44** (1977), No. 6, 71–74 (in Russian).

[2.35] Siegel J. I., Light microscope sees materials of low reflectivity, *Industrial Research and Development*, June 1981, 106–108.

[2.36] *Abbildende und beleuchtende Optik des Mikroskops. Objektive, Okulare, Kondensoren*, Brochure No. 512-99a VI/73/CX/g, edited by E. Leitz, Wetzlar 1973.

[2.37] Popielas M., Universal research microscopes—Unimat series, *Proc. of the Conference "Microscopy in Science and Practice"* (Polmic'76), SIMP-ZORPOT, Warsaw 1976, pp. 78–100 (in Polish).

[2.38] Hohn E., Leitz Laborlux 12 HL und Leitz Laborlux 12 ME—zwei neue Mikroskope für die industrielle Produktions—und Qualitätskontrolle, *Leitz-Mitt.*, **7** (1979), 195–198.

[2.39] Luthardt K., Metallux 3—ein neues Leitz-Mikroskop für die metallographische Werkstoff-Prüfung und Produktionskontrolle, *Leitz-Mitt.*, *Sonderheft, Achema Juni* 1985, 26–27.

[2.40] Hohn E., Leitz Metalloplan HL 6″ × 6″, das Grossfeld-Mikroskop für die neue Maskenund Wafergeneration in der Halbleiterindustrie, *Leitz-Mitt.*, **7** (1979), 174-177; Leitz Inspections- und Messmikroskop Ergolux für die Elektronik-Industrie, *Leitz-Mitt.*, **8** (1982), 68–72.

[2.41] Möllring F. K., IM 35 and ICM 405 inverted microscopes for Biology, Medicine and Metallography, *Opton Information*, No. 1/1979, 12–13.

[2.42] Knüpffer H., Das umgekehrte Auflichtmikroskop Metaval—ein modernes Gerät für die metallographische Routine und Forschung, *Jenaer Rundschau*, **23** (1978), 177–181.

[2.43] Beck R., Die technische Entwicklung umgekehrter Mikroskope, *Leitz-Mitt.*, **8** (1984),

179–189 (see also: Luthardt K., Neues umgekehrtes Leitz-Metallmikroskop Metallovert, *ibid*, 174–175).

[2.44] Gifkins R. C., *Optical Miscroscopy of Metals*, Sir Isaac Pitman and Sons Ltd., Melbourne and London 1970.

[2.45] Brandon D. G., *Modern Techniques in Metallography*, Butterworths, London 1966.

[2.46] Oettel W. O., *Grundlagen der Metallmikroskopie*, Akademische Verlagsgesellschaft Geest und Portig K.-G., Leipzig 1969.

[2.47] Greaves R. H. and Wrighton H., *Practical Microscopical Metallography*, Chapmann and Hall Ltd., London 1957.

[2.48] Richardson J. H., *Optical Microscopy for the Materials Sciences*, Marcel Dekker, Inc., New York 1971.

[2.49] Gräf I. and Draugelates U., Phasenidentifizierung in Nickelloten mit modernen metallographischen Techniken, *Leitz-Mitt.*, **7** (1978), 71–76.

[2.50] Grammatin A. P. and Kirichenko E. V., Problems in the calculation of an epi-illuminator with a paraboloidal mirror, *OMP*, **46** (1979), No. 4, 20–23 (in Russian).

[2.51] Beck J. L., A new reflecting microscope objective with two concentric spherical mirrors, *Appl. Optics*, **8** (1969), 1503—1504.

[2.52] Norris K. P., Development of reflecting microscopes, *Research*, **8** (1955), 94–101.

[2.53] Riesenberg H., Das Spiegelmikroskop und seine Anwendungen, *Jenaer Jahrbuch 1956*, 30–70.

[2.54] Claussen H. C., Mikroskope, in: Flügge S. (ed.), *Encyclopedia of Physics*, vol. 29, Springer-Verlag, Berlin–Heidelberg–New York 1967, pp. 343–425.

[2.55] Panov V. A., New catadioptric microscope objectives of long working distance, *OMP*, **36** (1969), No. 8, 19–22 (in Russian).

[2.56] Grammatin A. P. and Rybakov J. R., Objective with two concentric mirrors for infrared microscopy, *OMP*, **36** (1969), No. 10, 77–79 (in Russian).

[2.57] Panov V. A. and Zhidkova I. A., Microscope objectives with two reflecting ellipsoidal surfaces, *OMP*, **37** (1970), No. 11, 25–29 (in Russian).

[2.58] Panov V. A. Microscope objectives for infrared range of spectrum, *OMP*, **37** (1970), No. 2, 35 (in Russian).

[2.59] Burch C. R. and Murgatroyd P. N., Zonal coma of semiaplanat reflecting microscope, *J. Phys. E (Sci. Instrum.)*, **5** (1972), 1129–1130.

[2.60] Jurek B., *Optical Surfaces*, Academia, Prague 1976.

[2.61] Murie R. A., Microscopy at a distance, *International Laboratory*, **13** (1983), No. 8, 12–27.

[2.62] Maksutov D. D., Mirror objective, *USSR Patent* 40 859 (1932).

[2.63] Ahrberg H., Dialux 22, *Leitz-Mitt.*, **8** (1982), 73–75 (see also: Diaplan, *Leitz-Mitt.*, *Sonderheft*, *Achema Juni* 1985, 15–21).

[2.64] Baensch R., Leitz Laborlux 11 und Leitz Laborlux 12, zwei neue Mikroskope für das Mikroskopieren transparenter Objekte, *Leitz-Mitt.*, **7** (1979), 178–182.

[2.65] Baensch R., Leitz HM-Lux 3, ein Schul-, Kurs- und Amateurmikroskop zur Untresuchung transparenter biologischer Objekte, *Leitz-Mitt.*, **7** (1979), 184–186.

[2.66] Schindl K. P., The Reicher UnivaR—a new concept in microscopy, *The Microscope*, **22** (1974), 117–124.

[2.67] Michel K., Axiomat, ein Mikroskop mit neuen Konzept, *Sonderdruck aus den Opton Informationen*, No. 82 (1973).

[2.68] McCrone W. C., The new Zeiss Axiomat, *The Microscope*, **21** (1973), 167–176.

[2.69] Leman A. and Moritz P., Design concept for a new generation of microscopes, *International Laboratory*, **13** (1983), No. 8, 48–54.

[2.70] Keller H. E., Research microscope design: new concepts, *International Laboratory*, July/August 1986, 68–72 (see also *Proc-Roy. Micr. Soc.*, **22** (1987), 183).

[2.71] Malý M. and Veselý P., A new light microscopic method for the synchronous bidirectional illumination and viewing of living cells in different contrast modes, and/or different focal levels or magnifications, *J. Microscopy*, **117** (1979), 411–416.

[2.72] Haselmann H., Modern microscope design, *Proc. Roy. Micr. Soc.*, **17** (1982), 48–57; The future of light microscopy, *Proc. Roy. Micr. Soc.*, **18** (1983), 229–234.

[2.73] Klingenberg A., A new comparison microscope, *J. Microscopy*, **119** (1980), 257–265.

[2.74] Policard A., Bessis M. and Locquin M., *Traité de microscopie*, Masson et Cie, Paris 1957.

[2.75] Needham G., *The Practical Use of the Microscope, Including Photomicrography*, Blackwell Scientific Publications, Oxford 1958 (reprinted in 1977 by Ch. C. Thomas Publishers, Springfield 1977).

[2.76] Appelt H., *Einführung in die mikroskopischen Untersuchungsmethoden* (4. Auflage), Akademische Verlagsgesellschaft Geest und Portig K.-G., Leipzig 1959.

[2.77] Otto L., *Durchlicht-Mikroskopie, Geräte und Verfahren*, VEB Verlag Technik, Berlin 1959.

[2.78] Malies H. M., *Applied Miscroscopy and Photomicrography*, Fountain Press, London 1959.

[2.79] Barron A. L. E., *Using the Microscope*, Chapman and Hall, London 1965.

[2.80] Barer R., *Lecture Notes on the Use of the Microscope* (3rd ed.), Blackwell Scientific Publications, Oxford 1968.

[2.81] Gander R., *Rezepte zur Mikrophotographie für Mediziner und Biologen*, Urban und Schwarzenberg, München–Berlin–Wien 1968.

[2.82] Casartelli J. D., *Microscopy for Students* (2nd ed.), McGraw-Hill, London 1969.

[2.83] Zieler W., *The Optical Performance of the Light Microscope*, Parts I and II, Microscope Publications Ltd., London–Chicago 1973.

[2.84] Mc Laughin R. B., *Accessories for the Light Microscope*, Microscope Publications Ltd., London–Chicago 1975.

[2.85] James J., *Light Microscopic Techniques in Biology and Medicine*, Martinus Nijhoff BV Publishers (Medical Division), The Hague 1976.

[2.86] Gerlach D., *Das Lichtmikroskop*, Georg Thieme Verlag, Stuttgart 1976.

[2.87] Mc Laughin R. B., *Special Methods in Light Microscopy*, Microscope Publications Ltd., London–Chicago 1977.

[2.88] Rochow T. G., and Rochow E. E., *An Introduction to Microscopy by Means of Light,. Electrons, X-rays or Ultrasound*, Plenum Press, New York 1978.

[2.89] Marmasse C., *Microscopes and Their Uses*, Gordon and Breach Science Publishers, New York–London–Paris 1980.

[2.90] *Photography Through the Microscope*, Eastmann Kodak Company, Publication No. P-2; 7-80 (7th ed.), Rochester 1980.

[2.91] Locquin M. and Langeron M., *Manuel de Microscopie*, Masson, Paris 1978.

[2.92] Bradbury S., *An Introduction to the Optical Microscope*, Oxford University Press and RMS, Oxford 1984.

Chapter 3

[3.1]   Yzuel M. J. and Calvo F., A study of the possibility of image optimization by apodization filters in optical systems with residual aberrations, *Optica Acta*, **26** (1979), 1397–1406.

[3.2]   Wilkins J. E., Jr., Apodization for maximum Strehl ratio and specified Rayleigh limit of resolution, *J. Opt. Soc. Am.*, **67** (1977), 1027–1030; **69** (1979), 1526–1530.

[3.3]   Galbraith W. and Sanderson R. J., The energy distribution about the image of a point, *Microscopica Acta*, **83** (1980), 395–402.

[3.4]   Maréchal A. and Françon M., *Diffraction—Structure des images*, Éditions de la Revue d'Optique, Paris 1960.

[3.5]   Abbe E., Beiträge zur Theorie des Mikroskops und der mikroskopischen Wahrnehmung, *M. Schutzes Archiv f. mikr. Anatomie*, **9** (1873), 413–468.

[3.6]   Abbe E., *Gesammelte Abhandlungen*, Verlag Gustav Fischer, Jena 1904.

[3.7]   Beyer H., 100 Jahre Abbesche Mikroskoptheorie und ihre Bedeutung für die praktische Mikroskopie, *Jenaer Rundschau*, **18** (1973), 159–163.

[3.8]   Hofmann Ch., 100 Jahre Abbesche Sinusbedingung, *Jenaer Rundschau*, **18** (1973), 164–170.

[3.9]   Riesenberg H., Die Weiterentwicklung der Abbeschen Erkenntnisse in der Optikentwicklung moderner Mikroskope, *Jenaer Rundschau*, **18** (1973), 171–175.

[3.10] Wolfke M., Allgemeine Abbildungstheorie selbstleuchtender und nichtselbstleuchtender Objekte, *Annalen der Physik*, **39** (1912), 569–610.

[3.11] Wolfke M., Über die Abbildung eines Gitters ausserhalb der Einstellebene, *Annalen der Physik*, **40** (1913), 194–200.

[3.12] Michel K., *Die Grundzüge der Theorie des Mikroskops* (2. Auflage), Wissenschaftliche Verlagsgesellschaft, Stuttgart 1964.

[3.13] Reynolds G. O. and De Velis J. B., Review of optical coherence effects in instrument design, *Proc. of the SPIE*, **194** (1979), 2–33.

[3.14] Kintner E. C., Method for the calculation of partially coherent imagery, *Appl. Optics*, **17** (1978), 2747–2753.

[3.15] Rao K. P. *et al.*, Coherent imagery of straight edges with Straubel apodization filters, *Optik*, **50** (1978), 73–81.

[3.16] Zbralidze T. D., A representation of slit transfer function, *Optika i Spektroskopiya*,[3] **50** (1981), 1185–1186 (in Russian).

[3.17] Kirillovskii V. K. and Krynin L. I., Evaluation of imaging quality using line spread function, *OMP*, **47** (1980), 1–4 (in Russian).

[3.18] Duffieux P. M., *L'intégrale de Fourier et ses applications à l'optique*, F-té Sci., Besançon 1946 (2nd edition: Masson, Paris 1970); English edition: *The Fourier Transform and Its Application to Optics*, John Wiley and Sons, New York 1983.

[3.19] Bracewell R. N., *The Fourier Transform and Its Applications*, McGraw-Hill, Inc., New York 1970.

[3.20] Yu F. T. S., *Optical Information Processing*, John Wiley and Sons, Chichester 1983.

[3.21] Hesse G., Kohärenztheoretische Aspekte zur Abbeschen Theorie des mikroskopischen Abbildung, *Feingerätetechnik*, **22** (1973), 562–567.

[3] Translation into English: *Optics and Spectroscopy USSR*, published by Optical Society of America.

[3.22] Hopkins H. H., The concept of partial coherence in optics, *Proc. Roy. Soc. (London)*, **A208** (1951), 263–277.

[3.23] Hopkins H. H., Applications of coherence theory in microscopy and interferometry, *J. Opt. Soc. Am.*, **47** (1957), 508–526.

[3.24] Hopkins H. H. and Barham P. M., The influence of the condenser on microscopic resolution, *Proc. Phys. Soc.*, **B63** (1950), 737–744.

[3.25] Hopkins H. H., On the diffraction theory of optical images, *Proc. Roy. Soc.*, **A217** (1953), 408–432.

[3.26] Françon M., *Modern Applications of Physical Optics*, John Wiley Interscience, New York and London 1963.

[3.27] Lipson H. (ed.), *Optical Transforms*, Academic Press, London and New York 1972.

[3.28] Barakat R., Application of apodization to increase two-point resolution by the Sparrow criterion. I. Coherent illumination, *J. Opt. Soc. Am.*, **52** (1962), 276–283.

[3.29] Barakat R. and Levin E., Application of apodization to increase two-point resolution by the Sparrow criterion. II. Incoherent illumination, *J. Opt. Soc. Am.*, **53** (1963), 274–282.

[3.30] Arnulf A., Dupuy O. and Flamant F., Étude experimentale de la variation de la limite de résolution en fonction de la cohérence, *Rev. d'Optique*, **32** (1953), 529–552.

[3.31] Tsujichi J., Influence de l'aberration du condenseur sur l'image formée dans un microscope, *Rev. d'Optique*, **38** (1959), 57–74.

[3.32] Beran M. J. and Parrent G. B. (Jr), *Theory of Partial Coherence*, Prentice-Hall, Englewood Cliffs, N. J. 1964.

[3.33] Thompson B. J., Image formation with partially coherent light, in: E. Wolf (ed.), *Progress in Optics*, vol. 7. North Holland Publishing Co., Amsterdam and London 1969, pp. 169–230.

[3.34] Bhatnagar G. S. and Aggarwal A. K., On the limit of resolution of an achromatic microscope objective, *Atti Fond. G. Ronchi e Contr. Istit. Naz. Ottica*, **25** (1970), 457–462.

[3.35] Bhatnagar G. S. et al., Two-point resolution in partially coherent light, *Optics Com.*, **3** (1971), 269–271.

[3.36] Bhatnagar G. S. and Sirohi R. S., Effect of coherence on the resolution of a microscope. II. Annular aperture of condenser, *Optica Acta*, **18** (1971), 547–553.

[3.37] Sirohi R. S. and Bhatnagar G. S., Effect of partial coherence on the resolution of a microscope, *Optica Acta*, **17** (1970), 839–842.

[3.38] Som S. C., Influence of partially coherent illumination and aberration on microscope resolution, *J. Opt. Soc. Am.*, **61** (1971), 681.

[3.39] Som S. C., Influence of partially coherent illumination and spherical aberration on microscopic resolution, *Optica Acta*, **18** (1971), 597–608.

[3.40] De M. and Basuray A., Two-point resolution in partially coherent light. I. Ordinary microscopy, disc source, *Optica Acta*, **19** (1972), 307–318.

[3.41] De M. and Basuray A., Two-point resolution in partially coherent light. II. Ordinary microscopy, annular source, *Optica Acta*, **19** (1972), 523–532.

[3.42] Gehm U. and Wolter H., Überauflösung eines kleines Objektes, *Optik*, **33** (1971), 567–579.

[3.43] McKechnie T. S., The effect of condenser obstruction on the two-point resolution of a microscope, *Optica Acta*, **19** (1972), 729–738.

[3.44] McKechnie T. S., The effect of defocus on the resolution of two points, *Optica Acta*, **20** (1973), 253–262.

[3.45] Kintner E. C. and Sillitto R. M., Two-point resolution criteria in partially coherent imagery, *Optica Acta*, **20** (1973), 721–728.

[3.46] Mehta B. L., Effect of non-uniform illumination on the critical resolution in partially coherent light, *Optics Com.*, **9** (1973), 304–307 (see also: *Atti Fond. G. Ronchi e Contr. Istit. Naz. Ottica*, **30** (1975), 17–22).

[3.47] Nayyar V. P. and Verma N. K., Two-point resolution of Gaussian aperture operating in partially coherent light using various resolution criteria, *Appl Optics*, **17** (1978), 2176–2180.

[3.48] Nayyar V. P. and Verma N. K., Two-point resolution of a microscope: effect of non-uniform and nonsymmetric illumination using a semitransparent π-phase annular aperture, *Appl. Optics*, **16** (1977), 2460–2463.

[3.49] Rykhlov A. F., Effect of light coherence on resolution of point objects, *Zhurnal Nauchnoi i Prikladnoi Fotografii i Kinematografii*, **25** (1980), 424–426 (in Russian).

[3.50] Zieler H. W., What resolving power formula do you use? *The Miscroscope*, **17** (1969), 249–270.

[3.51] Becherer R. J., and Parrent G. B. (Jr), Nonlinearity in optical imaging systems, *J. Opt. Soc. Am.*, **57** (1967), 1479–1486.

[3.52] Swing R. E. and Clay J. R., Ambiguity of the transfer function with partially coherent illumination, *J. Opt. Soc. Am.*, **57** (1967), 1180–1189.

[3.53] Hewlett P. S., Visibility and resolution of periodic structures in oblique dark-ground illumination: images of diatoms with rectangulate and triangulate arrays of fine structure, *J. Microscopy*, **89** (1969), 349–357.

[3.54] Gupta S. D., Influence of apodized condenser in a microscope upon image of slit and bar objects, *Optik*, **41** (1974), 25–33.

[3.55] Pulvermacher H., Die Abbildung partiel kohärent beleuchteter ebener Gitter im Übergangsgebiet zwischen Köhlerschen und kritischer Beleuchtung, *Optik*, **39** (1974), 205–222.

[3.56] Fujiwara H., Effects of spatial coherence on Fourier imaging of periodic object, *Optica Acta*, **21** (1974), 861–869.

[3.57] Ichioka Y., Yamamoto K. and Suzuki T., Image of a sinusoidal complex object in a partially coherent optical system, *J. Opt. Soc. Am.*, **65** (1975), 892–902.

[3.58] Yamamoto K., Ichioka Y. and Suzuki T., Influence of light coherence at the exit pupil of the condenser on the image formation, *Optica Acta*, **23** (1976), 987–996.

[3.59] Jones F. T., Two apparent exceptions to Abbe's theory of resolution, *The Microscope*, **16** (1968), 4–11.

[3.60] Shimojima T. and Hayamizu Y., Analysis and synthesis of visual phenomena in microscopic vision—with particular reference to visual acuity, *Optica Acta*, **19** (1972), 455–458.

[3.61] Kingslake R., Influence of the eye on the performance of visual systems, *Proc. of the SPIE*, **39** (1973), 57–64.

[3.62] Lau E., Das Doppelmikroskop und Beispiele seiner Anwendung, *Feingerätetechnik*, **9** (1960), 112–118.

[3.63] Lau E., Untersuchungen von Phasen-Objekten mit dem Doppel-Mikroskop, *Optik*, **20** (1963), 333–346.

[3.64] Westheimer G., Optimal magnification in visual microscopy, *J. Opt. Soc. Am.*, **62** (1972), 1502–1504.

[3.65] Charman W. N., Optimal magnification for visual microscopy, *J. Opt. Soc. Am.*, **64** (1974), 102–104.

[3.66] Hanschke C. and Menzel E., Depth of field in incoherent image formation, *Optik*, **64** (1983), 67–72.

[3.67] Berek M., Grundlagen der Tiefenwahrnehmung im Mikroskop, *Sitzungsberichte der Gesellschaft zur Förderung der gesamten Naturwissenschaften zu Marburg*, **62** (1927), 189–223.

[3.68] Berek M., Zur Theorie der Abbildung im Mikroskop, *Optik*, **5** (1949), 1–30.

[3.69] Hopkins H. H., The frequency response of defocused optical systems, *Proc. Roy. Soc.*, **A231** (1955), 91.

[3.70] Schober H., *Das Sehen*, VEB Fachbuchverlag, Leipzig 1970.

[3.71] Valyus K. A., *Stereoscopy*, Focal Press, New York 1966.

[3.72] Muchel F. L., Essential optical features of stereomicroscopes demonstrated by a new instrument, the Zeis SV8, *Proc. Roy. Micr. Soc.*, **19** (1984), 89–97.

[3.73] Burke J. F., Indebetouw G., Nomarski G. and Stroke G. W., White-light three-dimensional microscopy using multiple-image storing and decoding, *Nature*, **231** (1971), 303–306.

[3.74] Häusler G., A method to increase the depth of focus by two stage image processing, *Optics Com.*, **6** (1972), 38–42.

[3.75] Courtney-Pratt J. S. and Gregory R. L., Microscope with enhanced depth of field and 3-D capability, *Appl. Optics*, **12** (1973), 2509–2519.

[3.76] Zając M. and Bilewicz Z., A simple method of the coherent and incoherent MTF visualization, *Optica Applicata* (Wrocław), **5** (1976), 33–40.

[3.77] Linfoot E. H., *Fourier Methods in Optical Image Evaluation*, Focal Press, London 1964.

[3.78] Smith D., *OTF*—quantitative image analysis, *Electro-Optical Systems Design*, **11** (1979), No. 12, 39–43.

[3.79] Rosenbruch K.-J. and Gerschler R., Die Bedeutung der Phasen-übertragungsfunktion und der Modulationsübertragungsfunktion bei der Benutzung der OTF als Bildgütekriterium, *Optik*, **55** (1980), 173–182.

[3.80] Maeda J. and Murata K., Retrieval of wave aberration from point spread function or optical transfer function data, *Appl. Optics*, **20** (1981), 274–279.

[3.81] Jaiswal A. K., Transfer functions of semitransparent and phase annuli, *Optics Com.*, **9** (1973), 161–164.

[3.82] Jaiswal A. K. and Bhogra R. K., Influence of condenser transmission upon the performance of microscopic systems, *Optica Acta*, **21** (1974), 819–834.

[3.83] Reddy G. R. Ch., Fourier imaging of sinusoidally periodic complex objects, *Atti Fond. G. Ronchi e Contr. Istit. Naz. Ottica*, **31** (1976), 402–404.

[3.84] Barton N. P., Application of the optical transfer function in visual instruments, *Optica Acta*, **19** (1972), 473–484.

[3.85] Kutter P., Modulation transfer functions of ground-glass screens, *Appl. Optics*, **11** (1972), 2024–2027.

[3.86] Maggo J. N. *et al.*, A method of improving the optical transfer function of an astigmatic circular aperture, *Optica Acta*, **21** (1974), 801–808.

[3.87] Hinds D., An *OTF* enhancement scheme for optical scanning microscopy, *The Microscope*, **22** (1974), 151–157.

[3.88] Van Leunen J. A. J., Problems of *OTF* standardization for non-perfect imaging devices, *Opt. Eng.,* **14** (1975), 169-171.

[3.89] Beyeler B. H. and Tiziani H. J., Die optische Übertragungsfunktion von dezentrierten optischen Systemen, *Optik,* **44** (1976), 317-328.

[3.90] Galpern D. Yu., Modulation transfer function of optical systems suffering from chromatism of magnification, *OMP,* **45** (1978), No. 4, 3-5 (in Russian).

Chapter 4

[4.1] Uhlig M., Prüfung der einzelnen Abbildungsfehler von Mikroobjektiven an verschiedenen Testplatten, *Mikroskopie,* **17** (1962), 273-284.

[4.2] Gander R., Hilfs- und Testpräparate in der Mikroskopie, *Mikroskopie,* **20** (1965), 117 –123.

[4.3] Göke G., Schöne und seltene Diatomen, *Mikrokosmos,* **68** (1979), 382-386.

[4.4] Kingslake R. (ed.), *Applied Optics and Optical Engineering,* vol. 3: *Optical Components,* Academic Press, New York 1965.

[4.5] Ivanova T. A. and Kirillovskii V. K., Performance testing of microscope objectives, *OMP,* **45** (1978), No. 3, 50-55 (in Russian).

[4.6] Stamnes J. J., Focusing of a perfect wave and the Airy pattern formula, *Optics Com.,* **37** (1981), 311-314.

[4.7] Garbunkov V. M. *et al.,* Calculation of energy distribution in the diffraction pattern of a point object imaged by means of a high aperture optical system, *OMP,* **46** (1979), No. 8, 56-57 (in Russian).

[4.8] Pluta M., Polarizing interferometer for testing microscope objectives, *Pomiary. Automatyka. Kontrola,* **8** (1962), 349-350 (in Polish).

[4.9] Pluta M., Polarizing interferometer with Wollaston prism for the study of microscope objectives, *Pomiary. Automatyka. Kontrola,* **9** (1963), 292-295 (in Polish).

[4.10] Pluta M., Polarizing interference method of the quality assessment of microscope objectives, *Optyka,* **2** (1967), 5-12, 51-61 (in Polish).

[4.11] Pluta M., Birefringent fibre as a secondary light source of interference devices, in: Bescos J. *et al.* (eds.), *Optica hoy y mañana (Proc. of the ICO-11),* Instituto de Optica, Madrid 1978, pp. 663-666.

[4.12] Pluta M., Double-refracting interferometer with variable direction of tilt of laterally sheared wavefronts, *Optica Applicata,* **15** (1985), 77-89.

[4.13] Roblin G. and Nomarski G., Localisation longitudinale des surfaces par interférometrie à modulation de phase, *Nouv. Rev. d'Optique,* **2** (1971), 105-113.

[4.14] Twyman F., *Prism and Lens Making,* Hilger and Watts, London 1952.

[4.15] Briers J. D., Interferometric testing of optical systems and components: a review, *Optics and Laser Technology,* **4** (1972), No. 1, 28-41.

[4.16] Steel W. H., A radial shear interferometer for testing microscope objectives, *J. Sci. Instrum.,* **42** (1965), 102-104.

[4.17] Freitag W. and Grossmann W., Interferometrische Prüfung der Abbildungsqualität an Jenaer Hochleistungsmikroskopobjektiven, *Jenaer Rundschau*, **25** (1980), 168–169.

[4.18] Ditchburn R. W., Focusing of instruments, *Optica Acta*, **27** (1980), 713–715.

[4.19] Murata K., Instruments for the measuring of optical transfer functions, in: Wolf E. (ed.), *Progress in Optics*, vol. 5, North Holland Publications Company, Amsterdam 1966, pp. 202–245.

[4.20] Sojecki A., Integral of squared modulation as a criterion of quality for the microscope objective, *Optyka*, **11** (1976), 171–173 (in Polish).

[4.21] Arnold W. and Rejman-Czosnyka M., Measurements of the modulation transfer function of microscope objectives (in incoherent light), *Optyka*, **13** (1978), 93–100 (in Polish).

[4.22] Arsenev A. A. *et al.*, An instrument for the measuring of modulation transfer functions of microscope objectives, *OMP*, **48** (1981), No. 1, 18–21 (in Russian).

[4.23] Lohmann A., Zur Messung des optischen Übertragungsfactors, *Optik*, **14** (1957), 510–518.

[4.24] Mallick S., Measurement of optical transfer function with polarization interferometer, *Optica Acta*, **13** (1966), 247–253.

[4.25] Steel W. H., A polarization interferometer for the measurement of transfer functions, *Optica Acta*, **11** (1964), 9–20.

[4.26] Merkel K., Common-Path-Interferometer zur Messung der Übertragungsfunktion, *Feingerätetechnik*, **30** (1981), 109–111.

[4.27] Hopkins H. H., Interferometric methods for the study of diffraction images, *Optica Acta*, **2** (1955), 23.

[4.28] Tsuruta T., A new type of shearing interferometer for the measurement of transfer functions of the microscope objective, *Appl. Optics*, **2** (1963), 371–378.

[4.29] Popielas M., Interferometric-photometric method for the assessment of contrast transfer of microscope objectives, *Third Czechoslovak-Polish Optical Conference*, Nové Mêsto na Moravach 1976.

[4.30] Gover C. P. and van Driel H. M., Autocorrelation method for measuring the transfer function of optical systems, *Appl. Optics*, **19** (1980), 900–904.

[4.31] Grammatin A. P., Occurrence of light reflections in microscope oculars, *OMP*, **37** (1970), No. 6, 21–23 (in Russian).

[4.32] Agroskin L. S. *et al.*, Experimental studies of the stray light produced by objectives for transmitted-light microscopy, *OMP*, **45** (1978), No. 9, 13–16 (in Russian).

[4.33] Arlevskii A. G. *et al.*, Determination of a stray light factor of the image plane of reflected-light microscopes, *OMP*, **47** (1980), 58 (in Russian).

[4.34] Magarill S. Ya. and Babak E. V., Light scattering in the objective, *Priborostroenie*, **24** (1981), No. 9, 54–63 (in Russian).

[4.35] Pluta M., Appearance and evaluation of the strain birefringence in microscope objectives, *Optyka*, **6** (1971), 6–11 (in Polish).

[4.36] Kozłowski T., Measures of quality of planachromatic microscope systems, *Optyka*, **6** (1971), 117–123 (in Polish).

[4.37] Popielas M., Measures of quality of planachromatic, apochromatic and planapochromatic microscope systems, *Optyka*, **8** (1973), 5–14 (in Polish).

[4.38] Ivanova T. A. and Kirillovskii V. K., *Design and Testing of Microscope Optical Systems*, Mashinostroenie, Leningrad 1984 (in Russian).

Epilogue to Volume 1

[E1.1] Abramowitz M., A new inverted research biological microscope, *International Clinical Products*, March/April 1986, 9–12.

[E1.2] Shotton D. M., Robert Day Allen (1927–1986), *Proc. Roy. Micr. Soc.*, **22** (1987), 62–63.

[E1.3] Brenner M., Computer enhancement of low-light microscopic images, *International Clinical Products*, September 1983, 30–34.

[E1.4] Shatton D., The current renaissance in light microscopy. I. Dynamic studies of living cells by video enhanced contrast microscopy, *Proc. Roy. Micr. Soc.*, **22** (1987), 37–44.

[E1.5] Nyman G. N., Real time video microscopy, *Proc. Roy. Micr. Soc.*, **22** (1987), Part 4, Supplement, S39.

[E1.6] De Brabander M. *et. al.*, Visualization of individual colloidal gold particles in living cells, *Proc. Roy. Micr. Soc.*, **22** (1987), Part 4, Supplement, S25.

[E1.7] Nyman G. N., Video enhanced microscopy in semiconductor industry, *Proc. Roy. Micr. Soc.*, **22** (1987), Part 4, Supplement, S15.

Index of Names

Subject Index

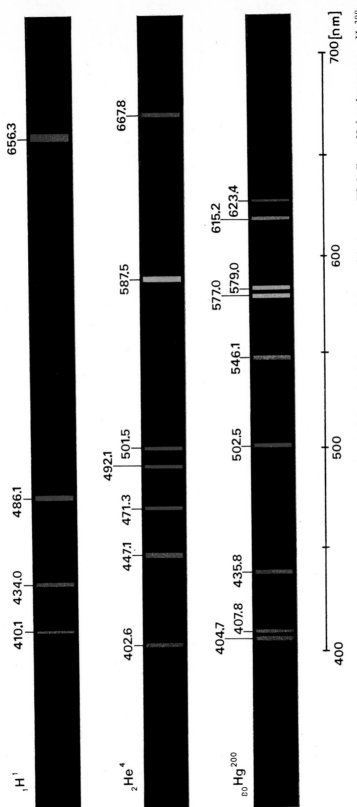

Fig. I. Visible spectral range of light microscopy: continuous spectrum of white light and emission spectral lines of hydrogen $_1H^1$, helium $_2He^4$ and mercury $_{80}Hg^{200}$ (cross references: Section 1.1 and Subsection 1.3.1).

458 PLATES

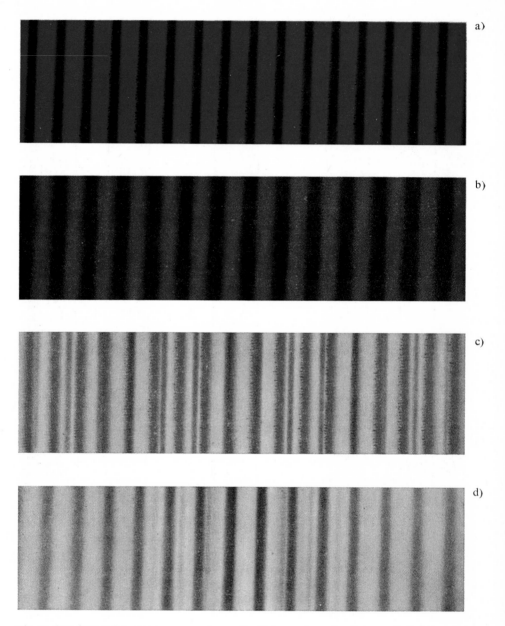

Fig. II. Interference between two monochromatic, bichromatic and polichromatic light waves: a) wavelength $\lambda = 546$ nm, b) $\lambda = 640$ nm, c) $\lambda_1 = 546$ nm together with $\lambda_2 = 640$ nm, d) $\lambda \approx 400$–700 nm, white light (cross refs.: Subsections 1.6.1 and 1.6.2).

Fig. III. Talbot's phenomenon (cross ref.: Subsection 1.7.3).

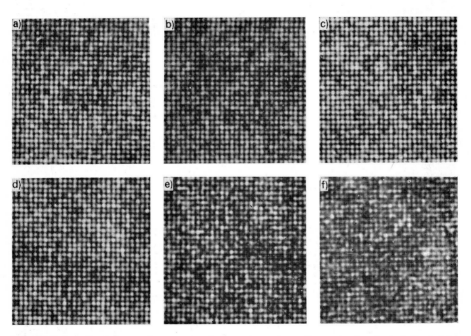

Fig. IV. Influence of the condenser numerical aperture A_c on the contrast and resolution of the microscopic image: a) A_c greater than the numerical aperture A of a microscope objective, b) $A_c = A$, c) $A_c/A = 0.65$, d) $A_c/A = 0.5$, e) $A_c/A = 0.3$, f) $A_c/A = 0.2$. Specimen: cross-line amplitude grating of period $p = 1.1$ μm; objective: planachromat $40 \times /0.65$ (cross refs.: Subsections 2.3.4, 3.6.4 and 3.8.4).

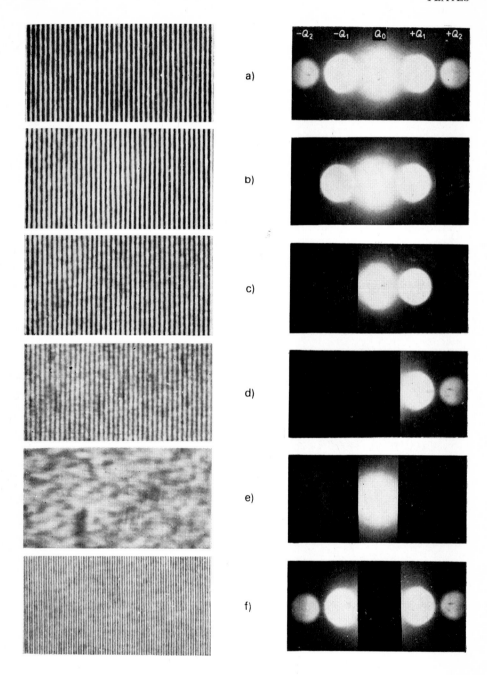

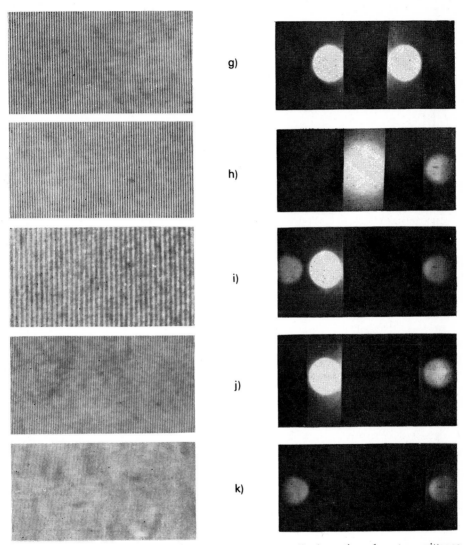

Fig. V. Spatial filtering of the Fourier spectrum of a line amplitude grating whose transmittance profile is nearly square: a) all diffraction maxima (spots) $-Q_2$, $-Q_1$, Q_0, $+Q_1$, $+Q_2$ accepted by the microscope objective are acting, b) diffraction spots $-Q_2$ and $+Q_2$ are masked, c) spots $-Q_2$, $-Q_1$ and $+Q_2$ are masked, d) all left-hand spots and Q_0 are masked, e) all spots except Q_0 are masked, f) only the zero order spot is masked, g) spots $-Q_2$, Q_0 and $+Q_2$ are masked, h) spots $-Q_2$, $-Q_1$ and $+Q_1$ are masked, i) spots Q_0 and $+Q_1$ are masked, j) spots $-Q_1$ and $+Q_2$ are only acting, k) spots $-Q_2$ and $+Q_2$ are acting (cross refs.: Subsections 3.3.1 and 3.5.6).

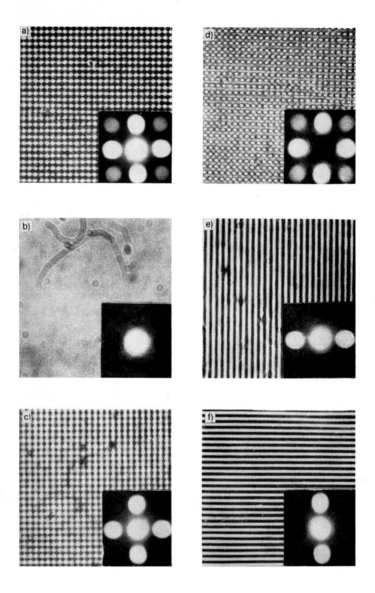

Fig. VI. Spatial filtering of the Fourier spectrum of a cross-line amplitude grating whose transmittance profile is nearly described by a function $\cos^2$ in x, y direction: a) all diffraction maxima accepted by the microscope objective form the grating image, b) all diffraction maxima with the exception of the zero order are masked, c) diagonal diffraction maxima are masked, d) maximum of zero order is only masked, e) all top and bottom maxima are masked, f) all left-hand

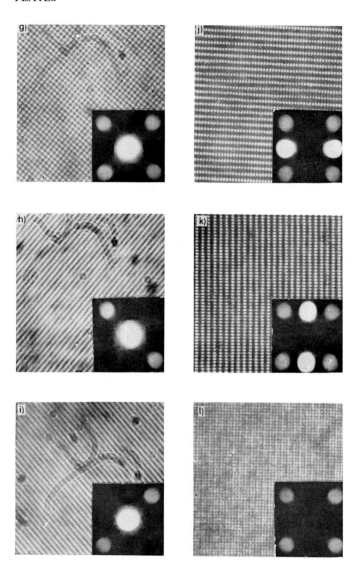

and right-hand maxima are masked, g) central and diagonal maxima form the grating image, h) central an two diagonal maxima form the grating image, i) as above, j) middle maxima at y-direction are masked, k) middle maxima at x-direction are masked, l) only diagonal maxima form the grating image (cross refs.: Subsections 3.3.1 and 3.5.6).

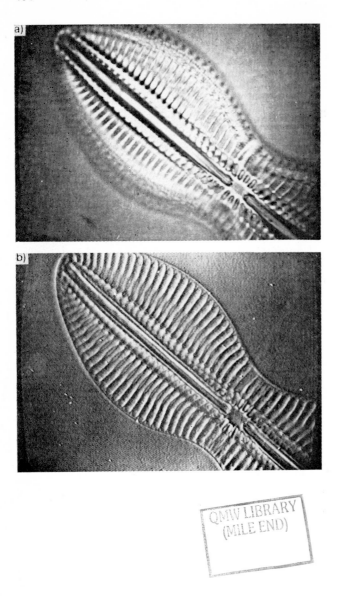

Fig. VII. Conventional photomicrograph of a diatom (a) and its counterpart with increased depth of field (b). Objective 100×/1.32, total magnification 1000×, depth of object 16 μm. Photos: by courtesy of G. Häusler (cross ref.: Subsections 3.7.2).